Aircraft Fuel Systems

Aerospace Series List

The Global Airline Industry	Belobaba	April 2009
Computational Modelling and Simulation of Aircraft and the Environment, Volume 1: Platform Kinematics and Synthetic Environment	Diston	April 2009
Aircraft Performance Theory and Practice for Pilots	Swatton	August 2008
Aircraft Systems, 3rd Edition	Moir & Seabridge	March 2008
Introduction to Aircraft Aeroelasticity and Loads	Wright & Cooper	December 2007
Stability and Control of Aircraft Systems	Langton	September 2006
Military Avionics Systems	Moir & Seabridge	February 2006
Design and Development of Aircraft Systems	Moir & Seabridge	June 2004
Aircraft Loading and Structural Layout	Howe	May 2004
Aircraft Display Systems	Jukes	December 2003
Civil Avionics Systems	Moir & Seabridge	December 2002

Aircraft Fuel Systems

Roy Langton

Retired Group VP Engineering, Parker Aerospace, USA

Chuck Clark

Retired Marketing Manager, Air & Fuel Division, Parker Aerospace, USA

Martin Hewitt

Retired Director of Marketing, Electronic Systems Division, Parker Aerospace, USA

Lonnie Richards

Senior Expert, Fuel Systems, Airbus UK, Filton, UK

A John Wiley and Sons, Ltd, Publication

This edition first published 2009

Registered office
John Wiley & Sons Ltd, The Atrium, Southern Gate, Chichester, West Sussex,
PO19 8SQ, United Kingdom

For details of our global editorial offices, for customer services and for information about how to apply for permission to reuse the copyright material in this book please see our website at www.wiley.com.

Library of Congress Cataloging-in-Publication Data

Aircraft fuel systems / Roy Langton . . . [et al.].
p. cm. — (Aerospace series)
Includes bibliographical references and index.
ISBN 978-0-470-05708-7 (cloth)
1. Airplanes—Fuel systems. I. Langton, Roy.
TL702.F8A35 2008b
629.134′351—dc22
2008036183

British Library Cataloguing in Publication Data

A catalogue record for this book is available from the British Library

ISBN 978-0-470-05708-7

Set in 10/12pt Times by Integra Software Services Pvt. Ltd. Pondicherry, India

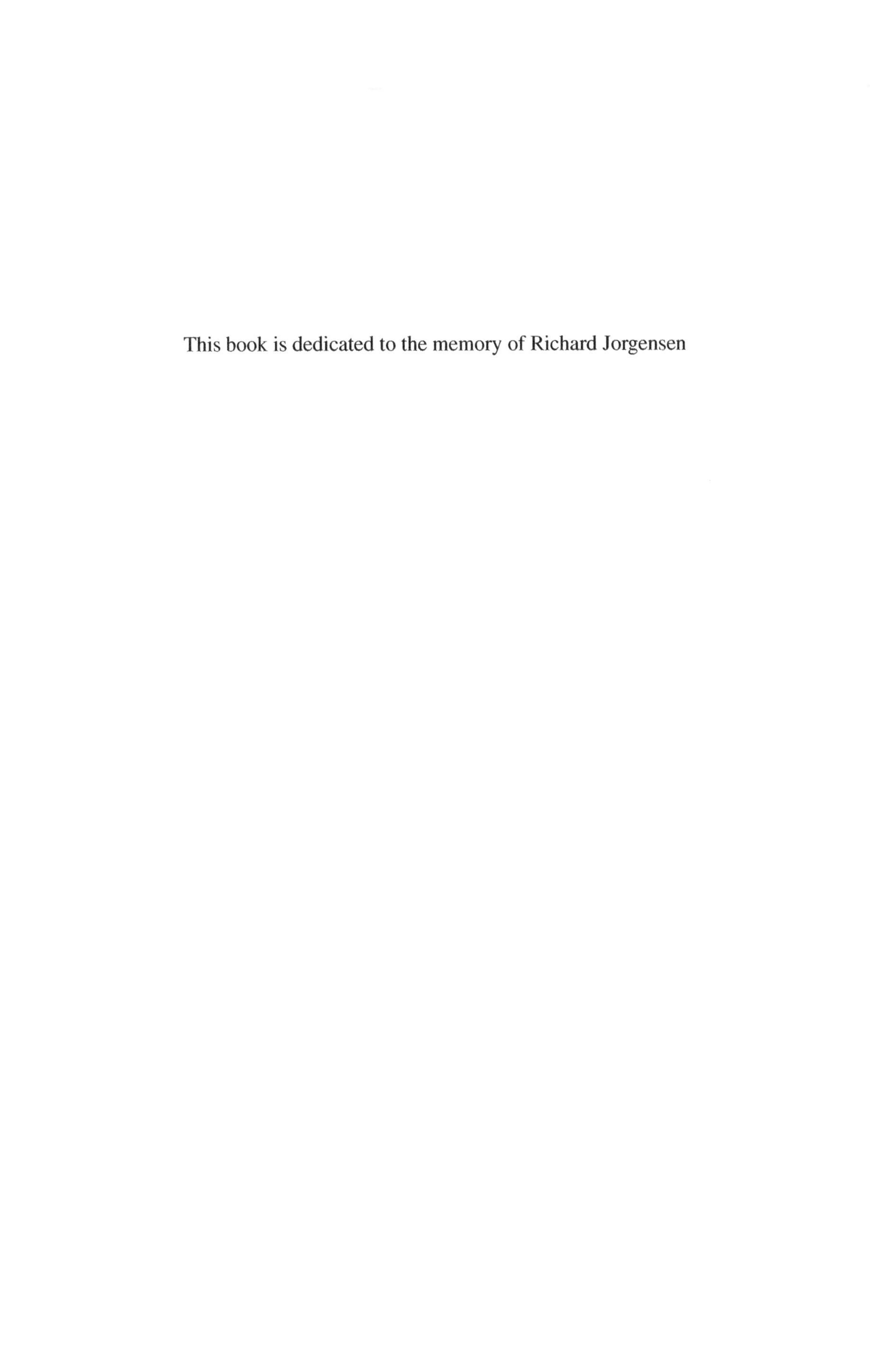

This book is dedicated to the memory of Richard Jorgensen

Contents

Acknowledgements

This book would not have been completed without the help and support of many colleagues and organizations who provided valuable information with enthusiasm. A large proportion of the design and equipment examples were provided by Parker Aerospace and Airbus by virtue of the fact that they are (or were) the employees of the authors for many years. Parker Aerospace has become a major source for fuel systems equipment to the industry and the Airbus fuel systems "Center-of-excellence" based in Filton UK has become one of Parker's major customers over the past decade.

The authors feel that it is appropriate to name specific individuals who gave much of their personal time in helping this project come to fruition:

Ron Bueter	Ray Bumpus
John Bunting	James Chu
Paul DiBella	Chris Horne
Alan Kocka	Joe Monaco
Mike Nolte	Candy Parker
David Sandy	Tim Pullen
John Passmore-Strong	Ira Rubel

In addition, the following organizations were an important source of information in support of the preparation of this book:

AMETEK, Inc
BAE SYSTEMS
Boeing
Bombardier
Embraer
GE Aviation

List of Acronyms

ACT	Additional Center Tank
ADCN	Air Data Communications Network
AFDX	Avionics Full DupleX
AGP	Alternate Gauging Processor
AIMS	Aircraft Information Management System
AIR	Aerospace Information Report
API	Armor Piercing Incendiary
APU	Auxiliary Power Unit
ARINC	Aeronautical Radio InCorporated
ARP	Aerospace Recommended Practice
ARSAG	Aerial Refueling Systems Advisory Group
ASCB	Avionics Standard Communications Bus
ASM	Air Separation Module
ASTM	American Society for Testing and Materials
AWG	American Wire Gauge
BDA	Boom-Drogue Adapter
BIT	Built In Test
BITE	Built In Test Equipment
CAD	Computer Aided Design
CCA	Common Cause Analysis
CDR	Critical Design Review
CDS	Central Display System
CENELEC	European Committee for Electrostandardisation
CG	Center of Gravity
CMCS	Central Maintenance Computer System
CMF	Central Maintenance Function
CPIOM	Central Processor Input/Output Module
CS	Certification Specification
DAC	Digital-to-Analog Converter
DEU	Data Entry Unit
DIN	Discrete INput

DMC Display Management Computer
DPUP Display Primary Micro-Processor
DSP Digital Signal Processor
DRUP Display Redundant Micro-Processor

EAP Experimental Aircraft Program
EASA European Aviation Safety Agency
ECAM Electronic Centralised Monitor
EICAS Engine Indication, Cautions and Advisories System
EIS Entry Into Service
ELMS Electrical Load Management System
EMI Electro Magnetic Interference
EMC Electro Magnetic Compatibility
ETOPS ExTended long range OPerationS

FAA Federal Aviation Authority
FAR Federal Airworthiness Regulation
FCU Fuel Conditioning Unit
FHA Functional Hazard Analysis
FMC Fuel Management Computer
FMECA Failure Modes Effects and Criticality Analysis
FMQGS Fuel Management and Quantity Gauging System
FMMS Fuel Measurement and Management System
FMS Flight Management System
FOD Foreign Object Damage
FPMU Fuel Properties Measurement Unit
FQDC Fuel Quantity Data Concentrator
FQGS Fuel Quantity Gauging System
FQIS Fuel Quantity Indication System
FQPU Fuel Quantity Processing Unit
FSAS Fuel Savings Advisory System
FWC Flight Warning Computer

HDU Hose-Drum Unit
HEI High Energy Incendiary
HEXFET Hexagonal Field Effect Transistor
HIRF High Incidence Radiated Frequencies

IDG Integrated Drive Generation
IFR Instrument Flight Rules
IFMCGS Integrated Fuel Measurement & Center-of-Gravity System
IFMP Integrated Fuel Management Panel
IGBT Insulated Gate Bi-Polar Transistor
IMA Integrated Modular Avionics
IML Inner Mold Line
INCOSE InterNational Council on Systems Engineering

INS	Inertial Navigation System
IRP	Integrated Refuel Panel
JAA	Joint Aviation Agency
JFGW	Jettison Fuel to Gross Weight
LCD	Liquid Crystal Display
LED	Light Emitting Diode
LP	Low Pressure
LROPS	Long Range OPerationS
LRU	Line Replaceable Unit
MAU	Modular Avionics Unit
MCU	Modular Concept Unit
MFD	Multi Function Display
MLI	Magnetic Level Indicator
MLW	Maximum Landing Weight
MMEL	Master Minimum Equipment List
MOSFET	Metal Oxide Semiconductor Field Effect Transistor
MTBF	Mean Time Between Failures
MTBUR	Mean Time Between Unscheduled Removals
MTOW	Maximum Take-Off Weight
NACA	National Advisory Committee for Aeronautics
NEA	Nitrogen Enriched Air
NPSH	Net Positive Suction Head
NPSHa	Net Positive Suction Head available
NPSHr	Net Positive Suction Head required
NTSB	National Transportation Safety Board
OBIGGS	On-Board Inert Gas Generation System
OEM	Original Equipment Manufacturer
OHP	OverHead Panel
O-O	Object-Oriented
PDR	Preliminary Design Review
PF	Power Factor
PSSA	Preliminary System Safety Analysis
PTFE	PolyTetreFluoroEthylene
PUP	Primary MicroProcessor
PWM	Pulse Width Modulation
RF	Radio Frequency
RTD	Resistance-Temperature Device
RTCA	Radio Technical Commission for Aeronautics
RUP	Redundant MicroProcessor

SAC	Strategic Air Command
SAE	Society of Automotive Engineers
SFAR	Special Federal Aviation Regulation
SSA	System Safety Analysis
STE	Special Test Equipment
TIU	Tank Interface Unit
TPU	Tank Processing Unit
TSP	Tank Signal Processor
TSU	Transient Suppression Unit
UAV	Uninhabited Air Vehicle
UARRSI	Universal Aerial Refueling Receptacle Slipway Installation
UML	Unified Modeling Language
USGPM	US Gallons Per Minute
VF	Variable Frequency
VOS	Velocity Of Sound
VSCF	Variable Speed Constant Frequency
WOW	Weight On Wheels

Series Preface

The field of aerospace is wide ranging and covers a variety of products, disciplines and domains, not merely in engineering but in many supporting activities. These combine to enable the aerospace industry to produce exciting and technologically challenging products. A wealth of knowledge is contained by practitioners and professionals in the industry in the aerospace fields that is of benefit to other practitioners in the industry, and to those entering the industry from university or other fields.

The Aerospace Series aims to be a practical and topical series of books for engineering professionals, operators and users and allied professions such as commercial and legal executives in the aerospace industry. The range of topics spans design and development, manufacture, operation and support of the aircraft as well as infrastructure operations, and developments in research and technology. The intention is to provide a source of relevant information that will be of interest and benefit to all those people working in aerospace.

This book co-authored by Roy Langton, Chuck Clark, Martin Hewitt and Lonnie Richards is a unique treatise on aircraft fuel systems, dealing with the design considerations for both commercial and military aircraft systems. As well as describing the system and its components in detail, *Aircraft Fuel Systems* also deals with the design process and examines the key systems drivers that a fuel system designer must take into account. This promises to be a standard work of reference for aircraft fuel systems designers.

Ian Moir, Allan Seabridge and Roy Langton

1

Introduction

While aircraft fuel systems are not generally regarded as the most glamorous feature of aircraft functionality they are an essential feature of all aircraft. Their implementation and functional characteristics play a critical role in the design, certification and operational aspects of both military and commercial (civil) aircraft. In fact the impact of fuel system design on aircraft operational capability encompasses a range of technologies that are much more significant than the nonspecialist would at first realize, particularly when considering the complexities of large transport and high speed military aircraft applications.

To illustrate this point, Figure 1.1 shows the power and intersystem information flow for a typical fuel system in a modern transport aircraft application. This 'aircraft perspective' demonstrates the interconnectivity of the fuel system with the overall aircraft and provides an indication of the role of the aircraft fuel system in the functionality of the aircraft as a whole.

This book brings together all of the issues associated with fuel systems design, development and operation from both an intersystem and intrasystem perspective covering the design, functional and environmental issues associated with the various technologies, subsystems and components.

The range of aircraft applications covered herein focuses on gas turbine powered aircraft from the small business jet to the largest transport aircraft including military applications such as fighter aircraft and helicopters.

The fuel systems associated with small internal combustion engine-powered aircraft used by the General Aviation community are not discussed in this publication since the system-level challenges in this case are minimized by the flight envelope which is confined to low altitudes and speeds and therefore the fuel system issues for these aircraft applications are relatively straightforward.

The scope of the material presented herein is focused on all areas of aircraft fuel systems from the refuel source to the delivery of fuel to the engine or engines of the aircraft. The engine fuel control system is only covered here at a high level since it is a separate and complex subject in its own right and will therefore be addressed in depth in a separate Aerospace Series publication addressing aircraft propulsion systems.

Aircraft Fuel Systems R. Langton, C. Clark, M. Hewitt, L. Richards

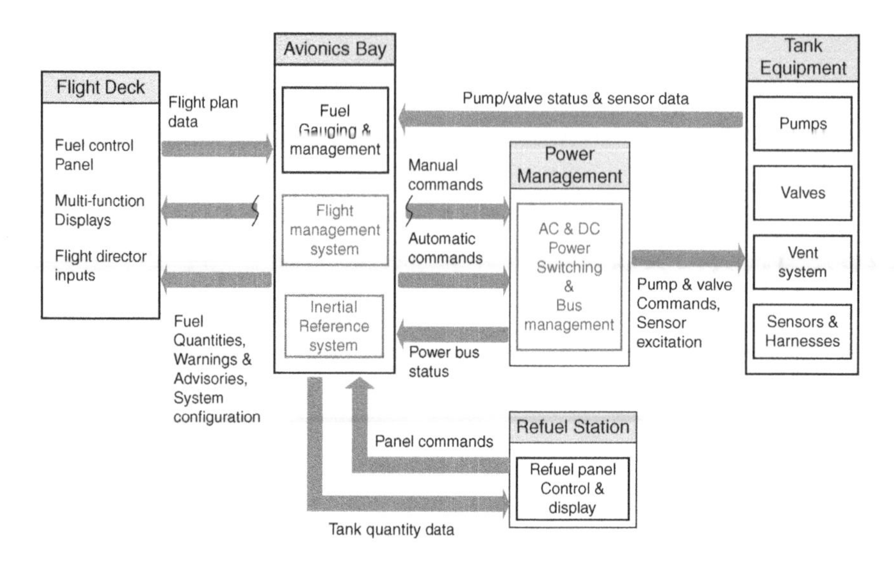

Figure 1.1 The fuel system from an aircraft perspective.

1.1 Review of Fuel Systems Issues

To introduce the subject of aircraft fuel systems the following paragraphs provide an overview of a number of the fundamental issues in an attempt to provide the reader with a feel for many of the key system and operational features that must be addressed routinely by the system design team. The comments offered in this introductory chapter are maintained at a fairly high level since a much more detailed treatment of every aspect of aircraft fuel systems is covered in the ensuing chapters.

1.1.1 Basic Fuel System Characteristics and Functions

To begin it must be appreciated that very large quantities of fuel (in terms of the fuel volume to aircraft volume ratio) must be stored aboard in order that the aircraft can meet its operating range requirements. This in turn demands a high refueling rate capability particularly in commercial transport applications where turnaround-time is a critical operational factor. While the introduction of pressure refueling goes a long way to solving this problem it does bring with it other related challenges such as the control of surge pressure following valve closure as the required tank quantities are reached. See Chapter 4 for more detail on this subject. Another pressure refueling-related issue concerns the prevention of electrostatic charge build-up resulting from fuel movement through piping at high velocities (see Chapter 9). Fuel spillage or structural damage must also be prevented through careful tank venting system design and rigorous control of the refueling process. This is addressed in Chapter 3 which discusses fuel storage and venting issues in detail.

Pressure refueling has become the standard used by all commercial and military aircraft where significant fuel quantities are involved (say 1000 gallons or more) although provision for gravity refueling is typically available on all but the largest transport aircraft where such a capability becomes impractical. The system must also make provision for defueling the aircraft for maintenance purposes and also in the, hopefully, rare event of an accident where it becomes necessary to remove the fuel from the aircraft before the aircraft can be safely moved. This process utilizes an external suction source. Frequently the on-board fuel pumps can be used to defuel the aircraft or to transfer fuel between tanks in support of ground maintenance needs.

An issue related to the refuel and defuel function is fuel jettison in flight. This function becomes an important procedure for large transport aircraft where take-off weight with a full fuel load can be substantially higher than the maximum landing weight. Therefore a major failure that takes place during or shortly after take-off can require jettison of fuel to reduce the weight of the aircraft to an acceptable level before an emergency landing can be made without exceeding the undercarriage/landing gear equipment design limits. The jettison system is required to move large quantities of fuel overboard as quickly as possible and to stop jettison before safe minimum fuel quantities are reached. This recognizes that in such emergency situations the crew will be very busy flying the airplane and would prefer not to have to spend valuable time monitoring the jettison process. Today's modern transports therefore have sophisticated jettison systems that must be prevented from uncommanded activation and be able to stop jettison automatically when some predetermined minimum fuel load or aircraft gross weight has been achieved.

Figure 1.2 shows a typical fuel tank layout for a commercial aircraft. Wing structure is a common location for fuel storage and in many commercial transports additional tanks are located in the area between the wings. Longer range aircraft and business jets may have tail tanks and/or additional fuselage tanks; however, in most cases the fuselage is primarily the place for passengers, cargo, flight deck (cockpit) and avionics equipment.

Military fighters are a special case and while the wing space is used for fuel storage in these applications, almost any available space in the fuselage is fair game for fuel since range

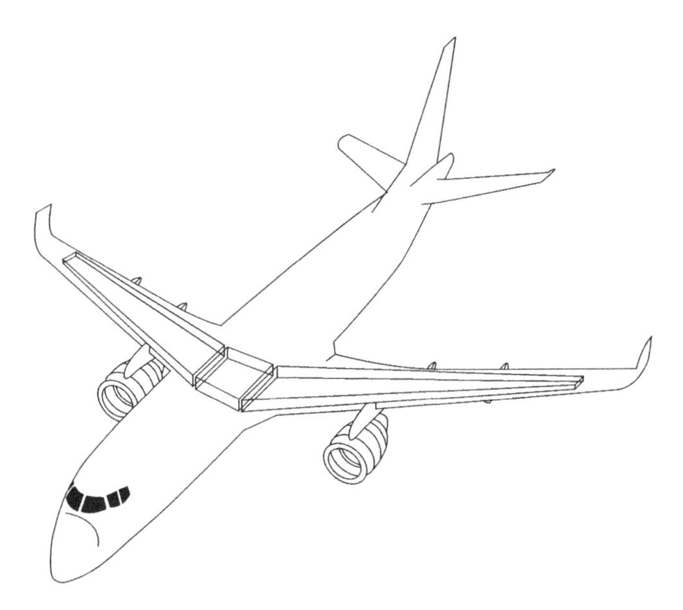

Figure 1.2 Typical transport aircraft fuel tank arrangement.

limitations are a perennial challenge for the military aircraft designer. Often the result is a number of fuselage tanks with complex shapes and a challenging fluid network design.

Fuel tank design for such large quantities of fuel on board an aircraft is a challenge for the aircraft structural designer who must also take into account the potential impact of an uncontained engine rotor burst. Such an event can generate high energy debris that can result in penetration of fuel tanks that are located in the path of the debris with subsequent loss of fuel. Consideration has to be made with regard to the ability of the aircraft to survive such an engine failure when establishing certification of the aircraft. This important issue is dealt with in more detail in Chapters 2 from an aircraft design and equipment location perspective.

In military applications battle damage with fuel tank penetration can result in loss of fuel and possible fuel tank explosion with an almost certain loss of the aircraft. For this reason, fuel tank inerting systems are commonly used to render the space above the fuel (ullage) safe from potential explosion. From a survivability perspective it is the overpressure resulting from a fuel tank explosion that can destroy an aircraft. Many different inerting techniques have been used by the military community over the past forty years including halon, stored liquid nitrogen, and reticulated foam installed in the tanks. More recently the On-Board Inert Gas Generation System (or OBIGGS) has become the standard approach to tank inerting because of the significant improvements in air-separation technology that have taken place in the past ten years. This system uses special-purpose fiber bundles to strip and dispose of a large percentage of the oxygen molecules from incoming air resulting in the generation of a source of nitrogen-enriched air (NEA). Engine bleed air is typically used as a source of air for separation. NEA output from the fiber bundles contains only a small percentage of oxygen and much less than what is required to sustain a fire or an explosion and therefore replacement of the ullage air with NEA will render a fuel tank inert and safe from potential explosion. A secondary but important issue concerns the air in solution within the fuel itself. Kerosene fuel can contain up to 14 % of air by volume at standard sea level conditions. Therefore as the aircraft climbs, this air, and more importantly the oxygen in the air, comes out of solution and can serve as a potential ignition source that must be dealt with in any effective inerting system solution.

Since the loss of TWA Flight 800 over Long Island in July 1996, the OBIGGS type of system, hitherto only used by the military, is becoming a commonly used subsystem in today's commercial aircraft.

This and other fuel tank safety issues are covered in detail in Chapter 10.

In many large transport aircraft the ratio of maximum fuel weight to total aircraft gross weight can be as much as much as 50 %. This can be compared to about 5 % for the typical automobile. This feature can in turn result in substantial variations in aircraft handling characteristics between the initial and final phases of flight.

Also, since fuel tanks are located in the wings, the effect of wing sweep is to change the longitudinal center of gravity (CG) of the aircraft as fuel is consumed causing a change in aircraft static stability and hence handling characteristics. In some aircraft the longitudinal CG is actively controlled by the fuel management system through the movement of fuel between fore and aft tanks automatically during the cruise phase.

The subject of CG control and other fuel management issues are described in depth in Chapters 4 and 5.

For commercial aircraft, optimizing the aircraft longitudinal CG during cruise minimizes profile drag which, in turn, maximizes the operating range of the aircraft.

In the case of the Concorde supersonic transport, the fuel system was used to keep the aircraft stable over the wide range of speeds involved by moving fuel aft during supersonic flight and pumping it forward as the aircraft decelerated at the end of the cruise phase. Thus the fuel system became a critical part of the aircraft's flight control system and its failure mode criticality played an important role in the fuel system design solution. A description of the Concorde fuel system is presented in Chapter 12 of this book. Even though this aircraft is no longer in service the fuel system design issues outlined would remain applicable to any future supersonic transport aircraft application.

Another frequent use of the fuel contained in the wings of larger aircraft is to provide wing load alleviation to minimize wing bending moment and thereby reduce long-term wing fatigue effects. This benefit is achieved by using inner wing tank fuel before outer tank fuel. This is discussed further in Chapters 3 and 4.

In military applications the CG variation issue can be further aggravated by the use of variable geometry (variable sweep) wings and by the use of afterburners (reheat) where very large fuel flow rates can cause fast changes in aircraft balance. The United States F-111, B-1 and F-14 and the Panavia Tornado are examples of the use of variable wing sweep technology. In these cases the fuel system must compensate for the aircraft CG variation that occurs during changes in wing sweep so that pilot workload and variations in aircraft handling characteristics are kept to a minimum.

A major fuel system issue regarding military aircraft applications is the ability to provide aerial (or in-flight) refueling. This critical need has become an essential function in modern military aircraft applications. For strike aircraft, take-off with a full weapons load followed by a climb to altitude can consume a large percentage of the fuel on board. The ability to top-off the fuel tanks after reaching operational altitude provides an essential extension of the aircraft's mission capability and is considered a key force multiplier. The aerial refueling function further complicates the fuel system design by having to provide an in-flight hook-up system with fluid-tight connections and appropriate safe disconnect capability in case of unforeseen emergencies.

Over the past 50 years standard aerial refueling equipment and procedures have been established by NATO countries ensure full interoperability between coalition forces. The US Air Force has developed a flying boom standard that provides a much higher flow rate capability than the probe and drogue standard adopted by the US Navy and NATO. A detailed description of all aerial refueling standards used by the United States and NATO is covered in Chapter 5.

A major requirement that presents a number of key operational issues to the fuel system designer is the need to provide venting of the ullage space in the fuel tanks. Wing tanks while large in volume remain relatively thin particularly at the more outboard sections. During flight these tanks bend and twist with aerodynamic loads as well as being subject to wide variations in both pitch and roll attitude. The challenge for the vent system designer is to ensure that air pockets cannot be trapped during any combination of tank quantity and aircraft attitude throughout the complete flight envelope of the aircraft. It is also critical that there is sufficient vent capacity to maintain a small differential pressure between the tank ullage and the outside ambient during maximum descent rate since only a small pressure differential between the outside ambient and the fuel ullage space can induce very large loads on the aircraft structure because of the large surface areas involved.

A challenging vent system related issue concerns the management of water as a fuel contaminant. This is most significant in large transport, long-range applications where substantial quantities of water can condense into the fuel tanks during a descent into a hot and humid destination following an extended cruise at high altitude. The high utilization rates of modern commercial transports often make the practice of routine water drainage impractical since there is seldom enough time between missions for the water in the fuel to separate out so that it can be drained from the fuel tank sump.

Water management, therefore, is a major operational issue facing today's transport aircraft and the designers of the next generation of long-range aircraft need solutions to this problem that can be effectively applied. A more in-depth discussion of this problem is presented in Chapter 3.

Military aircraft that operate at very high altitudes use 'Closed vent' systems to ensure that the ullage pressure in the fuel in the tanks remains above the fuel vapor pressure under all operational flight conditions. This adds considerable complexity to the vent system since the pressure in the ullage relative to the outside ambient conditions must be kept within safe limits by controlling the airflow in and out of this space as the aircraft climbs and descends.

During flight the fuel system must make sure that all of the fuel on board remains available to the engines through timely transfer of fuel from the auxiliary tanks (where applicable) to the engine feed tanks as the mission progresses. This process has flight-critical implications and therefore flight deck (or cockpit) displays typically provide a continuously updated display of the total fuel on board and its specific location. In order to ensure high integrity of the fuel transfer process the crew will usually have the ability to manually intervene if necessary, by selecting various pumps and valves, to provide continued safe flight in the event of a transfer system malfunction. The good news is that fuel system faults do not have an immediate impact on aircraft safety in the same way that a flight control system fault would have, because the effects on the aircraft performance of fuel system-related failures tend to develop slowly. If the fuel system develops a fault that results in a fuel transfer problem, it may be many minutes or even hours before the fault has any significant impact on the aircraft. In most cases warnings to the crew of fuel system functional faults do not have to be acted upon urgently (except perhaps a low level fuel warning which requires the crew to act or land immediately). While this situation is comforting it can also be a reason for overlooking potentially serious issues.

This was the case with Air Transat flight TS 236 from Toronto to Lisbon in August 2001 that ended in an emergency landing in the Azores after loss of both engines as a result of a fuel leak in the Starboard engine. The automatic fuel management system continued to compensate for the fuel leak on the right-hand side of the aircraft by transferring fuel from the good side of the aircraft to the leaky side of the aircraft so that fuel was eventually lost overboard to the point where first one and than then both engines lost power. With more vigilance during the early part of the flight instead of being overly dependent upon the automatic systems that had the effect of masking an ongoing problem, this event could probably been avoided. Fortunately the aircraft landed safely and no lives were lost.

1.1.2 Fuel Quantity Measurement

The challenge for the fuel quantity measurement system is to provide accurate information over a wide range of aircraft attitudes and variations in fuel properties that occur, even for a common fuel type, as a result of refueling from different locations around the world. A 1 % error in fuel quantity measurement for a commercial transport aircraft with a 100 tonnes fuel capacity is 1 tonne which is equivalent to some 10 passengers and their baggage. Also, as a result of the tank geometry, tank sumps and fuel transfer galleries on board, a small portion of the total fuel stored on board may be classified as either unusable or ungaugable. In either case this represents an operating burden for the aircraft.

Measurement of fuel quantity is accomplished by an array of in-tank sensors that are designed to detect the fuel surface at a number of locations within the tank from which volumetric information, and hence mass, can be calculated.

The most commonly used sensing technology in aircraft fuel quantity gauging systems today is that utilizing capacitance sensors, commonly referred to as probes or tank units. A capacitance probe typically comprises a pair of concentric tubes designed for near vertical mounting at a specific location within a fuel tank to act as an electronic 'dip stick'. The capacitance between each of the two concentric tubes varies with the wetted length due to the permittivity difference between fuel and air.

The number of probes required is determined by the accuracy requirements and a number of separate arrays may be required to provide adequate functionality in the presence of equipment failures.

Capacitance gauging has been the mainstay of aircraft fuel quantity measurement technology for decades. A key factor for reluctance within the industry to make changes in the fuel gauging technology is the cost of in-tank maintenance. The expectation of the airline operator is to never have to go inside a fuel tank to perform unscheduled maintenance and that in-tank hardware must continue to operate safely and without the need for maintenance for 20 years or so.

Nevertheless, alternative technologies have been tried and are continually being studied. Some of the new gauging technologies currently being evaluated are discussed in Chapter 13.

The Boeing 777 gauging system is a particularly good example of this point. Boeing made a decision on the 777 program to change from traditional capacitance gauging technology to ultrasonic gauging in an attempt to improve in-tank maintenance costs.

Ultrasonic gauging locates the fuel surface using a 'Sonar-like' technique wherein an ultrasonic wave is emitted and its echo from the surface detected. By knowledge of the speed of sound through the fuel, the fuel surface position can therefore be identified and, using a number of emitters, a surface plane can be defined and the fuel quantity computed. A detailed treatment of gauging system sensor technologies is presented in Chapter 7 and a description of the Boeing 777 fuel system and particularly the ultrasonic gauging system is described in Chapter 12.

It is interesting to note that the latest Boeing commercial transport aircraft program, the Boeing 787 Dreamliner reverted back to capacitance gauging technology for its fuel quantity measurement system.

The in-tank sensor arrays are excited electronically and the 'Fuel height' signals from the various probes are converted, using proprietary software algorithms, into tank quantity information for display to the flight crew.

Fuel quantity gauging systems are mass measuring rather than volumetric systems and they provide for continuous measurement over the full range of fuel quantity. Mass measurement is the most important parameter as it is a measure of stored energy related to fuel calorific content and, therefore, engine thrust. Nevertheless, discrete volumetric fuel measurement is also important and may be catered for by fuel level sensors which can either be integral to the gauging system or a separate system, depending on the requirements.

Examples of level sensing functions include high level sensing to ensure adequate fuel expansion space in a specific fuel tank by initiating fuel shut-off during refuel and transfer operations and low level sensing to provide warning to the flight crew of a low fuel tank quantity state.

In addition to the primary fuel gauging function a second measurement system called 'Secondary gauging' is required to ensure the integrity of this critical function and to permit safe aircraft dispatch in the presence of gauging system failures. The secondary gauge must use dissimilar technology to guard against common mode failures. A common type of secondary gauge is the Magnetic Level Indicator (MLI) where the position of the floating magnet can be read by the ground crew on a stick protruding from the base of the fuel tank. Several MLIs are usually provided. While this secondary measurement technique is significantly less accurate than the primary system it does serve to support the gauging system integrity requirement.

A detailed treatment fuel quantity measurement can be found in Chapters 4 and 7 from a system and equipment perspective respectively.

1.1.3 Fuel Properties and Environmental Issues

Perhaps the most significant issues to be recognized and dealt with regarding aircraft fuel systems are the wide variations in environmental conditions imposed by the flight envelope and the associated variations in local pressure and temperature that must be tolerated by the fuel and the equipment involved in its management. This is illustrated in qualitative terms in Figure 1.3.

A detailed treatment of fuel properties as they affect aircraft fuel systems can be found in Chapter 8.

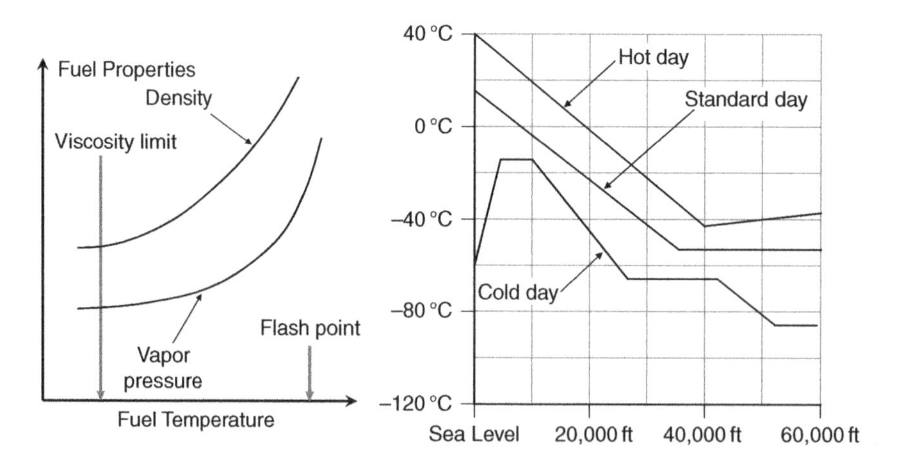

Figure 1.3 Fuel characteristics vs operating conditions.

Three highly significant characteristics of today's fuels are density (as it varies with temperature), vapor pressure and freeze point. The density variation means that an aircraft with full tanks at high temperature will have a significantly lower range (and gross weight) than an aircraft with full tanks at low temperature because the energy stored in the fuel is a function of its mass rather than its volume. This characteristic creates problems for the fuel quantity gauging system that must accommodate this variable either by widening the accuracy tolerance of the measurement system or by inferring density from dielectric constant or, preferably, measuring density directly thus compensating for this parameter in tank quantity computations.

Vapor pressure is a key factor in determining the limiting operational altitude for a given fuel (assuming an open vent system) since fuel vaporization (high evaporation rates and ultimately boiling) can occur at high altitudes, particularly with wide-cut fuels. For this reason, one critical certification test involves a maximum rate of climb with hot fuel in the tanks to identify or quantify any such limitations. In military aircraft with very high altitude ceilings a closed vent system is employed in order to maintain an adequate margin between tank ullage pressure and the fuel vapor pressure during operation at very high altitudes.

Freeze point is an important characteristic during long-range high altitude operations. Towards the end of a long flight when fuel quantities are lower, the fuel bulk temperature can approach the freeze point of commonly used jet fuels causing wax to precipitate out of solution. This wax can create obstructions and block filters and can as a result lead to engine shut-down. For this reason, the fuel bulk temperature is continuously monitored and safe operating margins maintained. If safe operating fuel temperature limits are approached the crew is required to take action (descend and/or increase Mach number) to alleviate the situation. This can be a serious problem when operating in Polar Regions if operating Mach number margins are small since descending may not necessarily locate any warmer air.

Each of these characteristics will be covered in detail in the ensuing chapters; however, they are presented here to illustrate some of the issues that must be addressed by the aircraft fuel system designer in the process of integrating the fuel system into the total aircraft design from a functional perspective.

As fuel is supplied to the engine it is pumped to high pressure and metered into the combustion chamber. The high pressures are necessary because combustion chamber pressure can be as high as 1000 psig (68 bars). Along the way, the fuel is used to cool the engine lubrication oil via a heat exchanger. The concern now becomes high temperature limitations for the fuel which, if fuel temperatures exceed 350 deg F (177 deg C), can lead to coking of the fuel nozzles with attendant loss of performance.

The engine thermal problem is largely due to the fact that the high pressure fuel pump is usually sized by the engine starting requirement when cranking speeds are low. Therefore during operation above idle speed there is typically an excess of high pressure pump capacity and this is spilled back to the pump inlet by the fuel metering control system (see Figure 1.4).

The thermal problem is exacerbated further when operating at high altitude cruise conditions where engine rotational speed (and hence pump speed) is high but fuel consumption is low resulting in a lot of undesirable heat generation. In some aircraft designs hot fuel from the engines is fed back to the aircraft fuel tanks as a cooling measure. There are issues related to this arrangement and these are covered in detail in Chapter 4 under 'Ancillary Systems'.

The most important functional requirement of the aircraft fuel system is to provide fuel to the propulsion engines (and to the Auxiliary Power Unit (APU) when fitted) within a predetermined range of acceptable pressures and temperatures as and when required throughout the specified operational envelope of the aircraft.

Providing an appropriate and reliable source of fuel to the propulsion system is fundamental to the need to keep the aircraft airborne by allowing the engine(s) to convert the fuel's chemical

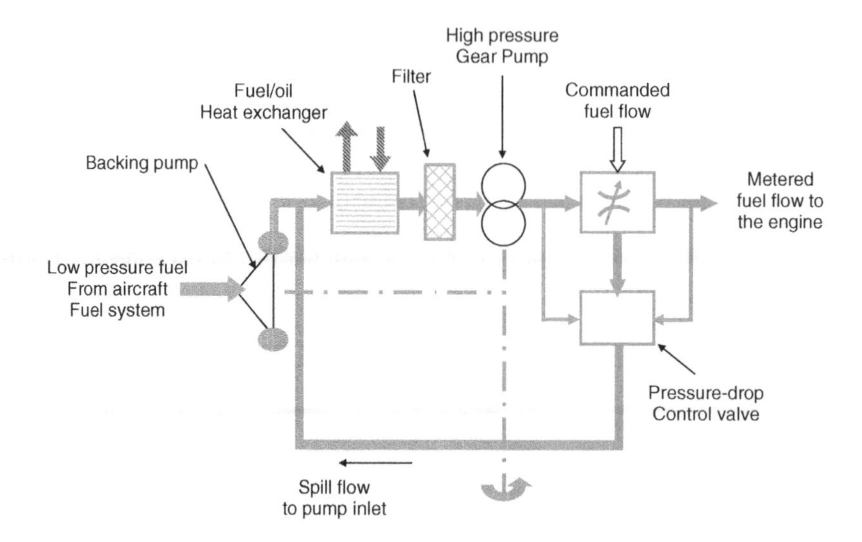

Figure 1.4 Simplified engine fuel metering control schematic.

energy into thrust continuously in accordance with aircraft's control system requirements. The criticality of this requirement therefore demands that the integrity of the fuel system from a functional perspective be equivalent to that of any of the flight critical systems on board since failure to provide this function would result in a catastrophic event, i.e. loss of aircraft.

It is also critical to the aircraft operation that the pilot and crew know how much fuel is on board and where it is located. The fuel measurement and fuel management system provides this function which must have extremely high integrity.

In summary, the primary aircraft fuel systems functions are as follows:

- The fuel system must make sure that the feed tanks associated with each engine are maintained full as long as possible by transferring fuel from the other (auxiliary) tanks into the feed tanks in accordance with a predetermined fuel burn schedule taking care to ensure that the balance of the aircraft laterally is maintained. Lateral imbalance can result from differences in fuel consumption between engines or inter-system failures that result in inadvertent fuel transfer. Fuel leakage overboard can also be a serious hazard resulting in imbalance and, more seriously a reduction in range. Detection and location of leaks is a major issue that has to be addressed by the fuel system designer.
- The system must also accommodate the effect of an engine failure by providing the ability to crossfeed between feed tanks so that the remaining engine(s) have access to the failed engine's fuel and that the aircraft does not become significantly unbalanced laterally
- In the event of an engine failure during or shortly after take-off the pilot may decide to return to the airfield. This situation can be problematic for the landing gear since maximum take-off weight can be significantly higher than the maximum landing weight. The fuel jettison system provides a means to quickly dump fuel overboard to achieve a safe landing weight before attempting a landing.

The above commentary is intended to set the scene for the more detailed discussions that are addressed in detail in the later chapters.

1.2 The Fuel System Design and Development Process

In many cases the fuel system function can be classified as a complex integrated process that involves major interactions between many aircraft systems. The process of designing, developing and certifying a modern aircraft fuel system is therefore a major undertaking and the demand for mature functionality at entry into service is, as with any major operational system, critical to both the aircraft manufacturer and the aircraft operator. Also the importance of addressing lessons learned from previous systems cannot be over-emphasized.

For this reason Chapter 11 provides an important insight into the technical and program management issues that must be addressed in the fuel system design and development process in order to maximize the potential for high system maturity at entry into service. The downside of this situation, where operational issues remain to be discovered in service by the airlines or the military operators, can be an enormous cost in both monitory and reputation terms to the aircraft manufacturer and to the equipment supplier.

Therefore the supplier/airframer combination that can develop a design and development process that guarantees a maximum probability of achieving maturity at entry into service offers an enormous operational benefit to both the aircraft manufacturer, the equipment supplier and to the user communities.

The design and development process can be best expressed by the well known 'V' Diagram (see Figure 1.5).

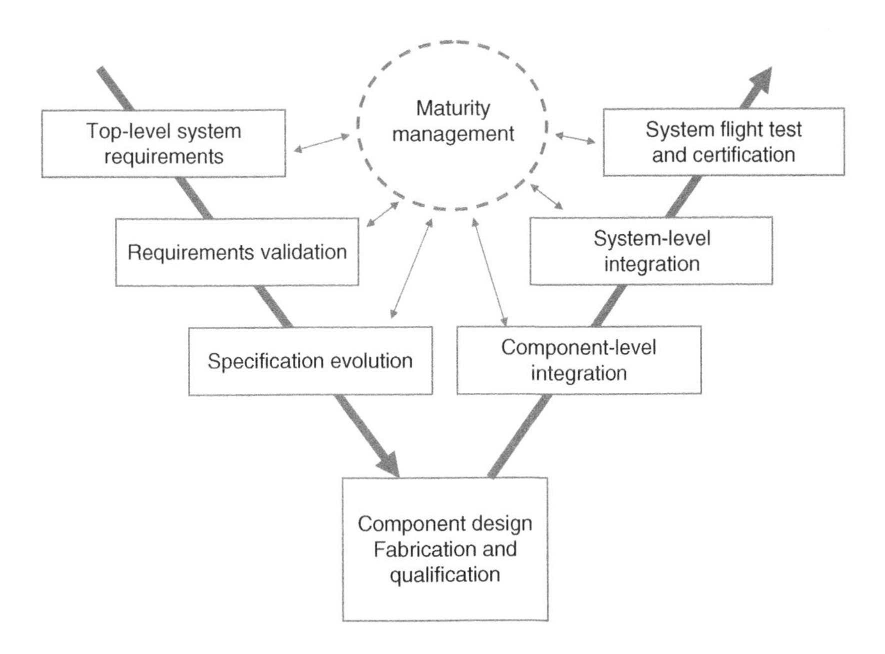

Figure 1.5 The 'V' diagram concept.

The process begins at the top left with top level system requirements. As the process moves down the left side of the 'V' the level of detail increases until the requirements of the major components of the system are defined. At each level, the applicable requirements are validated for completeness and correctness.

At the bottom of the 'V' the components are designed to meet the newly validated requirements followed by fabrication, testing and qualification at the component level.

The right side of the 'V' represents the integration, verification and certification phases of the program beginning with component integration, then system integration and finally integration with the aircraft for flight test and certification.

This is only a brief overview of this methodology which is developed in more detail in Chapter 11.

1.2.1 Program Management

Program management is a critical skill that can make an immense difference between success or failure of a project to meet the operational expectations of the system provider and end user. People who can demonstrate this capability remain in high demand within the industry because their potential economic benefit, in the long run, can prove to be substantial to both the manufacturing and operating communities.

In the past decade or so the industry has seen a major shift in the responsibility and role played by the typical equipment supplier who is today expected to be 'Systems smart' with the ability to contribute to the functional requirements definition of its products from an aircraft system perspective. In fuel system applications this issue is particularly important because of the complexities of many modern applications where there are a large number of significant functional interactions with other aircraft systems including:

- ground refuel station
- flight management system
- power management system
- flight warning and advisory system
- display management system
- central maintenance computer
- propulsion system
- tank inerting system
- on-board maintenance system.

Historically the equipment supplier was typically isolated from the operational problems seen by the operator community who had to pay the price (via the purchase of expensive spare parts) of functional immaturity. In this scenario there was no incentive to the supplier community to change its way of doing business. More problems in the field meant more sales of spare parts which were priced to provide good margins.

Today this situation is no longer viable and applies not only to complex fuel systems but to aircraft systems design and development in general. Today the equipment supplier is typically required to take responsibility for meeting direct maintenance guarantees and equipment removal rates that are part of the contract. The equipment supplier community is now required to participate in the system design, development and certification process and to take an active involvement in the performance of their components as an integrated entity within the aircraft itself.

The design and development process aspect of this book, presented in Chapter 11 describes some of the essential tools that are necessary in the successful certification of today's complex fuel systems. The importance of 'Joint working' between the supplier and aircraft manufacturer's design team is emphasized as a major contributor to a successful program.

Specific methodologies for system design and development are described in detail including the SAE standards ARP 4754 reference [1] and 4761 reference [2]. These advisory documents instill a discipline into design and development process that emphasizes the importance of safety from the earliest conceptual phase through final definition of the requirements both at the system and component levels. Examples include the System Safety Analysis (SSA) and the Functional Hazard Analyses (FHA) which is a relative of the Failure Modes Criticality and Effects Analysis (FMECA) used throughout the industry.

An important issue with the design of aircraft fuel systems is to ensure that there are no common failure modes that can eliminate the effectiveness of functional redundancy. For example, since fuel properties are a common factor in fuel system operation, consistency in fuel quality standards may become a critical factor since any single event that could impact this situation would be considered a common mode failure. Potential causes include excessive fuel contamination; say with water or ice or freezing (waxing) of fuel due to operation for extended periods at flight conditions with recovery temperatures below the fuel freeze point.

Risk management is also a key discipline that is designed to identify all potential risks to the program and to develop and manage mitigation plans in order of criticality. This aspect of program management is crucial in minimizing the possibility of late developing crises with the attendant schedule and development cost penalties.

1.2.2 Design and Development Support Tools

There are many powerful support tools that are typically utilized to support the design and development activities. These are general purpose and in some cases proprietary software tools that allow the design team to validate requirements and verify the functional behavior of the system at various levels of integration through to the flight test phase.

Requirements traceability is important in complex integrated systems such as a modern aircraft fuel system. Requirements management tools are available that can ensure that all requirements are traceable from a top level system requirement to an expanding number of requirements at each level down to the component level.

Modeling and simulation tools are used to analyze fluid network performance during the various modes of operation providing information on pipe sizing, pressure losses and surge pressures. Models are also necessary to facilitate early testing of the system functionality by providing a pseudo aircraft to exercise the equipment against well before the real system/aircraft becomes available.

Fuel tank geometry analysis tools are essential to the fuel system designer particularly when tank shapes are complex and aircraft attitudes and g force variations are considerable. The system designer needs to be able to define the quantity and location of fuel probes that are necessary to meet accuracy requirements at an early stage. Similarly the location of pumps to minimize unusable fuel must be established early so that structural penetrations and installation provisions can be made.

An important activity that must occur in parallel with the design and development of the fuel system components is the provision of the Special Test Equipment (STE) necessary to both verify the performance and achieve the hardware and/or software qualification of those components.

Finally as the program enters the flight test and certification phase, there is a critical need for the specialist systems engineering groups to access, review and analyze enormous amounts of test data from both ground and flight test situations. Special tools for data recording, classification and analysis are essential to the efficient synthesis and evaluation on a continuous basis as the certification process proceeds.

These are just a few examples of the importance and benefits of design support tools that are available and necessary for the successful execution of a fuel system development program.

1.2.3 Functional Maturity

Functional maturity at entry into service is the Holy Grail of successful program management because the benefits of such an accomplishment are so overwhelmingly powerful from an operational perspective.

The challenge to the program manager is to establish an effective process that can measure and react to the prevailing maturity status in an effective and program-beneficial manner.

Maturity management must be applied to all of the various phases of a design and development program at the system requirements level. Lessons learned form previous programs can be used to establish derived requirements and associated verification processes.

One of the most challenging areas for maturity management is in the area of software design and verification. Software maturity typically improves with exposure to the operating environment. The maturity manager must therefore strive to maximize software operating time on test rigs and flight test aircraft while providing a software design that can incorporate changes quickly and with minimal disruption to the program operating structure.

The prospects for the successful implementation of a maturity management system and the downside of paying lip service to this critical issue are discussed in Chapter 11.

1.2.4 Testing and Certification

The testing, integration and certification activities involve the right side of the 'V' Diagram. For the first time real hardware and flight operational software are tested in integrated test rigs with varying degrees of fidelity. The equipment suppliers are primarily interested in the sub-system and component level integration issues while the aircraft manufacturer is focused more on the aircraft level functionality of the system.

The aircraft fuel system is one of the most interactive systems in modern aircraft today.

It is critical, therefore, that the development phase provide the maximum possible exposure of the fuel system components and operational software to detect, isolate and correct system-level functional issues well before the system is exposed to the actual aircraft.

Testing and certification issues are also covered in detail in Chapter 11.

1.3 Fuel System Examples and Future Technologies

The penultimate chapter contains several real world examples of aircraft fuel systems all of which are concerning commercial (civil) aircraft applications because of the security constraints imposed by the military aircraft community. Included are several modern aircraft fuel systems including large transport aircraft such as the Boeing 777, the new Airbus super jumbo A380 and the Concorde which contains many unique functions related to the extensive operational flight envelope of this aircraft. Smaller regional aircraft fuel system examples are also included in order to provide a perspective of the difference between the system technologies and the component solutions.

These applications attempt to put into perspective the content of the book from the design, development, certification and operational aspects of aircraft fuel systems.

The final chapter provides a view of the future regarding aircraft fuel systems technology and where this may take us from a systems and component design and development perspective.

The need to develop a revolutionary gauging technology has occupied the engineering community for many years. Capacitance gauging in either its AC or DC form has been the accepted standard of the industry for the past fifty years or more notwithstanding the recent ventures into ultrasonic gauging technology adopted by Boeing for their 777 aircraft. Suggestions regarding the most likely prospective new gauging technologies currently being considered are presented and discussed in this chapter.

While fuel pumping and management products and methods represent well established and mature technologies, the increasing power of electronics and sensing capabilities over the past twenty years or so has led to the consideration of new integrated concepts to improve fuel system functionality particularly during the refuel process. These and other new ideas are described and discussed in this chapter.

1.4 Terminology

The fuel systems engineering community has established, over many years, a terminology associated with the system, functional and product aspects of aircraft fuel systems that is worth explaining here as an aid in the understanding of the principles and examples presented throughout this book.

The following therefore is a definition of some of the terms and expressions commonly used in the industry today that appear within the chapters that follow in an attempt to assist the reader in obtaining a more in-depth understanding of fuel systems, their functions and the equipment involved. It is recommended that the following definitions be used as a reference to support the discussion that follows in the ensuing chapters:

Brick wall architecture: This architecture, used extensively by Boeing, requires that the gauging of each tank is independent of the other tanks. Thus no failure in one tank can propagate into the gauging of the remaining tanks.

Closed vent: As the term implies a closed vent system connects the ullage to the outside air via a control valve or valves (often referred to as 'Climb and dive' valves) in order to control ullage pressure. This is necessary to prevent high levels of fuel evaporation or even boiling in aircraft fuel tanks where flight at very high altitudes is involved.

Compensator: The compensator is a capacitor mounted low down in the fuel tank to provide a standard measure of fuel permittivity that allows the capacitance fuel probes to provide a fuel immersion coefficient that is ratiometric and independent of the prevailing fuel permittivity. A fully immersed tank probe can also serve as a compensator.

Crossfeed: This applies to multi engined aircraft and is the supply of fuel to an engine from the opposite side of the aircraft fuel system typically during operation with one engine shut down.

Defueling: The process of off-loading fuel from the aircraft via suction or by the use of on-board pumps.

Densitometer: The densitometer is typically an in-tank fuel density measurement sensor. The most common implementation uses the spring-mass resonance concept where the mass term is represented by the fuel. These sensors provide a frequency output that is proportional to fuel density; however, some sensor characterization is usually required (temperature compensation for example) to achieve the best accuracy performance.

Dual channel: Fuel quantity gauging and management systems typically require some level of redundancy to deliver the integrity requirements dictated by the airworthiness authorities. The dual channel approach is in common use. Here there are two independent channels, one declared as the 'Master' and the other as the 'Slave' or 'Hot spare'. During normal operation the Master Channel is in control while the Slave executes identical software. If a fault should develop in the Master Channel, the Spare Channel takes over.

Dual-dual channel: The Dual-Dual architecture is essentially the same as the Dual Channel; however, here each channel has two computers; one as a command computer and the other as a monitor. This arrangement significantly improves achievable fault coverage.

Engine feed pressure: This is the pressure in the feed line to the engine which must be maintained above fuel vapor pressure with some margin. The term boost pressure is synonymous with feed pressure, as are boost pump and feed pump.

Fuel gauging and management: The Fuel Quantity Gauging System (FQGS) calculates the fuel quantities in each fuel tank for display to the flight crew and to the refueling station. In some applications this system is referred to as the Fuel Quantity Indication Systems (FQIS) or, when implemented as an integral part of the fuel management system, it may be referred to as the Fuel Measurement and Management System (FMMS) or the Fuel Management and Quantity and Gauging System (FMQGS). Other similar acronyms are also in use today.

Fuel-no air valve: A fuel no-air valve is normally located at the bottom of a fuel tank and is designed to close (stop transfer) when air enters the valve.

Fuel stratification: Fuel stratification can occur when an aircraft is turned around after a long flight. Any residual fuel may be very cold (say –30 degrees C). If the uplifted fuel for the continuing flight leg is relatively warm (say 10 degrees C) this fuel can lie on top of the colder, denser fuel causing gauging system errors.

Fuel transfer: Moving fuel from one location to another in the aircraft.

Non-modulating level control valve: A level control valve that provides a discrete deadband between the shut-off level and the re-opening level.

Open vent: Here the ullage is continually open to the outside air via the vent system piping and is typically of most commercial aircraft.

Pilot valve: A float operated valve that controls the state of a separately located valve. The float may be positioned to sense a high or low fuel level condition

Pre-check: This is the process of verifying the integrity of the refuel shut-off system by simulating a full tank condition via fluidic or electrical means prior to actually reaching the full tank condition.

Pressure refueling: The use of a high pressure source to facilitate fast refueling of aircraft

Scavenge: This involves moving the last vestiges of fuel from a fuel tank to a more accessible location (e.g. a feed tank or collector tank). Scavenged fuel would otherwise be trapped and/or unusable.

Sensing level: The level at which the pilot valve is set to operate by closing the pilot valve.

Suction feed: This is the supply of fuel to the engine created by suction from the engine. To accomplish suction feed the pressure at the engine must be above the fuel vapor pressure for the prevailing operating condition.

Surge pressure: This is the pressure rise in the refuel system upstream of the shut-off valve caused by the closure of the valve and is related to the phenomenon known as 'Water hammer'.

Tank pressurization: This is the provision of air pressure to the fuel tank ullage (usually from engine bleed air) to assist with high altitude and/or high temperature performance. Typically this is provided in conjunction with a closed vent system.

Tank probe/tank unit: Tank units or tank probes typically refer to fuel gauging probes that measure the wetted length or depth of immersion. Capacitance probes use the difference in permittivity between air and fuel.

Ullage: This is the space above the fuel surface within a fuel tank.

Valve overshoot: The volume of fuel that passes through a shutoff valve after the instant the valve is selected closed either by a pilot valve or other means.

The above list is just a snapshot of the many terms used within the industry and while they are mostly covered within each chapter to some extent, it may be useful for the reader to refer back to this section if confusion of terms occurs.

2

Fuel System Design Drivers

The most important phase in the design and development of a new aircraft fuel system is the initial concept phase when the basic aircraft design is still somewhat fluid as design trade studies are done in order to arrive at the optimum solution with regard to market needs and regulatory requirements.

It is at this stage that major functional aspects of other interfacing or interacting aircraft systems need to be understood and discussed with top-level aircraft design specialists. The most significant issues driving this activity are safety and economics since the solution that finally evolves must be safe and capable of satisfying regulatory authorities from a type certification perspective. Secondarily the aircraft must be operationally competitive to the end user operators in a highly challenging marketplace.

Even in military aircraft design, while mission effectiveness may, in some cases, be considered second to acquisition and sustainment costs, the concept of affordability is becoming a major driver in the weapons procurement process thus bringing the early conceptual phase more in line with traditional commercial operational methods.

Figure 2.1 shows that the conceptual phase relates to the overall design and development process as the top level and hence the most influential stage of the program.

The system concept definition phase is both the most critical and the most fluid phase on the system design process and therefore warrants the most careful study as the preferred system solution evolves. During this conceptual phase it is important that representatives of all of the interfacing systems and subsystems have the opportunity to represent their opinions and recommendations in the review of system alternatives that evolve and that everyone understands the benefits and issues at the aircraft level of all of the key system design decisions and implications that arise.

Figure 2.2 illustrates the concept definition phase schematically.

From the initial concept studies, a number of possible system design solutions typically emerge for further detailed evaluation reference [3]. During this trade study exercise both technical and program risks are identified and mitigation strategies considered. Certification strategies including the need for special purpose test rigs may be developed. One or two of the most favorable approaches are then reviewed by the aircraft systems design team before

Aircraft Fuel Systems R. Langton, C. Clark, M. Hewitt, L. Richards

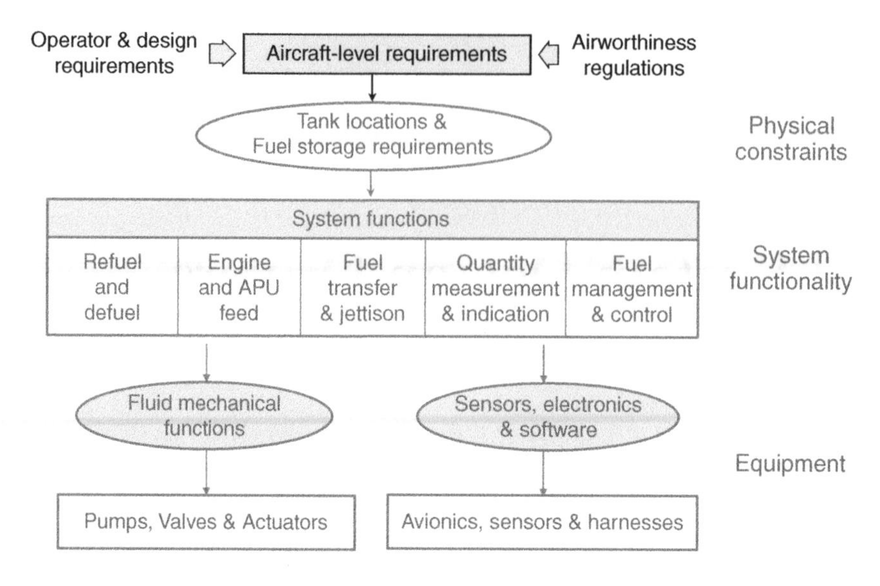

Figure 2.1 Fuel system design process overview – conceptual phase.

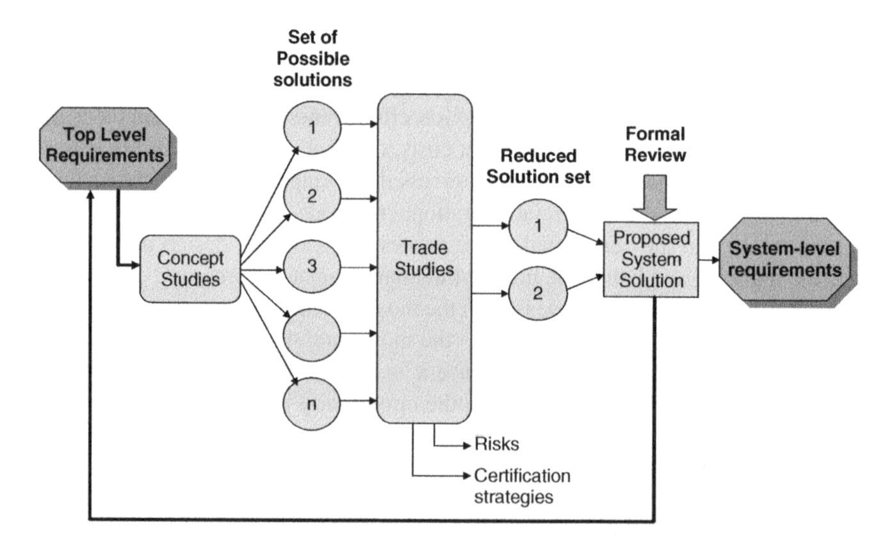

Figure 2.2 System concept definition phase activities.

a decision as to the preferred system configuration is made. As indicated in the figure, this process is an iterative process that may require several passes.

The following paragraphs summarize many of the key drivers that influence the design of the typical aircraft fuel system and how these drivers relate to major aspects of aircraft functionality.

2.1 Design Drivers

The most important fuel system design drivers are dependent upon many fundamental and operational aspects of the aircraft design that may not necessarily be related to the fuel system yet they must be accommodated within any design solution. It is critical, therefore, that the fuel system designer be intimately involved in the evolution of the basic aircraft design so that key performance, operational and cost drivers can be identified and included in the early trade studies.

The following list addresses some of these key aspects of the aircraft system preliminary design phase that can have a substantial impact on the final fuel system design implementation:

- the intended aircraft mission
- dispatch reliability goals and system operational availability
- fuel tank boundaries and location issues
- measurement and management system functional requirements
- electrical power technology and management system architecture.

Each of the above issues is discussed in the following paragraphs.

2.1.1 Intended Aircraft Mission

In commercial aircraft the fuel system requirements for a long-range aircraft requiring regulatory authority approval for ETOPS (Extended Twin OPerationS) routes may drive the fuel gauging system integrity and hence the system design architecture. For example, a traditional dual channel/duel sensor array design has an established integrity (defined as the probability of the display to the crew of an incorrect but believable fuel quantity) of 10^{-7} fault occurrences per flight hour. This may be inadequate for ETOPS certification thus driving the need for a higher integrity solution with, perhaps a third independent channel with dissimilar hardware and/or software implementation. A more detailed discussion of this issue is included in Chapter 4.

In a similar manner, the engine crossfeed design implementation using a single shut-off valve to connect the left and right feed tanks in a twin engined transport aircraft (in the event of an engine shut-down) may be deemed inadequate due to the probability of loss of function due to, say valve icing. In this case the recommended approach may require the implementation of a dual crossfeed valve arrangement perhaps with significantly different locations to guard against a potential common mode failure.

In military aircraft the operational flight envelope is often much more demanding with maximum operational altitudes of 60,000 ft and higher together with Mach numbers up to and beyond Mach 2. The provision of afterburning engines adds an additional dimension of complexity to the fuel system design solution. Control of the aircraft CG during the very large fuel flows involved with afterburner operation may be critical to safe operation of the aircraft thus imposing special considerations in the fuel system design solution. For example, flow proportioning valves may be employed to ensure that flow from fore and aft tanks is managed to maintain a fairly constant longitudinal CG.

2.1.2 Dispatch Reliability Goals

Dispatch reliability or system functional availability is a major operational objective in commercial aircraft operations and customer expectations/requirements make this issue particularly challenging. These requirements typically determine the extent of functional redundancy required in order to allow aircraft dispatch in the presence of equipment failures. In this instance, the design regulatory authorities do not play a major role since their primary concern is safety. For example, a design solution that has relatively inferior dispatch capability yet has a safe operational solution is totally acceptable as a certifiable design. Here, therefore operational and economic issues may become the main drivers in this specific aspect of the fuel system design solution. This issue is best illustrated by Figure 2.3 which shows how the major system design drivers are driven either by safety regulations, by mission effectiveness or by operational economics.

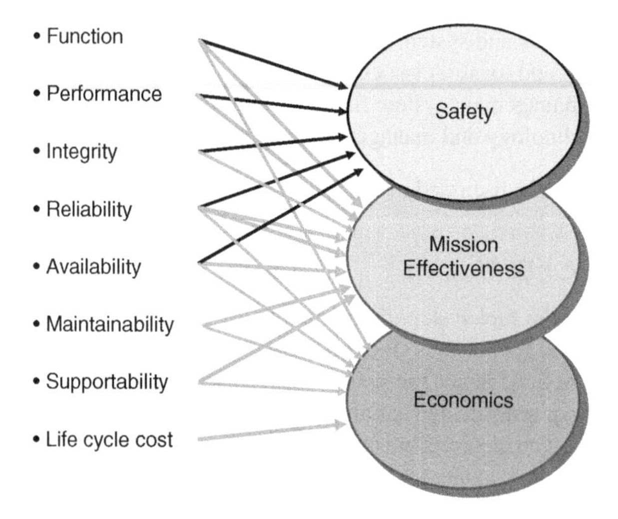

Figure 2.3 Top-level system design drivers.

2.1.3 Fuel Tank Boundaries and Tank Location Issues

Because of the large quantities of fuel carried by most aircraft today, the location and geometry of the fuel storage tanks plays a critical role in the aircraft design and its operational capabilities. The fuel storage issues associated with normal operation are addressed in Chapter 3; however there is a potentially catastrophic (albeit extremely unlikely) failure mode that must be addressed very early in the aircraft conceptual design phase. This failure mode is the 'Uncontained rotor burst' associated with the engines, and, to a lesser extent the Auxiliary Power Unit. These devices contain high speed rotating spools containing enormous amounts of kinetic energy, which if released due to some mechanical or control system failure, have the potential to cause major structural damage to the aircraft and in the process could result in penetration of the fuel tanks with the resultant loss of fuel overboard.

In conducting an uncontained rotor burst evaluation, it is required to assume that an uncontained rotor burst will result in the emission of debris of infinite energy normal to the engine rotational axis. The debris envelope expands along a five degree plane as indicated in Figure 2.4 which is an example of rotor burst implications for a wing-mounted engine configuration. In the example shown the left feed tank and center tank boundaries are maintained outside the

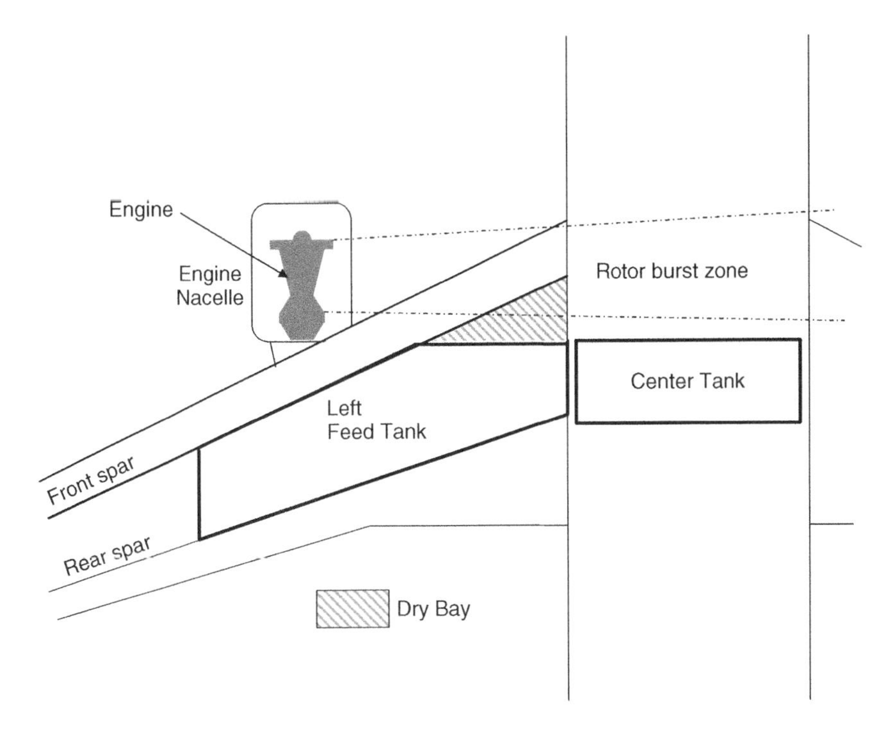

Figure 2.4 Uncontained rotor burst example for a wing-mounted engine.

rotor burst zone by providing a dry bay in the forward inboard corner of the wing tank and keeping the center tank forward boundary aft of the rotor burst zone.

The outcome of the rotor burst analysis can have far-reaching implications on tank structural boundaries as well as on system architecture and fuel handling equipment installation. This also involves the routing of electrical supplies to key fuel system equipment in order to allow operation after the event and hence continued feed and transfer as needed to the remaining good engine(s).

The effects of CG shift following a rotor burst due to loss of fuel overboard is also an important consideration and this may be more critical with a rear mounted engine configuration as indicated in Figure 2.5 which shows the rotor bust zone impacting the wing tip of a swept wing, twin engine aircraft with rear mounted engines.

In this situation it would be desirable to ensure that the wing fuel tanks stay out of the rotor burst zone. If this is not possible, then any potential fuel loss must not generate a rolling moment in excess of that available from the flight control system even when taking into account any loss of aileron control due the rotor burst event itself. This may require the introduction of a sealed rib within the wing tank to provide a fuel tank outboard compartment that is sufficiently small to ensure that roll control is not compromised following an engine rotor burst.

The fuel system designer must ensure that continued safe flight for several hours on the remaining operational engine or engines is achievable to allow safe diversion and landing following a rotor burst event. In this context, the rotor burst issue is not just a tank location issue but a broader system design issue since potential damage to critical equipment, fuel lines, electrical power supplies etc must also be taken into account.

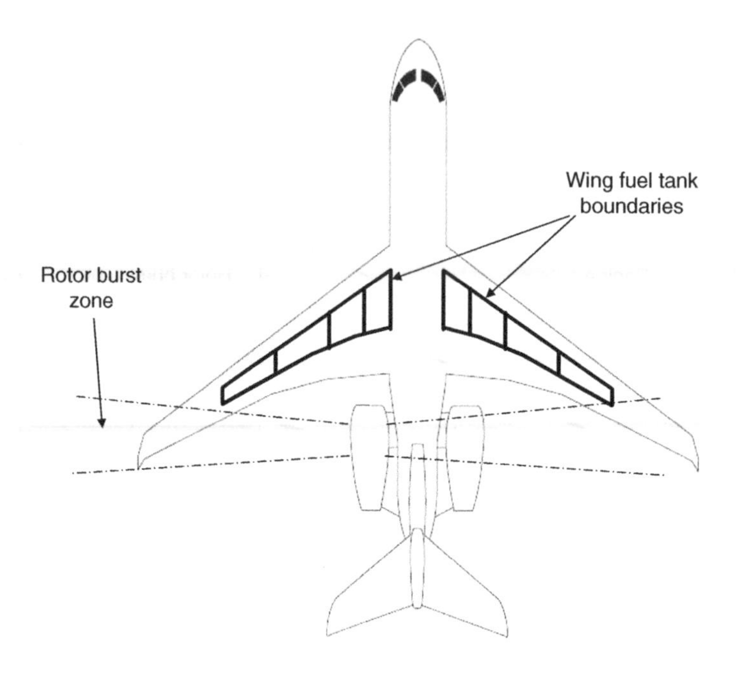

Figure 2.5 Rotor burst example for a rear-mounted engine application.

The possibility of the initiation of a fuel fire is also a consideration regarding fuel tank location that should be addressed, for example tank boundaries should remain well away from potentially hot sections near the engine mounting. To accomplish this it may be desirable to introduce a dry bay area around the engine mount region.

While under-wing engine installations are clearly the most preferable from a fuel system design perspective the close proximity to the fuel tanks presents a significant rotor burst failure concern as shown above. With wing-mounted engines there is already a positive fuel head between the fuel tanks and the engine inlet which helps to ensure a positive fuel feed head at the engine interface even with feed pumps inoperative. Rear-mounted engines, however, present a much more challenging situation to the aircraft fuel system designer. In this situation, feed lines are typically much longer and hence line losses are more substantial. The pitch-down situation is now much more challenging from a feed system perspective since the feed tanks (and hence the feed/boost pumps) are now below the engines and this additional head plus line losses must be provided by the feed/boost pumps during the worst case operating condition.

The influence of the large quantities of fuel on aircraft CG location also plays a key role in aircraft stability, balance and handling and the implications that these parameters have on trim drag and hence operating range may be significant.

In-flight stability is an issue that must be considered with regard to the location and CG of the stored fuel. This point is illustrated by Figure 2.6 which shows that for longitudinal static stability, the aircraft longitudinal CG must be located forward of the aircraft center of lift.

Referring to the stable aircraft in the figure, in steady level flight, a pitch-up trim moment is required (from the horizontal stabilizer) to counteract the pitch down moment due to the forward CG. If now the aircraft experiences a pitch-up disturbance, the increased lift generated

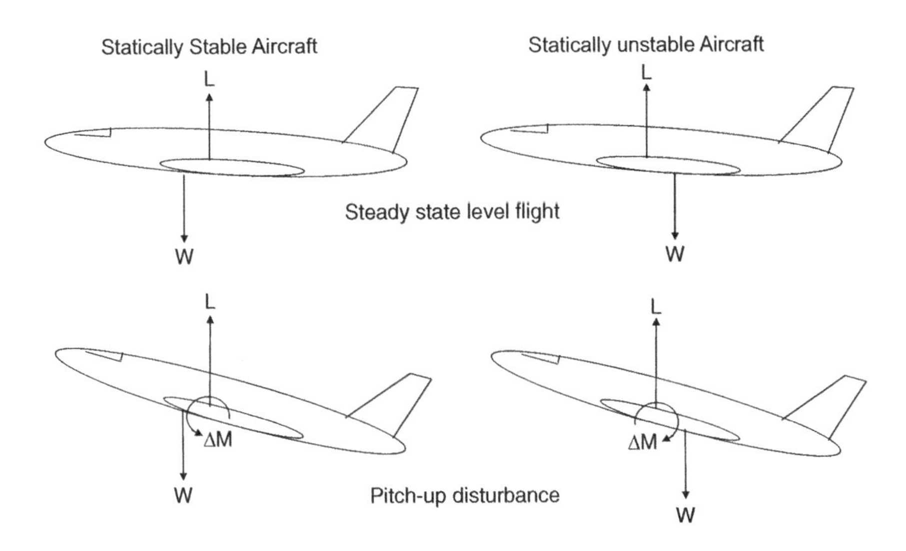

Figure 2.6 Definition of aircraft static stability.

by the wing, due to the increased angle of attack, produces a restoring pitching moment ($\Delta \boldsymbol{M}$) to the aircraft.

The opposite is true for the statically unstable aircraft. Here for steady level flight a pitch-down trim is required in steady state level flight because the weight acts behind the lift force. Now if a pitch-up disturbance takes place the increase in lift from the wing tends to increase pitch attitude further which is clearly destabilizing.

It is important therefore, to ensure that the fuel system must not have potential failure modes that would cause an unstable aircraft CG location. This is important in many of today's aircraft designs with aft located fuel tanks for minimization of trim drag in cruise. In these applications trapped fuel in the aft-most tank due to equipment failure, icing or any other potential cause, must not result in a longitudinally unstable aircraft.

One notable aircraft application with special longitudinal stability issues is the Concorde where supersonic flight results in large aft shifts in the aircraft center of lift requiring an equivalent movement in aircraft CG to optimize trim drag during the high Mach number cruise phase. This is accomplished by moving fuel aft during transonic acceleration and reversing the process during deceleration at the end of cruise. In this application, the need to transfer fuel at the end of cruise is critical to ensure that the aircraft is stable during subsonic flight. As a result the Concorde fuel management system is required to have the functional integrity equivalent to that of the flight control system. See Chapter 12 for a description of the Concorde fuel system.

From a lateral stability point of view, the ability of the lateral flight controls to offset any fuel load induced rolling moment due to a lateral CG shift must be understood at the outset. This issue must also cover the complete flight envelope and all aircraft configurations. If roll control power is not sufficient to support the worst case lateral CG shift then a flight-critical fuel lateral balance system must be provided. It is therefore the responsibility of the fuel system design team to ensure that critical design issues such as these are recognized by the aircraft design team at the earliest stages in the design.

2.1.4 Measurement and Management System Functional Requirements

As transport aircraft grow in size and operational range, the accuracy of the fuel gauging system becomes more critical since even small errors in gauged quantity can represent a significant payload penalty. Airworthiness requirements also demand that worst case gauging errors in addition to any unusable or ungaugable fuel be used in the determination of the fuel load for a particular flight.

The availability of inertial reference information may also be a critical factor in the design and complexity of the fuel gauging system. This system is often used to provide fuel surface attitude information; however, it is often unavailable on the ground during the refuel process where gauging accuracy is most critical. Fortunately, pitch and roll attitudes during ground refueling operations is usually limited to ± 2 degrees or so and therefore the number of fuel gauging probes required for fuel surface attitude calculation over the full quantity range is usually not excessive. In flight, however, significant variations in both pitch and roll fuel surface attitudes can take place and by using supplementary information from the inertial reference system significant savings in in-tank gauging equipment (probe-count) can be made. If the fuel system is to rely on the inertial reference system in flight then sufficient information system redundancy must be available.

In large long-range transport aircraft, fuel management system function can be complex and with the traditional two person flight deck crew that is typical of today's modern aircraft, it becomes crucial that crew workload be minimized so that they can concentrate on flying the aircraft.

In order to provide a level of functional integrity that ensures a fully automated fuel management function that meets flight critical performance would demand redundancy levels similar to a fly-by-wire flight control with attendant cost and complexity.

A more practical solution typically employed is to provide a lower level of functional integrity for the automated function while providing a manual back-up capability requiring crew intervention in the event of loss of function. This approach is acceptable provided that the integrity of the automated system is adequate. An industry standard for this approach is to have the automated fuel management function have a probability of loss of function of less than once per ten million flight hours (i.e. one failure per 10^7 flight hours).

2.1.5 Electrical Power Management Architecture and Capacity

The availability of electrical power to engine boost pumps can be a critical issue particularly when operating a very high altitudes when loss of boost pressure could result in engine flame out. This situation may dictate the level of redundancy required for the electrical generation system and the need to communicate power system status to the fuel management system so that auxiliary dc pumps can be powered on following the loss of ac power redundancy.

Another key issue related to the electrical power generation system is the type of ac power provided. Today there are three major types of ac power generation as indicated on Figure 2.7, namely reference [4]:

- the Integrated Drive Generator (IDG)
- the Variable Speed Constant Frequency (VSCF) system
- the variable frequency generation system.

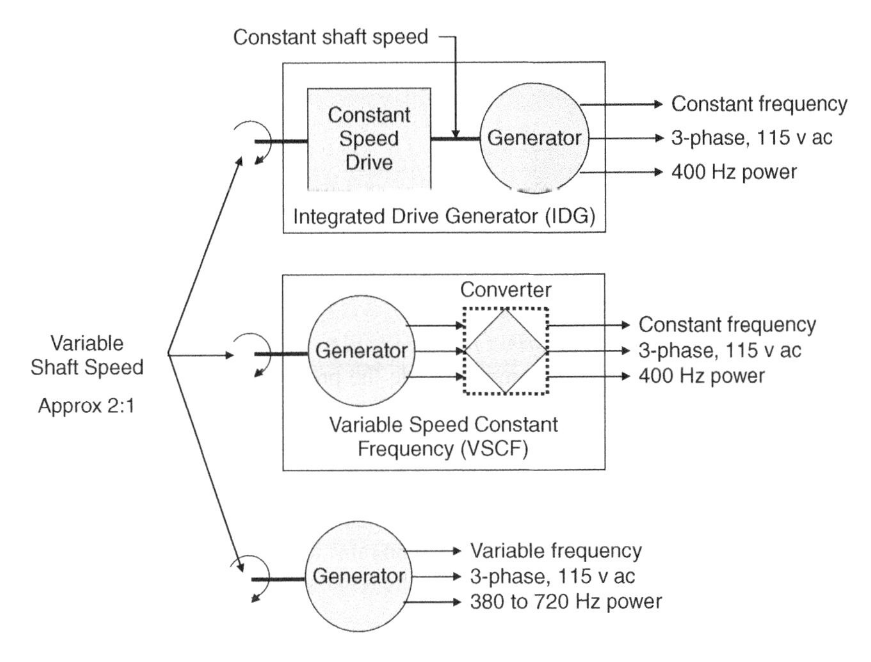

Figure 2.7 Power generation systems.

The first two systems provide the aircraft with constant frequency ac power (three phase 115 v ac, 400 Hertz) while the third system delivers three phase 115 v ac at a frequency dependent upon the rotational speed of the engine.

Variable frequency ac power is becoming more common today with most new aircraft adopting this standard. The main issue with the fuel system relates to the ac motor-driven pumps. Traditional induction-motors must be designed to accommodate the frequency variation resulting in a significant speed variation between minimum and maximum frequency. Therefore, both transfer and boost pumps must be sized to meet minimum flow and pressure at the lowest available ac frequency (and hence the lowest rotational speed).

Two significant issues result from this situation:

1. Boost pumps are typically oversized for the cruise flight condition where engine speeds are high and hence pump rotational speeds are high.
2. Power factors are much lower than for constant frequency power systems.

Both of the above features impose a weight penalty. Pumps become heavier to accommodate the specified performance at low frequencies and more current is required to deliver the required power.

2.2 Identification and Mitigation of Safety Risks

The initial concept phase must also focus on the identification, prioritization and mitigation of system safety risks. Mitigation of these risks is typically required to be addressed in the certification documentation.

Summarized below is a list of fuel system risks that are typically identified from which a certification methodology must be identified.

2.2.1 Fuel System Risks

The following risks need consideration and the establishment of mitigation strategies in fuel system design:

Identification of all potential thermal risks that can lead to the development of smoke and/or fire: A mitigation strategy for this risk may include the provision of adequate ventilation in areas that may accumulate fumes. Materials and equipment located adjacent to the fuel tanks should also be considered.

Uncontained rotor burst associated with APU and engines: This issue has been addressed in some detail primarily from considerations of tank boundaries as well as location and routing or key fuel lines and electrical harnesses. Consideration must also be taken regarding the impact of fuel loss as it affects the availability of fuel for the remaining engines as well as the impact on aircraft weight and balance.

The APU rotor burst may be less of a risk; however, in aircraft with fuel storage in the horizontal stabilizer the implications of tank penetration and fuel loss should be evaluated and an appropriate mitigation strategy developed that will support certification.

The effects of a fan blade out failure: This event will result in sustained engine imbalance with the potential for high vibration levels. Appropriate steps must be taken to accommodate this situation within the equipment installation design.

Flailing shaft: This is a similar event to the fan blade out failure and should be accommodated in a similar manner.

Ram air turbine rotor burst: Even though the likelihood of this event is extremely remote, it must be addressed through prudent location of equipment to circumvent potential damage from such an event.

Hydraulic accumulator burst: Consideration should be made regarding location of equipment, piping and power lines for potential damage from this event.

Wheel and tire failure: The main gear is typically located in close proximity to wing fuel tanks and therefore it is important to consider the potential damage to local structure and equipment due to the high energies associated with this failure.

Fuel ignition risk: Fuel ignition risk is addressed in detail in Chapter 9 'Intrinsic safety, lightning, EMI and HIRF' and Chapter 10 'Fuel tank Inerting', however it is important in the early stages of fuel system design to consider the potential for fuel ignition resulting from high current and/or stored energies associated with in-tank equipment as well as the installation of electrical harnesses in or adjacent to the fuel tanks. Bleed air duct rupture is also an important potential cause of fuel ignition if high temperature air is allowed to impinge of fuel tank surfaces and mitigation of this risk should be considered early in the aircraft design phase.

Crashworthiness: Consideration should be made of the installation security of all major equipment including the extension-ability of pipe installations. Frictional heating of fuel tank surfaces due to rubbing contact with the ground should also be considered and evaluated.

Tank overpressure: This can occur due to refuel and/or transfer system failures and appropriate protection against this occurrence should be included within the fuel system design in order to provide acceptable mitigation of the risk. Over pressure conditions can also result from restricted venting area and icing of flame arrestors.

Rapid decompression and aft pressure bulkhead rupture: Consideration of potential damage to fuel lines and electrical harnesses routed through the fuselage following rapid decompression and any resulting structural; distortion should be considered.

There are a number of other potential risks that must be addressed in the formal application for type certification that must be addressed and are listed below for completeness:

- nose wheel imbalance
- bird strike
- ditching
- weather – hail rain ice and snow
- tail strike
- collision-ground and air.

As in all aircraft designs, the first priority must always be safety. The comments and recommendations presented in this chapter serve to emphasize the importance of designing for safety and functional integrity at the outset.

Also the discipline of risk identification and mitigation imposed by the airworthiness authorities is designed to ensure that acceptable levels of operational safety will be achieved in the final aircraft configuration.

3

Fuel Storage

Storing large quantities of fuel aboard an aircraft is a challenge to the aircraft designer and while fuel storage is not in itself a 'Function' in the traditional sense it nevertheless imposes a number of functional constraints and requirements on the design and certification of both the fuel system and the aircraft as a whole.

Referring to Figure 3.1 it can be seen that the fuel storage aspect of the fuel system design is the second logical step in the process (the first being the operational and safety design drivers covered in the previous chapter). In this chapter the fuel storage requirements to be addressed involve the definition of the fuel volume required which in turn results from the primary aircraft mission requirements which include, for example range, speed and payload targets. These physical constraints must be accommodated in the aircraft design taking into account how this volume and mass and its location can meet critical weight and balance limitations.

It must be recognized at this point that the design process is not as straightforward as indicated by the figure but becomes an iterative process with each iteration cycle following the logical flow shown in the figure.

A major factor in the design of the fuel storage system is associated with management of the ullage above the fuel. For this reason, the system-level issues associated with fuel tank venting are covered in this chapter.

Two types of fuel tanks used in today's aircraft will be addressed in this chapter, namely 'Integral tanks' and 'Bladder tanks', together with the structural issues associated with penetrations for equipment installation and access. The impact on the aircraft and fuel tank design of tank location, tank sealing, fuel slosh, plus the effects of bending and twisting of the tank structure will also be addressed here.

A separate subsection is devoted to the fuel storage issues associated with military aircraft applications. This will include a discussion of the issues associated with closed vent systems, drop tanks and conformal tanks.

Finally aircraft operator maintenance issues associated with water contamination management are discussed. Micro-biological growth in fuel tanks is addressed in Chapter 8 on Fuel Properties.

Aircraft Fuel Systems R. Langton, C. Clark, M. Hewitt, L. Richards

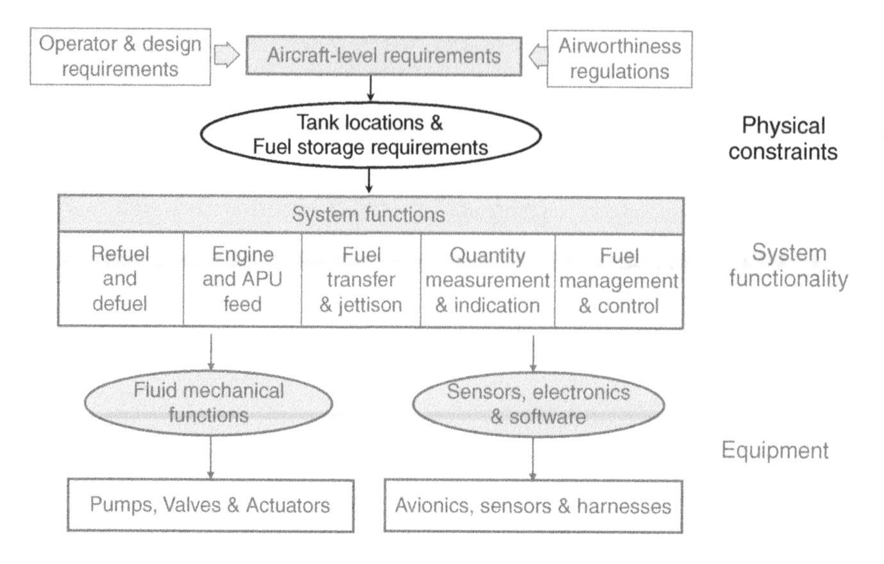

Figure 3.1 Fuel system design process overview.

3.1 Tank Geometry and Location Issues for Commercial Aircraft

The most common location used for fuel storage is within the wing structure typically between the forward and aft wing spars. This space is not readily usable for other aircraft functions due to geometry and access considerations.

From an aircraft and fuel system perspective, there are advantages and disadvantages associated with the use of wing structure for fuel storage. On the plus side, the considerable fuel weight which acts in opposition to the wing lift force results in a lower wing bending moment during flight than would be the case if the same mass of fuel were stored in the fuselage. In aluminum wing structures, this can provide a significant positive contribution to the fatigue life of the aircraft structure by reducing the magnitude of the stress cycle associated with each take-off and landing. In some applications, fuel is kept in outboard tanks until the end of the cruise phase before transferring the fuel contents to the engine feed tanks in order to maximize the fatigue life of the wing structure.

On the negative side of the equation, since the wing internal cavity is typically very thin (especially towards the wing tip) the challenge to the fuel quantity gauging system designer is considerable since a large number of measurement probes are required to provide adequate coverage over the full range of fuel quantities and aircraft attitudes for accurate computation of fuel quantity. This is further complicated by the bending and twisting of the wing structure due to aerodynamic loading.

Wing bending and twisting together with the presence of internal rib structures provide a challenge to not only the gauging system designer but also to the vent system designer since trapped air pockets within the fuel tank can seriously impair the performance of the engine feed pumps and could result in an engine shut-down through starvation of the boost pump fuel inlet if the ullage is not allowed to 'breathe' as the aircraft climbs and descends and as fuel is consumed. Vent system design issues are addressed in detail later in this chapter.

Since the wing tank structure itself bounds the fuel storage area and this type of tank is called an 'Integral tank'. A major issue with integral tanks is the provision of a reliable, flexible sealing method to ensure the integrity of the tank structure throughout the operational flight envelope. Fasteners associated with the wing structural assembly and penetrations for equipment and for access must be reliably sealed. Sealant materials must be compatible with the fuels to be used.

Figure 3.2 is a photograph of the inside of an integral fuel tank for a large commercial transport aircraft. The photograph shows internal equipment, specifically a spar-mounted transfer pump and piping associated with the fluid mechanical function of the system.

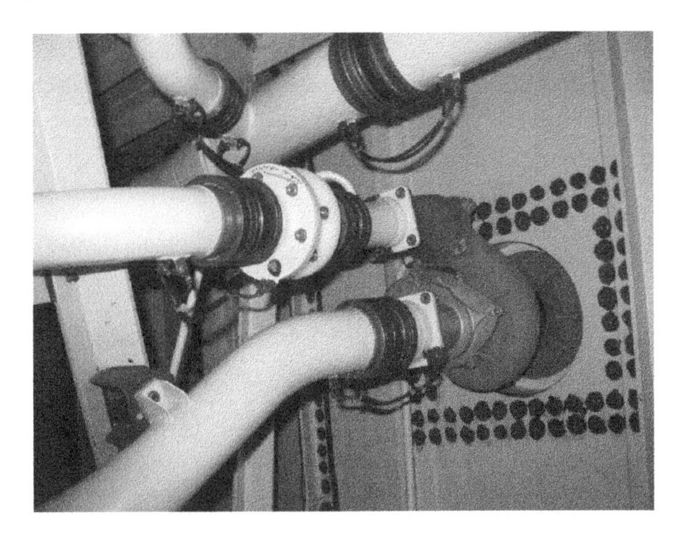

Figure 3.2 Photograph of a typical integral wing tank (courtesy of Airbus).

Also shown in the photograph are the bonding straps across all of the pipe connectors which are critical features that are necessary to prevent arcing resulting from the large currents that can flow through piping following a lightning strike. This is discussed in more detail in Chapter 9. Ribs and stringers within the integral tank structure must allow fuel and air migration within the fuel tank boundary and inertia loads resulting from fuel slosh during aircraft acceleration must be accommodated by this structure.

An even more critical requirement for fuel tank structure is the accommodation of fuel inertia-induced forces during worst case hard landings. This requirement is particularly critical in helicopter fuel tank design where the accommodation of an auto-rotation landing can impose very large g loads upon impact and therefore the structural design of the fuel tanks must ensure that fuel leakage, which would bring with it an obvious fire risk, is prevented. The structural designer must therefore accommodate the specified maximum g forces in each of the three orthogonal axes taking into account the maximum mass of fuel within each tank or compartment.

Fuel tanks can also be provided within the fuselage and are typically confined to the following locations:

- the section between the wings below the passenger compartment
- the empennage behind the pressure bulkhead.

Here the tank shape is typically rectilinear and therefore presents a much simpler challenge to the fuel quantity gauging system designer with regard to the number and location of tank probes than for wing tanks.

In special applications where extra long range is required, as in some business jet conversions, additional tanks may be installed within the cargo space.

In many Airbus transports provision is made within the baseline design to install one or more Additional Center Tanks (ACT's) within the cargo space. Figure 3.3 shows a simplified schematic of an Airbus ACT installation.

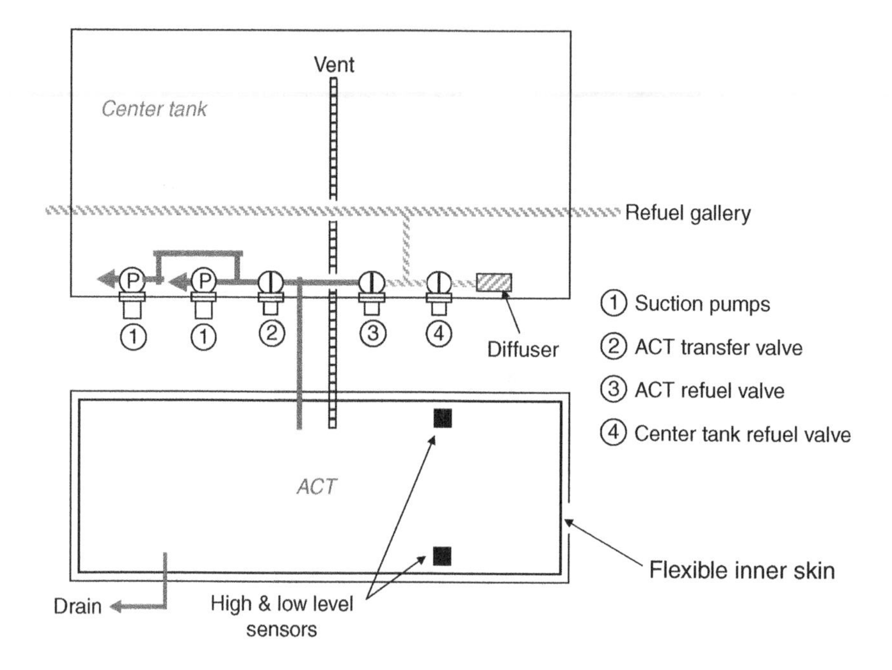

Figure 3.3 Airbus ACT functional schematic.

This is a bladder tank assembly with provision for easy installation of plumbing and electrical connections to integrate into the aircraft the refuel, defuel, transfer and quantity indication functions for the added tank. In operation, the bladder fuel contents are transferred into the center tank using redundant suction pumps mounted in the center tank. An alternative transfer means where cabin air pressure provides the primary transfer means is also used in some applications

Fuel can also be stored inside the horizontal stabilizer. The MD-11 and all Airbus long-range aircraft use the horizontal stabilizer for fuel storage. The -400 version of the Boeing 747 has additional fuel storage in the base of the vertical fin.

Airbus uses aft fuel storage to provide active CG control by transferring fuel between the forward tanks (wing and/or center tanks) and the horizontal stabilizer tank (referred to as the trim tank) as fuel is consumed, in order to maintain an optimum longitudinal aircraft CG throughout the cruise segment of the flight.

For reference purposes, the following two figures show plan-forms of several in- service commercial aircraft showing the different fuel tank storage configurations. Figure 3.4 shows two large transports, specifically the Boeing 767-200 and the Airbus A340-500.

The Boeing aircraft shown in the figure is a medium-range wide bodied aircraft which entered service in the early 1980s. This aircraft continues to be used extensively on intercontinental routes in the USA. In this early version, the center tank consists of 'Cheek tanks' formed at the wing root which are interconnected via large diameter tubes between each wing tank. This allows these tanks to be utilized as a single center tank.

Longer range versions of the Boeing 767 which include the 767-300, 767-300ER and 767-400 versions use the additional center tank dry bay space to add fuel capacity.

The Airbus A340-500 also shown in this figure was introduced into revenue service in the 2002 as a long-range four-engined transport aimed at fulfilling specific long-range routes including Singapore-to-Los Angeles.

This aircraft has a much more complex fuel storage arrangement, each wing having two feed tanks (one per engine) and an outer wing tank. The center tank is the largest of the auxiliary tanks. An aft center tank provides additional fuel to support the long-range mission goals while keeping a more aft longitudinal CG location. Finally the horizontal stabilizer tank is used by the fuel management system to optimize longitudinal CG and hence trim drag during the aircraft cruise phase.

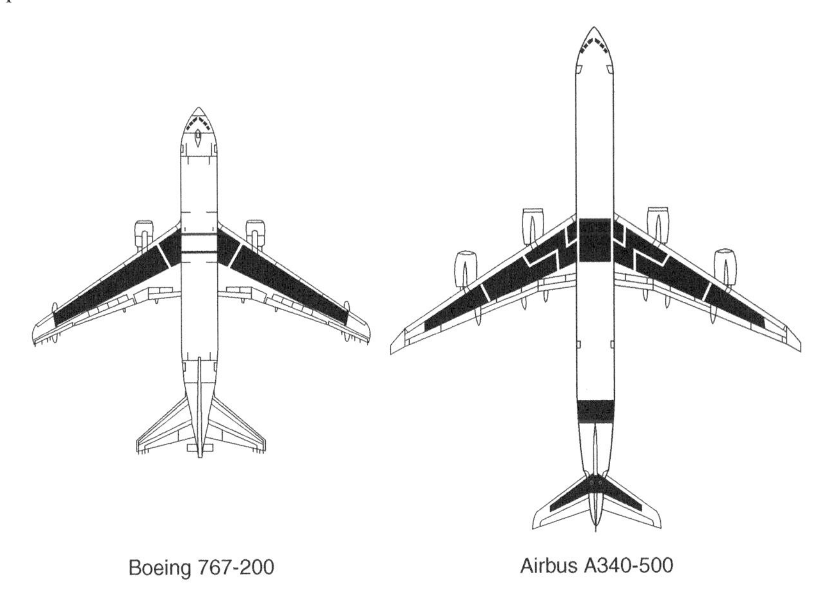

Figure 3.4 Fuel tank arrangements for large transport aircraft.

Figure 3.5 shows plan forms of two regional jets showing the fuel tank configurations. These aircraft are not shown to scale relative to the previous figure however; they are shown in relative scale to each other. For perspective, the wingspan of the Airbus A340-500 is approximately 208 ft versus the Embraer 190 which is slightly more than 94 ft.

The regional aircraft have much simpler missions and hence much simpler fuel storage arrangements. The Embraer configuration is a continuation of their two tank design used on

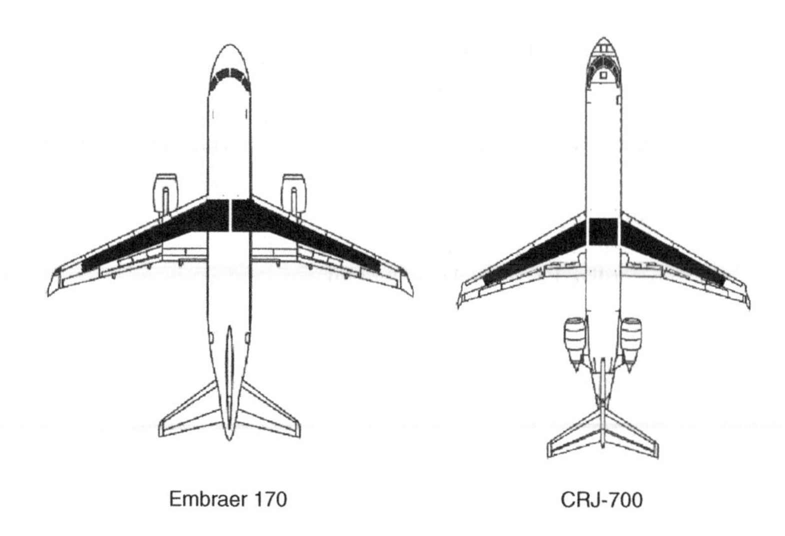

Figure 3.5 Fuel tank arrangements for regional aircraft.

the previous ERJ 145 series aircraft where the wing tanks continue below the cabin joining at the aircraft centerline.

This approach simplifies the fuel management requirements and in some designer opinions, may offer some structural weight benefits over the 'Wing-center-wing' fuel tank arrangement adopted by the Bombardier CRJ-700 example shown.

3.2 Operational Considerations

The uncontained rotor-burst issues affecting fuel storage are addressed in Chapter 2 since they are major safety and system design drivers. The following operational issues relate to normal aircraft operation.

3.2.1 CG Shift due to Fuel Storage

An important issue associated with fuel storage tanks and particularly wing tanks is concerned with the CG shift that can occur as a result of fuel migration during changes in pitch attitude. Fuel movement associated with lateral accelerations tends to be less critical because of their transient nature.

In a swept wing aircraft fuel will migrate aft during the climb phase due to the positive pitch attitude thus moving the longitudinal aircraft CG aft and reducing the aircraft static stability margin. This effect can be reduced by dividing the wing tank into a number of semi-sealed compartments that allow easy fuel migration inboard while minimizing fuel migration outboard.

To facilitate inboard fuel movement, a number of check valves (non-return valves) are installed along the ribs that form the boundaries between compartments. These check valves are installed using a flexible material allowing the fuel to move easily inboard while preventing fuel movement outboard (see Figure 3.6). These check valve devices are often referred to as 'Baffle check valves', 'Flapper-check valves' and sometimes as 'Clack valves'.

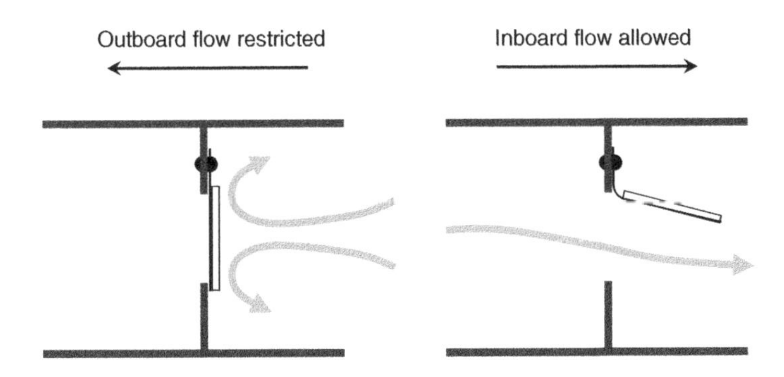

Figure 3.6 Illustration of baffle check-valve function.

In addition to the baffle check valves, small bleed holes located at the upper and lower tank rib boundaries of each compartment allow fuel and air to migrate between compartments hence the term 'Semi sealed'. Since these bleed holes are small, outboard fuel migration from this source can be essentially neglected for the short duration of the climb phase which is typically less than thirty minutes.

To further reinforce this concept consider a twin engine, swept wing aircraft example which has integral wing tanks designed to contain most of the fuel on board as indicated by Figure 3.7. Let us assume that this aircraft has a low wing design with significant, say five degrees, of dihedral. With this configuration, fuel would migrate naturally inboard during level flight.

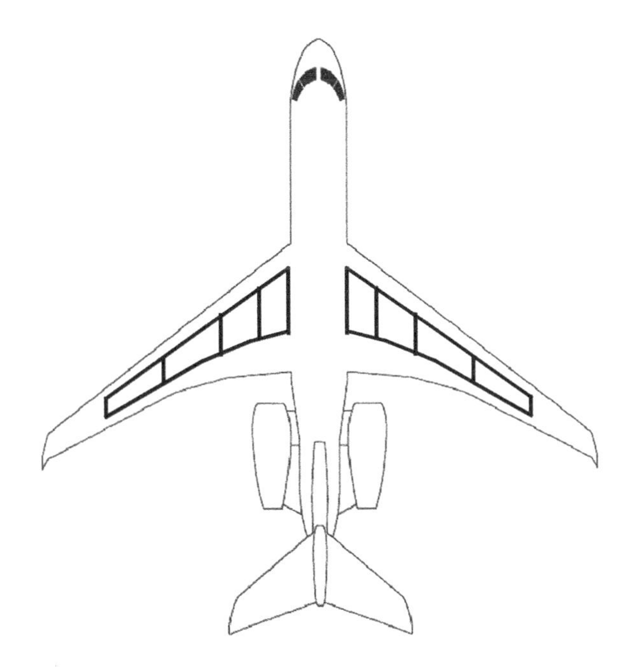

Figure 3.7 Swept wing transport aircraft wing tank arrangement.

As shown in the figure, the wing tanks are located between the forward and aft main spars and between the wing root and an outboard rib close to the wing tip. As previously discussed, these tanks can be considered as a single compartment per wing or as a number of separate compartments separated by semi-sealed ribs.

Figure 3.8 compares the single compartment wing tank and four compartment configurations in terms of fuel contents location during climb and dive maneuvers.

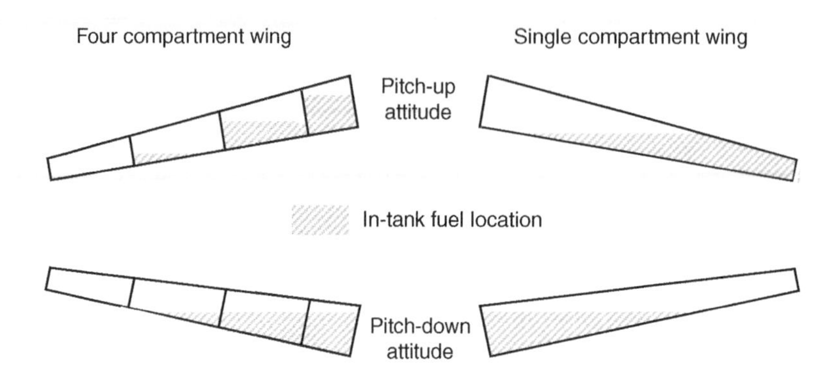

Figure 3.8 Comparison of single and four compartment configurations.

In the four-compartment version, baffle check valves located at each compartment boundary limit outboard fuel migration thus minimizing the aft-ward movement of aircraft longitudinal CG. The single compartment approach on the other hand allows all of the fuel within the tank to move aft during a climb creating a large aft shift in aircraft CG.

It is fairly intuitive that the single compartment solution is unacceptable and that a multiple compartment solution is necessary to ensure acceptable levels of CG shift during pitch angle changes. With the multiple compartment approach, the fuel boost pumps supplying fuel to the engines can be safely located in the inboard compartment of each wing. Under level flight or pitch down conditions fuel will migrate naturally towards the wing root under the action of gravity because of the wing dihedral. During a climb, fuel is held in the inboard compartment under the action of the baffle check valves thus maintaining coverage of the engine boost pumps.

One disadvantage of the four-compartment solution worthy of note is that each compartment must now be gauged as a separate tank each with its own fuel surface. This will require more gauging probes (tank units) than for the single-compartment design with a single fuel surface.

We can calculate the change in aircraft longitudinal CG as the aircraft climbs and descends for different tank compartment configurations and different fuel tank quantities as part of the fuel storage design trade study.

Figure 3.9 indicates qualitatively how longitudinal CG would vary with tank fuel quantity for a single compartment wing and a four compartment wing. In the four compartment application, baffle check valves are installed at the compartment rib boundaries to prevent outboard fuel migration during pitch-up attitudes. As indicated in the figure, the single compartment wing shows a much larger excursion in aircraft CG than the four compartment version.

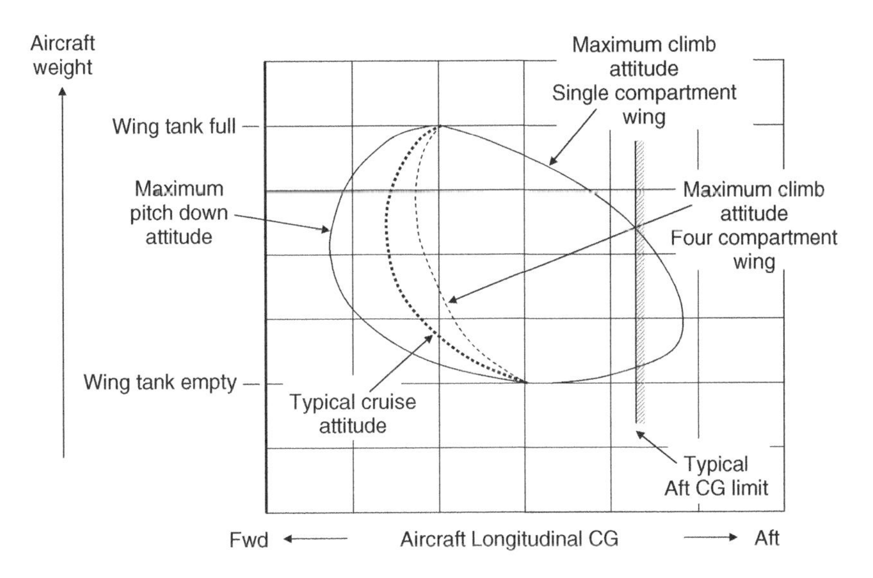

Figure 3.9 Climb and descent CG shift plots.

As indicated in the figure, cruise and pitch-down attitudes are common to single and multiple-compartment arrangements; however, the single compartment wing shows a large aft CG excursion which is likely to exceed the aft CG limit for positive aircraft static stability.

Thus the introduction of a multi-compartment wing has a favorable impact on longitudinal CG by keeping the fuel inboard via the action of the inter-compartment baffle check valves.

In today's CAD environment, it is relatively easy to develop a computer program which, when using the tank structural geometry CAD database, can calculate the fuel CG for combinations of tank quantity and aircraft attitude in order to obtain a design-specific plot similar to Figure 3.9 above.

3.2.2 Unusable Fuel

Since aircraft store large amounts of fuel in large tanks, even a fraction of an inch spread over the bottom of a large fuel tank if inaccessible to the fuel pumps feeding the engines can represent a significant amount of unusable fuel. Unusable fuel represents a direct weight penalty to the aircraft since credit for this quantity cannot be used in calculating the fuel required to perform the mission.

Center tanks can present a particularly challenging problem because these tanks typically have a flat bottom with a large area plan form. Wing tanks on the other hand typically have a significant amount of dihedral causing the fuel to accumulate in the inboard section during normal flight. Location of the feed pumps in the inboard section of the wing tank is common practice therefore in order to minimize the amount of unusable fuel.

Consideration must be made, however, for variations in fuel surface attitude due to aircraft attitude and acceleration forces. Aircraft pitch angle is the most critical issue here because both positive and negative sustained pitch angles can be applied during climb and descent

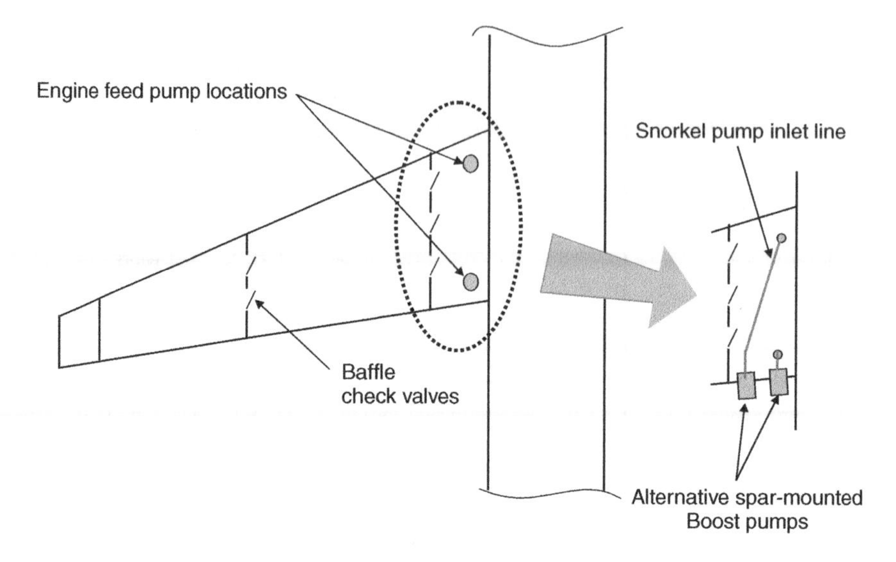

Figure 3.10 Feed pump location concepts.

respectively. Roll maneuvers affect the fuel surface only transiently since turns are usually coordinated.

In wing tanks therefore, unusable fuel can be minimized by using two fuel feed pumps per tank with one pump at an inboard-forward location and the other inboard-aft as shown diagrammatically in Figure 3.10. This design technique has been used successfully in many of today's commercial transport aircraft. From an installation point of view, the pumps may be installed remotely and 'snorkel' inlet lines used to pick up the fuel from the tank fore and aft corners. For example, the fuel feed pumps may be spar mounted on the rare spar of the feed tank as indicated in Figure 3.10. Here the advantage is that there are no lower skin penetrations requiring fairings to accommodate access to pump motor wiring although there will be a loss of suction pressure performance, which could be significant during high altitude performance.

With this arrangement, the aft pump will support the climb phase of flight as fuel migrates towards the rear of the tank and the forward pump will be covered during descent.

3.2.2.1 Collector Cells

A design approach also in common use as one means to minimize unusable fuel is to use the collector cell principle. Here a small section of the engine feed tank is assigned as a collector cell. This is a separate compartment within the feed tank which is kept full at all times during flight using scavenge pumps to continually top up the collector cell from the main feed tank section. Figure 3.11 shows this concept schematically. Scavenge pumps are ejector devices that use the high pressure discharge from the main feed pumps to 'Suck' fuel from the bottom of the tank.

The ejector pump concept is illustrated in Figure 3.12. Ejector pumps, also known as jet pumps, are low efficiency devices however they are extremely reliable since they have no moving parts. The only realistic failure mode is as a result of blockage of the discharge nozzle

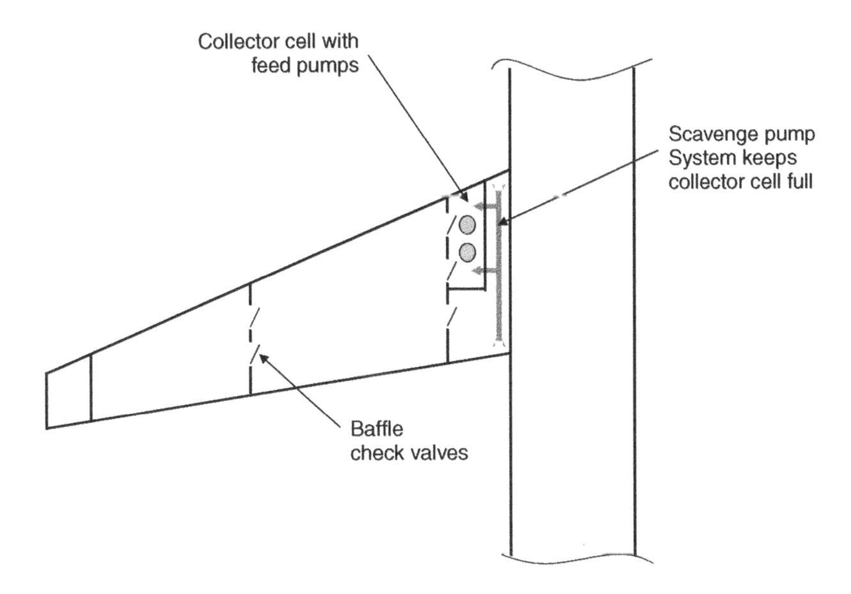

Figure 3.11 Collector cell example.

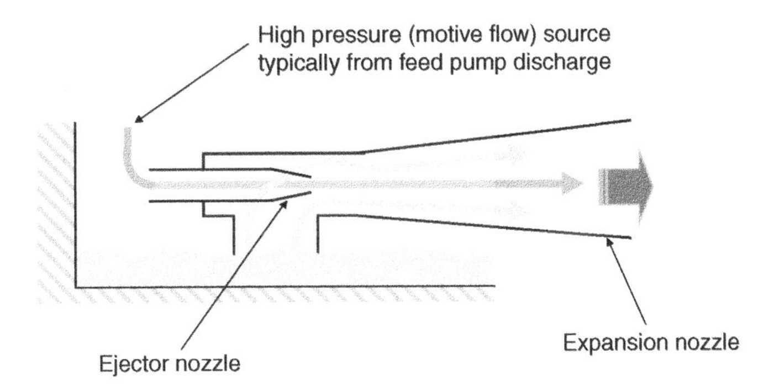

Figure 3.12 Ejector (scavenge pump) concept.

due to contamination or ice since the nozzle diameters can be small and are often required to be protected by screens.

The design specifics of the ejector pump will be addressed in detail in Chapter 6.

3.3 Fuel Tank Venting

Commercial aircraft use what is termed an 'Open vent system' to connect the ullage above the fuel in each tank to the outside air. The provision of adequate fuel tank venting throughout the aircraft operational flight envelope is critical in that it allows the tanks to 'Breathe' as the aircraft climbs and descends. Without this provision large pressure differences would develop between the ullage and outside air resulting in very large forces on the tank structure. It is impractical to accommodate these forces via the wing structural design because of the resultant weight penalty; therefore the design of the vent system plays a critical role in protecting the tank structure from structural failure as the aircraft transitions between ground and cruise altitudes.

During the refuel process, the uplifted fuel displaces air in the fuel tanks. For safety reasons, spillage of fuel to the outside must be avoided. To accomplish this consistently and reliably, a vent box (sometimes referred to as a surge tank) is provided to capture any fuel that may enter the vent lines.

For most commercial aircraft the vent box is located towards the wing tip. For a low-wing aircraft design which is typical of most commercial transports, the wing will have a significant dihedral so that the vent box conveniently becomes the high point during flight. However, even aircraft with anhedral the vent box often located outboard. Figure 3.13 shows two different aircraft configurations, one with a low wing installation with a significant amount of dihedral typical of most commercial transports and the other with a high wing installation with anhedral which is more typical of military transport aircraft.

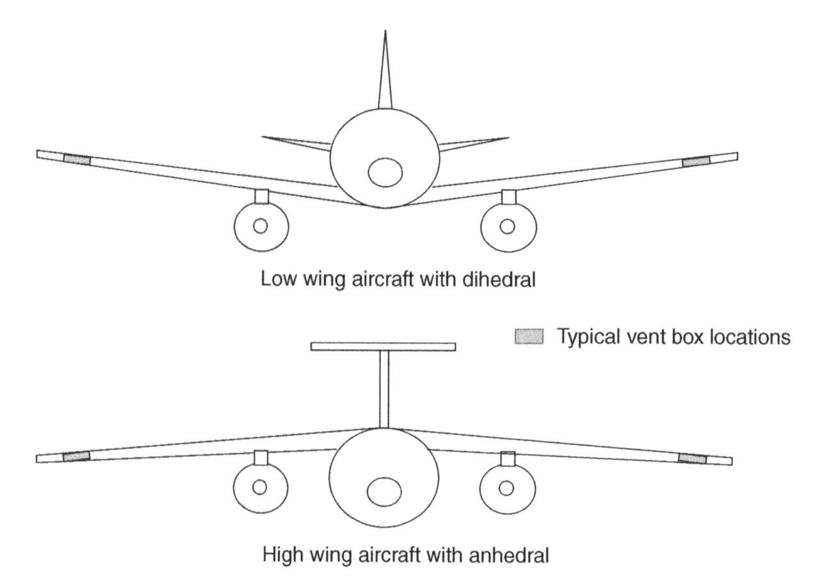

Figure 3.13 Vent box locations in typical transport aircraft.

From the figure it can be seen that the vent box location in the high wing aircraft may not be the high point during flight and therefore the vent system will typically employ scavenge pumps to prevent any build-up of fuel in the vent box from the tank vent lines. The reason for keeping the vent box outboard is to locate the air scoop (the connection of the vent system to the outside air) well away from the fuselage in order to maximize the ram air recovery. This point is discussed in more detail later.

In spite of the dihedral, the low wing design must also deal with fuel accessing the vent box. For example, when the aircraft is taxiing on the ground, any fuel in the vent lines may be forced into the vent box by centrifugal force. The dihedral is also reduced (or eliminated) on the ground due to the weight of wing fuel plus the weight of wing-mounted engines. This is worse for large aircraft with large wing spans and typically the size of the vent box is based on the accommodation of a predetermined number of taxi turns assuming vent lines full of fuel.

For the low wing/dihedral design, there is usually no need for scavenge pumps in the vent box since any fuel that has been transferred into the vent box while on the ground can be arranged to drain back into the fuel tanks via a check valves after take-off.

The positioning of the vent openings within each fuel tank must account for the varying attitudes and acceleration forces that can occur both on the ground and in flight. Forward acceleration and pitch-up attitudes will force fuel aft and with swept wings fuel will be forced outboard. A vent must therefore be located at the wing root near the forward tank boundary.

Similarly forward deceleration and pitch down attitudes will force fuel forward and for swept wings inboard. To accommodate this situation a vent must be positioned outboard and aft in the fuel tank. Figure 3.14 shows a simple schematic of a vent system for a three tank aircraft

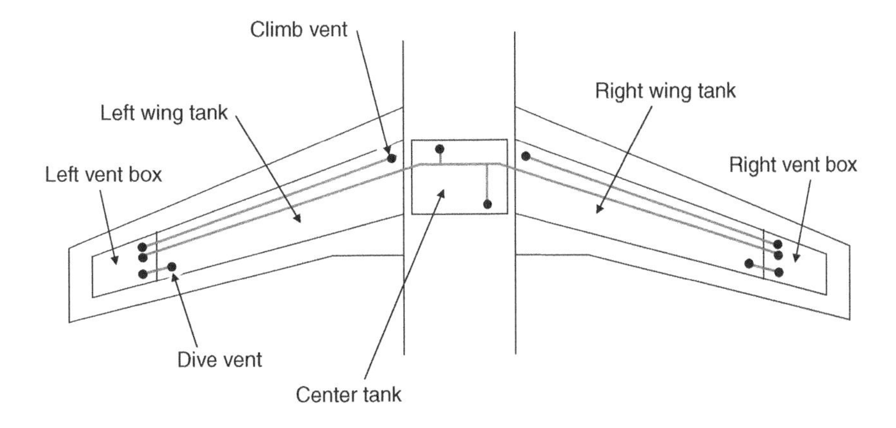

Figure 3.14 Simple three tank vent system schematic.

In the example shown, each tank is vented such that there is a breathing path to the outside air for all long-term flight attitudes that occur during normal flight operations. In order to prevent or minimize the opportunity for the vent lines to become filled with fuel, float actuated vent valves are often installed that close the vent lines when fuel is present.

This concept is illustrated by the schematic diagram of Figure 3.15 which shows a dihedral wing and how the vent line is shut off when there is fuel present via a float-actuated vent valve.

This figure shows the vent box and its connection to the outside air-stream. Also shown is the flame arrestor which prevents flame propagation into the vent system that could result from a direct lightning strike to the air inlet igniting fuel vapors escaping from the vent system. Flame arrestors must be designed to minimize the probability of icing-up since this would block the vent system. For this reason, in order to ensure adequate integrity of the vent system function, a secondary pressure relief provision is typically provided such as a burst disk or relief valve so that any single event causing a vent line blockage will not result in structural damage.

The outside air inlet to the vent box typically consists of a specially shaped scoop optimized by the National Advisory Committee for Aeronautics (NACA) in 1945. This air scoop provides the optimum combination of dynamic air pressure recovery and drag. Dynamic pressure recovery is important in improving boost pump performance margins particularly in hot and high operating conditions where vapor evolution from the fuel can impair pump performance.

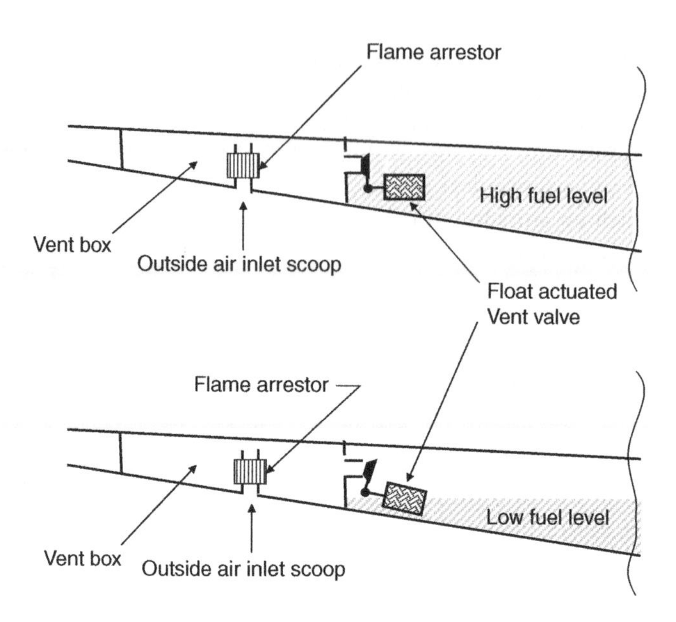

Figure 3.15 Float actuated vent valve schematic.

The forward speed of the aircraft is 'Recovered' in the form of increased temperature and pressure according to the following equations for adiabatic (constant energy) flow of an ideal fluid:

$$T_T = T_O\left(1 + \left(\frac{\gamma - 1}{2}\right)M^2\right)$$

and

$$P_T = P_O\left(1 + \left(\frac{\gamma - 1}{2}\right)M^2\right)^{\frac{\gamma}{(\gamma-1)}}$$

where T_T and P_T are total temperature and pressure respectively and
T_O and P_O are free stream temperature and pressure respectively
γ is the ratio of specific heats equal to 1.4 for air.

The NACA scoop (also referred to as a 'Submerged duct entrance') is shown in Figure 3.16 together with a graph of pressure recovery ratio for ideal fluid flow and the typical NACA scoop.

From the graph of Figure 3.16 it can be determined that an aircraft cruising at Mach 0.8 at 35,000 ft with a static air pressure of 3.46 psia will have a vent system ullage pressure of between 4.5 and 4.9 psia which represents a pressure recovery of between one and one and a half psi.

Figure 3.17 shows the wing and center tank vent system arrangement of the A340-600 aircraft illustrating the complexity required to provide the venting function of a modern long-range transport aircraft. In this case the wing tank comprises two inboard feed tanks and an outboard tank, each of which must be independently vented. The center tank has its own dedicated vent pipe and over pressure protectors which in this case are burst disks.

A contributor to vent system complexity is the flexibility of the wing structure. When on the ground the weight of fuel and wing mounted engines may move the high point in the ullage to some mid span position. This situation must be accommodated by the vent system for the

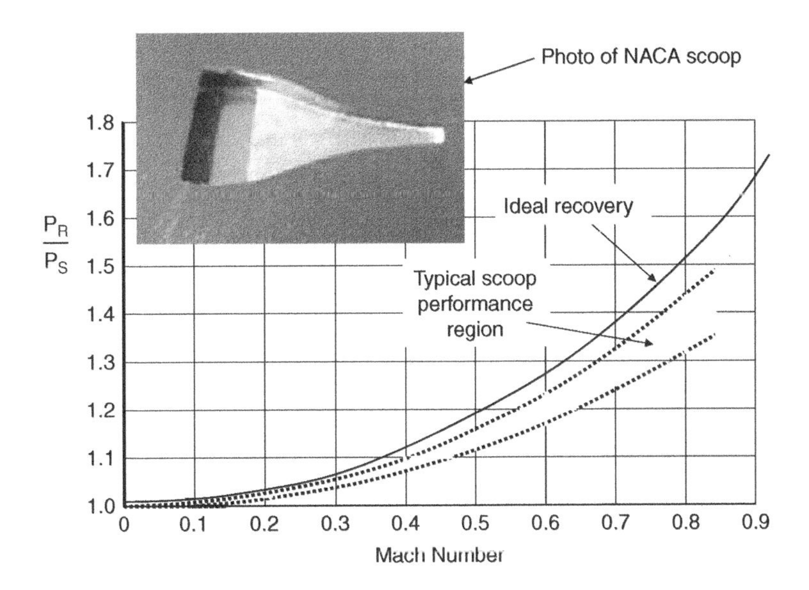

Figure 3.16 NACA scoop recovery pressure.

ground refueling case. Bubble studies are conducted as a routine to define the location of the ullage bubble for full or nearly full tanks for various ground and flight conditions.

Frequently, scale models of wing tanks are used to provide a qualitative evaluation of vent system operation for various attitudes and fuel tank quantities. These models contain representative rib structures and in some cases can be loaded to represent operational flight loads.

Figure 3.18 shows a quarter scale model of the Global Express™ wing and center tank. This model was used to evaluate vent system performance during the refuel process well before the real aircraft wing was available.

3.3.1 Vent System Sizing

The design requirements for sizing the vent system for commercial aircraft applications are usually defined by the emergency descent case. Here the aircraft has to descend as quickly as possible from maximum cruise altitude to below 10,000 ft following a loss of cabin pressurization. The worst case fuel state is for close to empty conditions when the ullage volume is near maximum and hence vent system inflow will be the highest. The design requirement typically specifies a pressure difference between the outside ambient air pressure and the ullage pressure must not be exceeded.

To satisfy this requirement, simulation techniques are used to provide evidence that the vent system design is compliant. These simulation studies together with model validation evidence are typically used to support the aircraft certification process.

3.4 Military Aircraft Fuel Storage Issues

Fuel storage in military aircraft, particularly fighter and attack aircraft, is typically much more complex than for commercial/civil aircraft. Fuel is stored in wing tanks as in their commercial counterparts; however, in high speed aircraft wings are small and thin with little

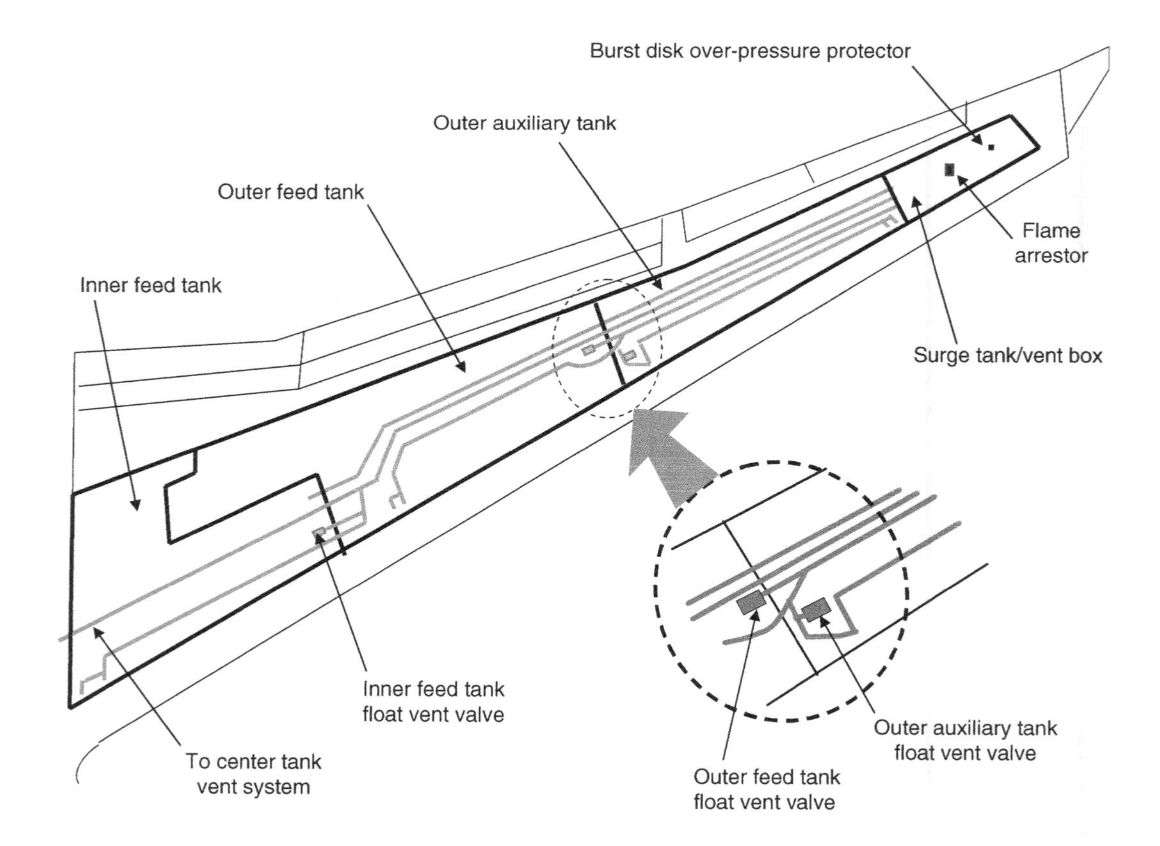

Figure 3.17 A340-600 wing vent arrangement.

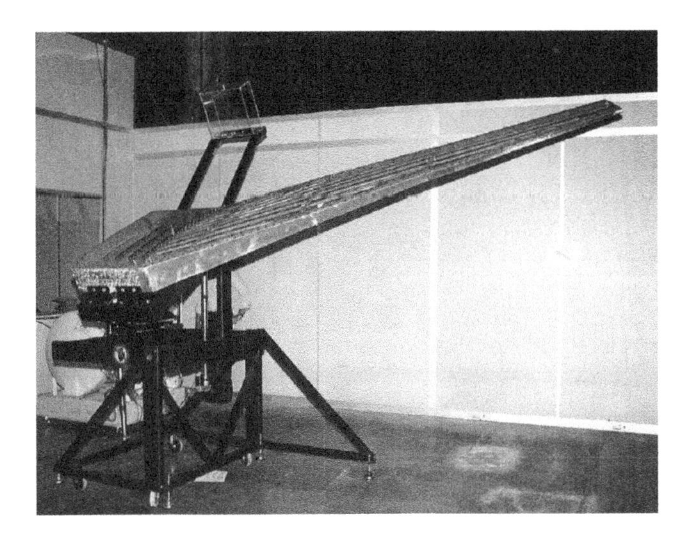

Figure 3.18 Global Express quarter scale wing tank model (courtesy of Parker Aerospace).

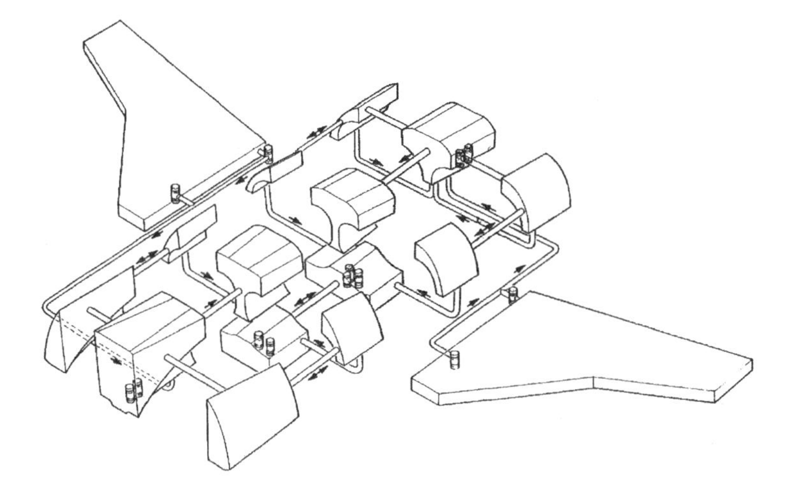

Figure 3.19 Fuel tank arrangement for a military fighter aircraft (courtesy of BAE Systems).

internal volume. The fuselage is therefore often used as a place to store fuel but since this location is filled with engines and weapon storage, the volume remaining for fuel storage often involves complex shapes as indicated in Figure 3.19 which shows the fuel tank arrangement for the Experimental Aircraft Programme (EAP) aircraft developed by BAE Systems as the fore-runner to the Eurofighter now known as the Typhoon.

It can be seen from the figure that the fuselage tanks of which there are 13 are wrapped around the two engines that are embedded within the fuselage. This situation clearly complicates the fluid network of piping, pumps and valves that support the refuel, transfer and feed functions of the fuel system. The complexity of the fuel quantity gauging system is also seriously impacted and the requirement for aerial refueling which is mandatory in all modern military aircraft adds an additional dimension of complexity to this already demanding design challenge.

Figure 3.20 F-16 with drop tanks (courtesy of US Air Force/Master Sgt. Andy Dunaway).

3.4.1 Drop Tanks and Conformal Tanks

To augment the operational range of military aircraft, external tanks are used. These tanks can be mounted under the wing and connected to the aircraft's fuel system via quick disconnects. This allows the tanks to be dropped during flight after the fuel has been used. Fuel is usually transferred into the aircraft integral tanks using compressed air.

A more recent development is the conformal tank which is designed to conform aerodynamically to the shape of the aircraft. These tanks are not discarded during flight but can be removed or installed depending upon the mission requirements. Conformal tanks are designed to have minimal impact on the aircraft's aerodynamic or stealth performance.

Figure 3.20 shows a photograph of an F-16 Fighting Falcon with drop tanks installed and Figure 3.21 shows an F-15 Eagle with conformal tanks. The latter photograph clearly illustrates the aerodynamic superiority of the conformal tank arrangement.

3.4.2 Closed Vent Systems

The flight envelopes of most military aircraft involve operation at extremely high altitudes (e.g. above 60,000 ft) where local ambient pressures are below 1 psi absolute. Even the best recovery pressure in the fuel tank ullage for an open vent tank of an aircraft cruising at Mach 0.8 at 60,000 ft will be less than 1.5 psi absolute.

This situation is aggravated by the fact that ambient temperatures (and recovery temperatures during subsonic operation) can be extremely low at these altitudes causing fuel waxing concerns for traditional commercial fuels.

As a result, military aircraft fuels often have vapor pressures that can be equal to or higher than the ullage pressures at these operating conditions making it impossible to pump the fuel to the engines. A closed vent system arrangement is therefore employed to provide a means of increasing the ullage pressure above the local ambient pressure when operating at very high altitudes.

Figure 3.21 F-15 with conformal fuel tanks (courtesy of US Air Force/Senior Airman Miranda Moorer).

Typically ullage pressure is controlled during high altitude operation using a pressurization and vent system comprising climb and dive valves and a bleed air regulator. During a climb the climb vent allows ullage air to vent overboard in order to maintain a nominally constant pressure differential between the ullage and the outside air. During descent, the dive valve opens to allow outside air into the ullage to maintain the same nominal pressure differential. Conditioned bleed air is also regulated into the ullage to maintain the same pressure schedule as fuel is consumed and ullage volume increases. Typically there will be a hysteresis band between the climb and dive pressure differential set point.

Implementation of the pressurization and vent function can be pneumatic-mechanical or electro-mechanical however the functional integrity required for this system must be adequate to ensure that unsafe ullage pressures cannot occur following probable equipment failures.

Pressure differentials used are typically in the range 2.0 to 5.0 psi and in some cases the control system reverts to an absolute pressure schedule above a predetermined altitude.

When aircraft are fitted with a fuel tank inerting system, this system must operate in conjunction with the pressurization and vent system. This issue is discussed in detail in Chapter 10 which addresses historical and current fuel tank inerting technologies.

3.5 Maintenance Considerations

3.5.1 Access

Access to the fuel tanks is typically via access panels located on both the upper and lower surfaces of the wing. Clearly structural penetrations into load bearing structures are a major concern to the stress engineers and the design decisions are usually required early in the design and development program.

3.5.2 Contamination

Within the aircraft fuel storage environment, fuel contamination is a major issue since it must be considered as a common mode failure prospect. Fuel contamination can result in loss of all propulsive power since the problem source affects all engines. Perhaps the most common source of fuel contamination is water. Even though water in ground-based hydrant sources is aggressively controlled via coalescing filtration systems, dissolved water can be present and undetectable at concentration levels of up to 80 ppm at typical ground ambient conditions. Also there remains a source of water contamination that occurs during normal operation that must be recognized and managed. In the operational environment, both the fuel and wing tank structure can become extremely cold during the high altitude cruise phase. During descent large quantities of outside air come into the ullage as the pressure difference between the outside air and the ullage is equalized. During descent into tropical climates, this air can be particularly humid and as a result water condenses onto the cold structure.

Another common source of fuel tank contamination involves microbial growth. This occurs as a result of spores in the air from the vent system that can grow when the fuel tank environmental conditions are favorable. This situation is exacerbated by the fact that fuel tanks are expected to operate without inspection for long periods.

The issue of microbiological growth is addressed in detail in Chapter 8 on Fuel Properties. The following paragraph discusses the management of water contamination which remains a serious operational issue with today's commercial aircraft particularly in long-range applications.

3.5.2.1 Water Management

Water can be present in fuel in three different forms, specifically:

- Water is readily absorbed from the atmosphere by kerosene fuels where it becomes "In solution" i.e. dissolved in the fuel. This form of water contamination is difficult to control and typically exists in uplifted fuel at levels of 50 to 100 ppm at typical ground ambient temperatures. At this level of contamination, kerosene fuel appears clear.
- Water in suspension. This occurs following the cooling of relatively warm fuel which releases water as fine droplets. This form of water dispersion is undesirable, since as fuel cools below freezing, the water can form super-cooled droplets that can turn into ice when coming into contact with pump inlet strainers and engine filters.
- Free water coming primarily from water condensing from air entering the tank from the vent system during aircraft descent. Due to the viscosity and density of kerosene fuels it can take a long time for this water to settle to the bottom of the tank where it can be drained from the tank sump.

It is standard design practice to provide water drain valves at tank low points allowing maintenance personnel to drain off any water that has settled out of the fuel. The problem however, is that it takes a long time (perhaps a day or more) for this settling process to be effective and with the high utilization rates of today's commercial fleets the standard maintenance practices for water management may often be ineffective.

One approach to circumvent this problem is to use small ejector pumps to suck fuel from the tank sump and to deliver the contents, ideally in small droplet form, to the inlet of the main boost pumps.

If any water is present in the area of the ejector pump inlet it will be entrained with the fuel and discharged in small droplets. This way, small amounts of water can be consumed in the engine which can handle small amounts of water (up to about 600 PPM) without any significant impact on engine performance.

Water detection sensors are also used in some aircraft applications that set a maintenance flag when unacceptably high levels of free water are present in the fuel tank near to the engine boost pumps. Water contamination issues are also discussed in Chapter 8 which covers fuel properties. The Certification section of Chapter 11 also addresses the water contamination issues associated with icing of dissolved and free water in fuel systems.

There is also the possibility that the introduction of fuel tank inerting systems in long-range transport aircraft, that is now becoming more typical of modern fuel system designs, may have a beneficial side effect resulting from the reduction of water vapor that accesses the fuel tanks during the descent phase due to the action of the tank inerting system which continues to deliver large quantities of Nitrogen Enriched Air (NEA) into the tank ullage. Fuel tank inerting is a topic that warrants its own dedicated chapter particularly in view of the major technology developments in fuel tank safety and inerting systems that have taken place over the past 50 years involving both military and commercial applications. The reader is therefore referred to Chapter 10 which addresses the issue of fuel tank inerting in detail.

4

Fuel System Functions of Commercial Aircraft

This chapter describes in detail the fuel system functions associated with modern commercial aircraft from the business jet up to the much larger wide body transports. The topics covered include both the ground refuel and defuel processes as well as the 'In flight' functions and their system-level interactions with the aircraft and flight crew.

The 'in flight' functions include fuel transfer, engine feed, fuel measurement, fuel management and fuel jettison tasks that must be safely executed and controlled between take-off and landing.

The schematic diagram of Figure 4.1 is an overview of the fuel system design process which emphasizes the system-level functions that are associated with the contents of this chapter and how they relate sequentially to the overall design and development process.

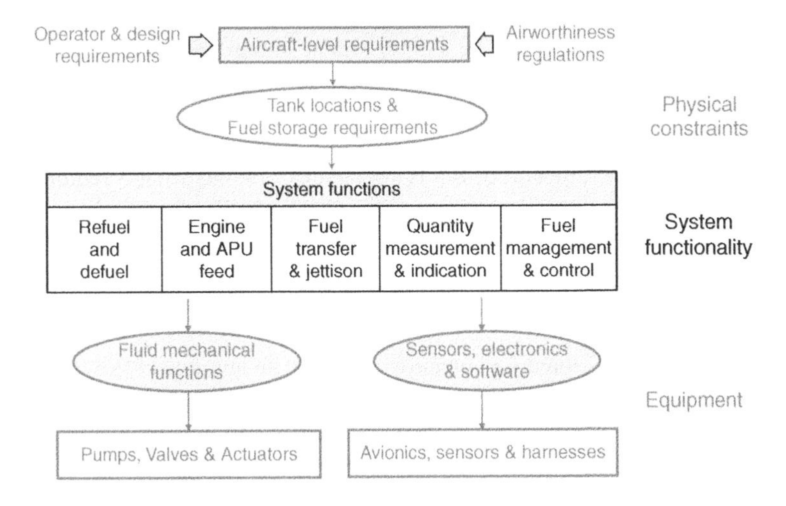

Figure 4.1 Fuel system design overview – system functionality.

Aircraft Fuel Systems R. Langton, C. Clark, M. Hewitt, L. Richards

In addition to the system functions outlined in Figure 4.1, ancillary functions are also addressed; these functions depend upon the fuel system as a heat sink which can lead to significant fuel system design and operational issues that must be taken into account early in the design of the fuel system and the aircraft. The following sections cover each of the major fuel system functions separately.

4.1 Refueling and Defueling

Refueling and defueling of aircraft is most commonly performed by connecting the aircraft refueling system to an airport based ground refueling system, sometimes referred to as a 'Hydrant' system. The airport ground refueling system supplies fuel to the aircraft at high enough flows and pressures to allow refueling of the aircraft in a short period of time. Aircraft refueling is performed prior to almost every flight whereas defueling is primarily performed during maintenance actions which, ideally, are only required on an extended schedule basis. Gravity, or over-the-wing refueling is limited to very small general aviation aircraft where the time to refuel the aircraft is not critical. All current day jet aircraft, both commercial and military, use pressure refueling with the possible exception of very small unmanned military aircraft. Provision for gravity refueling is normally provided on all aircraft as a backup to the pressure refueling system even though it is unlikely that it will ever need to be used.

4.1.1 Pressure Refueling

The fundamental need for pressure refueling of an aircraft is to provide a safe, quick aircraft turn-around-time. A prime example of an airline that relies heavily on the speed benefits of pressure refueling to increase profitability and provide superior customer service is Southwest Airlines who can unload passengers and luggage and refuel the aircraft in less than thirty minutes. Although not all airlines require a thirty minute turnaround time, there is normally a contracted refueling time the airlines have with the aircraft supplier. This contracted refueling time will always require pressure refueling, and in some cases, because of the large size and fuel quantity, may require multiple ground connection points for pressure refueling. When observing an aircraft in the refuel process it is not directly obvious what is required to ensure that the process is completed safely and expeditiously. Typically the refueling system must provide:

- fast refuel times, usually between 15 and 40 minutes depending upon the size and mission of the aircraft;
- accurate loading of the required fuel quantity, often via an automated system on board the aircraft that allows the refuel operator to preset the total fuel load required at the refuel station; this facility allows the airline to select a fuel load that matches the upcoming flight requirements thus avoiding the operational penalty associated with carrying the extra weight of unneeded fuel;
- accurate location of the fuel on board to ensure compliance with aircraft CG limits;
- protection against overboard spillage of fuel out of the tank vent system;
- protection against over-pressurizing of the tank structure (since the refuel source pressures needed (typically 35–55 psi) to provide fast fuel loading cannot be withstood by the aircraft tank structure).

The refueling system aboard the aircraft must also be compatible with airport facilities throughout the world where ground system operating conditions and fuel characteristics can vary substantially from airport to airport. The complexity of the refuel system is complicated further where a number of separate aircraft fuel tanks are involved.

While wing tanks are the most common location for fuel storage, fuel tanks may also be located in the fuselage and empennage (see Chapter 3 for more detail). In this case the refuel system will provide a predetermined fuel loading schedule so that the uploaded fuel is distributed to the optimum location in the aircraft based on the residual fuel (fuel remaining from the previous flight) and the fuel load required for the next flight.

Fortunately, there has been a long-standing standardized interface between the airport ground refueling facilities and the aircraft pressure refueling station that has been adopted by the commercial airline community. This standard which originated as United States Military Standard MS24484 reference [5] provides a common interface between all aircraft and airport facilities throughout the world.

The ground fueling station comprising ground refueling pumps, controls and monitoring equipment uses a hose that connects the fueling nozzle, frequently referred to as a 'D-1 nozzle' to the aircraft ground refueling adapter.

A clockwise twist of the nozzle once inserted over the aircraft adapter locks the nozzle and adapter together. A handle on the side of the nozzle is then rotated to open the flow path between the ground refueling system and the aircraft refueling system. The refueling operator controls the aircraft refueling process by pre-selecting the quantity of fuel to be loaded onto the aircraft via the Refuel Panel located next to the refueling adapter.

Figure 4.2 is photograph of the Embraer 190 refuel station showing the standard adapter, the D1 Nozzle and a typical Refuel Panel.

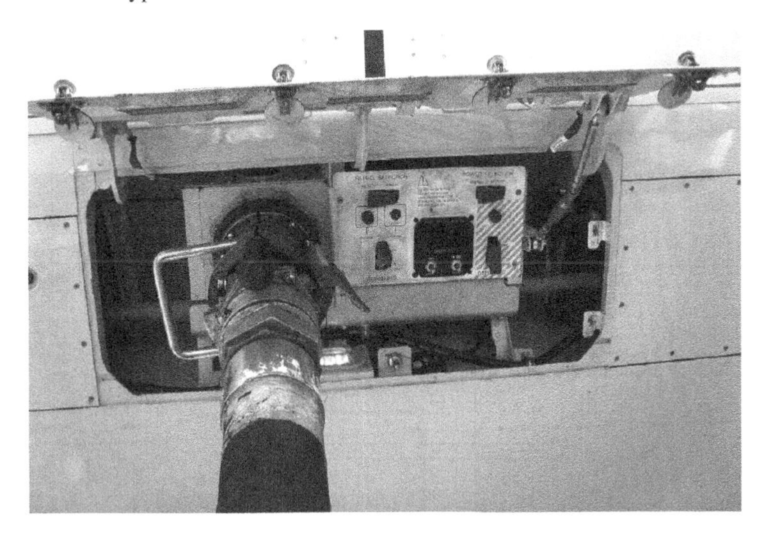

Figure 4.2 Embraer 190 refuel station (courtesy of Chris Horne).

This refueling panel interfaces with the aircraft fuel quantity gauging and management systems. These on-board systems automatically close the tank fill valves in the aircraft when the required fuel quantity has been achieved.

In the event of system faults, the operator can revert to a manual refuel mode where he controls the tank fill valves directly from the Refuel Panel.

4.1.1.1 System Architecture Considerations

Most aircraft only require a single pressure refueling point since the flow rate required to meet the refueling time is easily achievable. Larger commercial aircraft may require multiple refueling points to meet the contracted refueling time. When multiple refueling points are required, they are typically located on both sides of the aircraft. An aircraft with multiple refueling points can, of course be refueled from a single point provided that the resulting additional refuel time is acceptable.

The number of fuel storage tanks determines the number of refuel shutoff valves and the size of each tank determines the flow rate required into that tank which, in turn, determines line and valve sizing. Tank configuration (e.g. sealed/baffled ribs, low/high points, vent space location – see Chapter 3) will determine the location of the fuel entry points and location of high level sensors. The configuration of the tank will also determine if 'balance tubes' are required to properly refuel the aircraft. Balance tubes can be an essential feature where wing storage is divided into a number of semi-sealed compartments interconnected with baffle check valves. It may become impractical to load into the outboard compartment particularly when the thin outboard section results in a small volume and, as a result the migration of the uplifted fuel to the inboard compartments cannot occur fast enough to prevent a high fuel pressure developing in the outboard section.

The solution is to load the fuel into an intermediate compartment and connect this compartment to the outboard section via one or more balance tubes.

Figure 4.3 illustrates how balance tubes can be used to assist the refuel process for multi-compartment wing storage tanks.

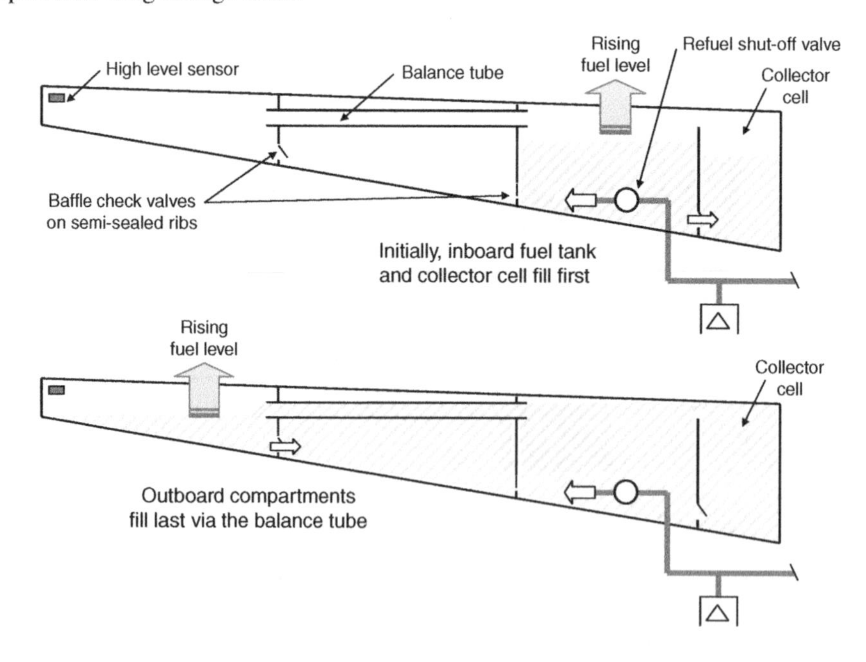

Figure 4.3 Example of the use of balance tubes.

As demonstrated by the figure the wing tank fill process can be optimized by allowing the larger inboard compartments to fill first and spill outboard through the balance tube as the maximum fuel volume is approached.

Dedicated Gallery vs Combined Refuel/Transfer Systems

Refueling systems in the majority of commercial aircraft are dedicated to ground refueling and defueling of the aircraft which means the refuel system is completely passive when in flight. Some of the larger commercial aircraft such as the DC-10/MD11 family of aircraft have combined the refueling system with the in-flight fuel transfer system. This has the advantage of reducing system cost and weight by combining system functions but this does introduce other significant design considerations such as:

- shut-off valve failure during in-flight transfer;
- the high number of open-close cycles for the shut-off valves compared to valves used for ground refueling only;
- the valves sized for fast refueling may be much larger than required for fuel transfer;
- the feed/transfer gallery must be drained to minimize unusable fuel with subsequent refueling of a completely empty galley.

Figure 4.4 is a schematic of a combined refuel and transfer system.

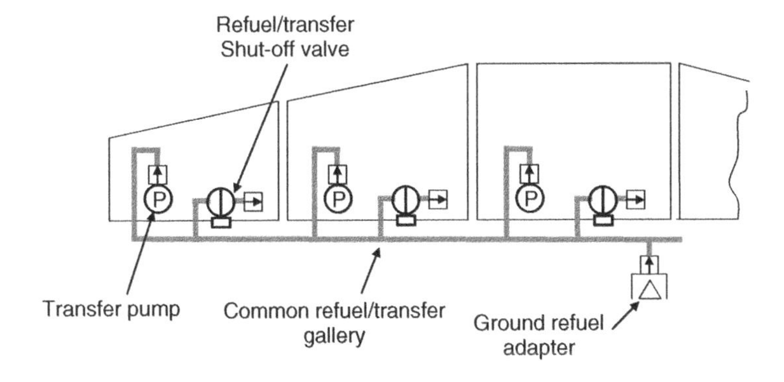

Figure 4.4 Combined refuel and transfer system.

As shown, the shut-off valves in each tank support both the refuel and transfer functions. In flight fuel can be transferred between tanks by selection of the appropriate pumps and valves.

Failure Modes

The potential modes of failure need to be considered in the design of the refuel system. For example a shut-off valve that is likely to fail in the closed state may represent a fail-safe situation but will not support aircraft dispatch requirements. A refuel system that has the potential to fail open may require special system venting configurations to prevent over-pressurization of the fuel tank structure. Although there may be a desired ‘most likely’ failure mode which influences the refuel system design, typically both failure modes must be accommodated in the aircraft design. The preferred system failure mode typically determines the type of refuel shutoff valves (fluid mechanical or motor operated) that will be used. As a means to detect a pending system failure, the refuel system will utilize a 'pre-check' function selectable at the refuel panel to cause the system to shutoff flow prior to reaching the selected fuel tank quantity. If the system shuts off as it should, the refuel operation can then be completed in the automatic mode. If the system fails to shutoff properly the system can be refueled in a manual mode (including gravity filling).

4.1.1.2 Refueling system design requirements

Typically during the design process of a new aircraft, technical requirements documents are created for each of the major subsystems. The fuel system, including the refuel/defuel requirement documents are typically derived from top level aircraft configuration and operational requirements. These requirements typically establish system design requirements which influence the following:

- line and component sizing – derived from required refuel time, and available ground refuel 'hydrant' characteristics;
- control of surge pressure and overshoot – derived from allowable maximum system pressures and final shutoff level;
- allowable line and exit velocities
- use of electrical or fluid-mechanical equipment – derived from availability of electrical power and preferred equipment failure state;
- fail-safe provisions:
  - prevention of ignition sources;
  - tank over-pressure protection – reference Chapter 3;
  - precheck function – electrical and fluid-mechanical methods.

4.1.2 Defueling

Defueling the aircraft is normally required only for maintenance of the aircraft although defueling of an in-service aircraft may be required to reduce the amount of on-board fuel. A possible example of this would be where a large aircraft such as a B747 or A380 had been over-fueled in error resulting in an overweight condition for the aircraft or for the runways available at that particular airport.

Defueling is normally performed by suction applied at the aircraft's ground refueling adapter or by using on-board transfer and engine feed pumps to pump the fuel off the aircraft. Suction defueling provides the fastest off load rates whereas the use of on aircraft pumps will get more of the fuel off the aircraft but at a much slower rate.

Defueling is also a necessary function following an accident where the aircraft is damaged and fuel must be removed before it can be safely moved for repair. For this reason, consideration must be taken during the design phase regarding the location of the refuel/defuel points to allow access in the unlikely event of a wheels-up landing.

Factors affecting the defuel system design include:

- allowed defuel time
- operation with world wide facilities
- suction or pressure defueling or both
- defuel flow shutoff when empty (suction)
- allowable remaining fuel
- engine feed and transfer pumps powered for pressure defuel
- final defuel through in-tank drain valves.

4.2 Engine and APU Feed

The fuel feed system provides fuel under pressure to the engines and to the Auxiliary Power Unit (APU) where fitted. Most modern aircraft have APUs installed whose primary function is to provide electrical power to the aircraft when the aircraft is on the ground with the main engines shut down. In some applications the APU may be required to operate in flight to provide an additional source of ac electrical power in case an engine-driven generator becomes inoperative. The APU is also a source of compressed air that is used to provide cabin air conditioning on the ground and to drive the air turbine starter that cranks the engine during the engine start sequence.

Compared with the propulsion engines the APU is much smaller and can be started via an aircraft battery-powered starter motor. During APU start the fuel feed system uses a small dc motor driven fuel pump again using battery power to provide fuel boost pressure.

During flight the feed system must ensure that fuel pressure at the engine interface is maintained above the fuel vapor pressure (i.e. above the fuel boiling point) by a predefined margin throughout the operational envelope of the aircraft for all possible combinations of fuel types and fuel temperatures. The system must also ensure that contamination of the fuel with air or water does not exceed the limits set by the engine manufacturer.

4.2.1 Feed Tank and Engine Location Effects

Airworthiness regulations require that each engine has its own dedicated feed tank. The location of the feed tank relative to the engines can have significant impact on the design of the feed system. For example, an aircraft with rear-mounted engines having feed tanks located at the inboard section of the wing will result in the following:

- long feed lines to the engines;
- double-walled fuel lines through the pressurized section of the aircraft in compliance with safety regulations;
- large variations in fuel pressure at the engine interface during aircraft pitch attitude excursions (see Figure 4.5).

The traditional location of the engines below the wing is therefore preferred by the fuel system designer since feed line lengths can be kept to a minimum and the effects of aircraft pitch variations will be minimal. The location of the feed tank above the engines will also provide a small but beneficial pressure head.

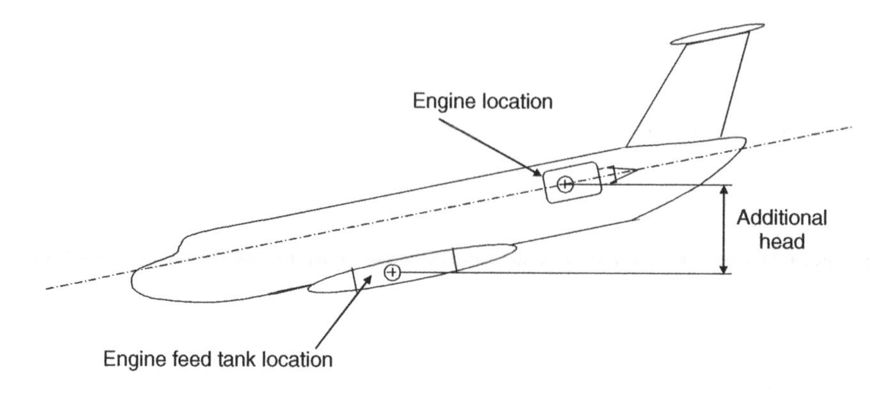

Figure 4.5 Rear engine aircraft with negative pitch angle.

4.2.2 Feed Pumping Systems

Feed pumps, also referred to as 'Boost pumps' come in many different designs (see Chapter 6 for equipment performance analysis and design details), however the feed pump designs currently used in today's aircraft fall into two main categories:

- motor-driven pumps
- ejector pumps (also referred to a 'jet pumps').

Motor-driven pumps comprise two main elements, an electric motor and a pumping element as shown in the schematic diagram of Figure 4.6. The centrifugal pumping element shown in the figure is used almost exclusively in aircraft fuel boost pump applications since it is ideally suited to the task which requires high volume flow, low pressure rise and high reliability.

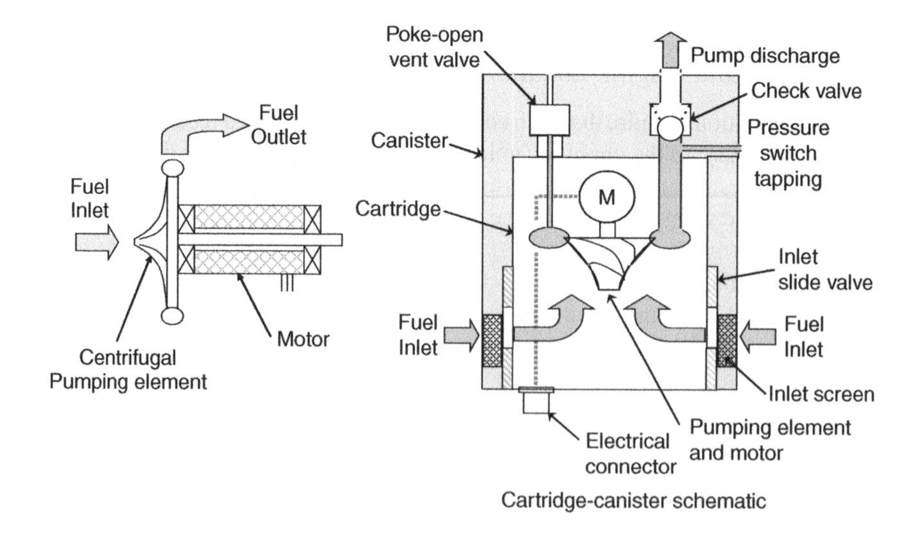

Figure 4.6 Motor-driven fuel pump schematic.

Motor-driven feed (boost) pumps are usually mounted on the lower boundary of the feed tank to support minimum unusable fuel objectives. Pump installations often use the cartridge-in-canister concept (shown schematically on Figure 4.6) to facilitate easy removal and replacement of the pumps without having to drain the tank. Feed pumps may also be spar-mounted however, the inlet to the pumping element is now above the bottom of the tank and a snorkel inlet must be used to allow the feed pump to 'Lift' the fuel in the tank to the feed pump inlet. There are performance penalties with this arrangement which become more significant as the operating altitude increases. Once again see Chapter 6 for a detailed treatment of this issue.

There are many attributes associated with the motor-driven fuel pumps that must be taken into account in feed system design. These attributes include:

- the pump's ability to continue to operate in the presence of air (or vapor) at the inlet;
- the ability of the pump to handle vapor evolution as operating conditions come close to the fuel vapor pressure;
- the ability to continue pumping fuel with little or no inlet fuel head;
- the ability to recognize and safely accommodate a locked rotor failure condition (that can be caused by fuel tank contamination or other internal failures) so that high drive currents and hence motor winding over-temperatures cannot occur.

The accommodation of entrained air in the fuel is a major issue in motor-driven fuel pump design and this is again covered in detail in Chapter 6. Since fresh kerosene fuel can contain as much as 14 % of air by volume at sea-level standard conditions (and even more for wide cut fuels), fuel pumps must be able to continue to operate effectively as the aircraft climbs and the dissolved air within the fuel bubbles-out creating an effect like opening a soda can. Similarly as the operating conditions approach the fuel vapor pressure, vapor formation will rapidly accelerate and the resulting vapor must be dispersed, or somehow managed, by the fuel boost pump in order to continue to support the engine feed pressure needs. Clearly this situation can only be accommodated to a point because boiling fuel cannot be pumped.

The third listed feature of motor-driven pump design is its ability to pump down to a minimum level. Most pumps will lose prime (i.e. stop pumping) when the inlet head falls below some minimum inlet head condition. This limiting condition will be a function of the operating point (pressure and temperature) and type of fuel. Once the prime is lost, it is typically not easily recoverable until some higher fuel inlet level is provided.

Accommodation of the locked rotor failure usually requires the installation of an isolation device within the pump motor winding. The high current and resulting high temperature within the pump winding will cause the isolator device to operate thus breaking the motor winding circuit thus preventing an unsafe situation from developing. Reliance on the power supply circuit breaker to provide this function is usually unacceptable because of the time delay involved.

Figure 4.7 shows generic performance curves for a typical centrifugal fuel pump in terms of pressure rise, mass flow and the operating altitude that defines the pump inlet condition.

This simplified presentation does not show the effects of fuel temperature variation which becomes dominant when the fuel inlet operating condition is close to the fuel vapor pressure.

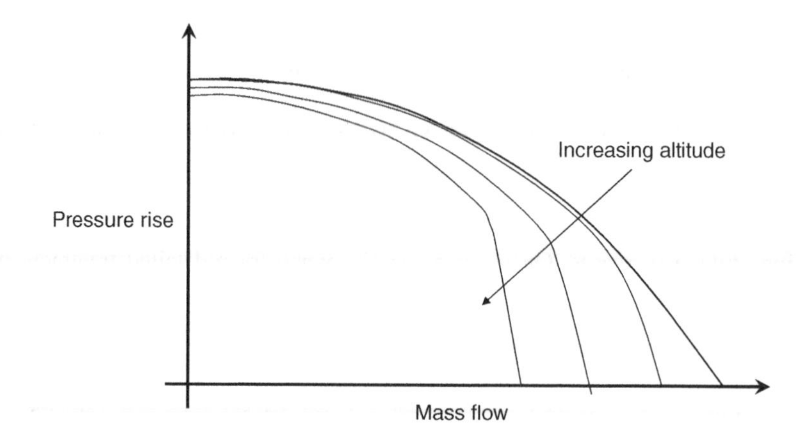

Figure 4.7 Generic centrifugal pump characteristics.

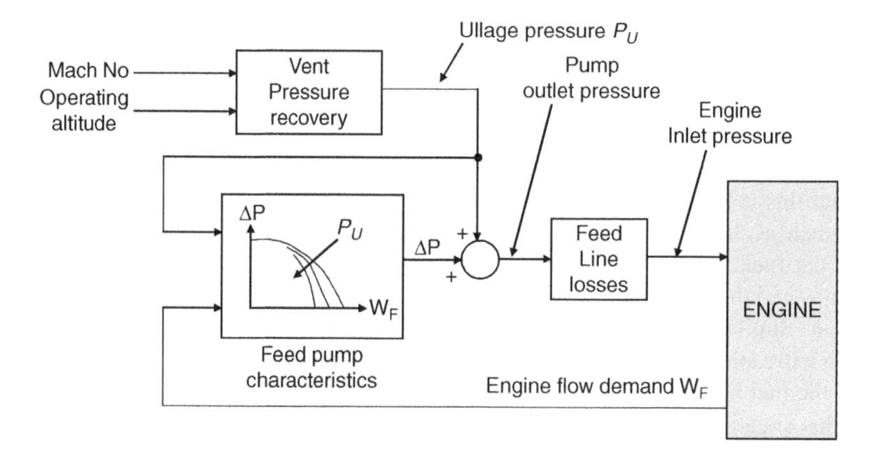

Figure 4.8 Feed system block diagram.

As indicated by the figure, the pump outlet pressure will depend upon the mass flow delivered by the pump, which is determined by the engine demand, and the pump inlet pressure, which is, in turn, a function of the prevailing flight condition.

This point is illustrated by the block diagram of Figure 4.8. This diagram shows that the inlet pressure (i.e. the tank ullage pressure P_U) is a function of the operating altitude and the pressure recovery associated with the aircraft flight Mach number. Pump pressure rise ΔP and hence pump outlet pressure can than then be determined from the engine flow demand.

The fuel pressure at the engine interface is the pump outlet pressure minus the feed line flow losses.

Ejector pumps (see Figure 4.9) are often used to provide the primary fuel boost function in smaller transport aircraft applications such as business jets and regional aircraft. As already mentioned in Chapter 3 regarding the use of ejector pumps for tank scavenging, the attraction of this pump type is the fact that the ejector pump has no moving parts and hence the reliability of the device is extremely good.

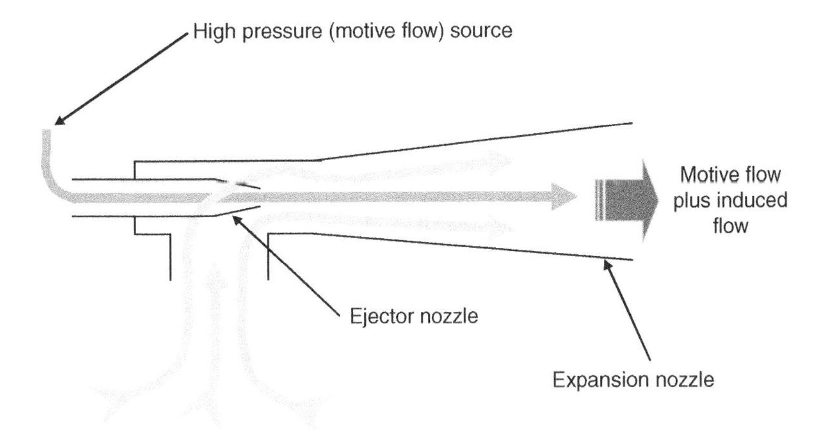

Figure 4.9 The ejector as a boost pump.

The ejector pump is also inherently very good at pumping down to very low fuel inlet levels, however, should the ejector pump lose prime its ability to re-prime will be dependent upon the prevailing operating condition and fuel type in a similar manner to the motor-driven pump.

In ejector feed pump applications, the motive flow high pressure source must come from the engine and therefore a separate motor-driven pump is required to provide boost pressure for engine starting. Once the engine is running, and motive flow is established, this 'Boot-strap' type of feed circuit will be sustainable. The motor-driven start-up pump can then be shut down.

Motive flow can be provided from two alternative sources:

- a dedicated pump mounted on the engine gearbox
- a motive flow outlet on the engine fuel metering unit.

The provision of a dedicated motive flow pump is a simple approach but does impose additional hardware with its associated cost and weight. Use of the engine fuel metering unit for this function comes for free since there is usually plenty of excess high pressure fuel available from this source. The difficulty with using this latter approach is the need for very close coordination with the engine manufacturer. Since there is typically a long lead time associated with engine design, development and certification, the engine fuel metering unit design is usually frozen by the time the aircraft fuel system design is addressed. The aircraft fuel system designer must therefore take what's available and any adaptation to the specific needs of the aircraft fuel system may not be practical at this stage of the development program.

An operational limitation associated with the ejector pump is its efficiency which is typically about 30 % meaning that the feed line must be sized to accommodate substantially higher flow than the specified maximum engine fuel flow. This is why the ejector feed approach is not viable for the larger transport aircraft.

Figure 4.10 shows, schematically, how a feed ejector pump would interface with a typical gas turbine engine. In the example shown, the source of motive pressure is via a dedicated motive pump which is typical of many applications in service today.

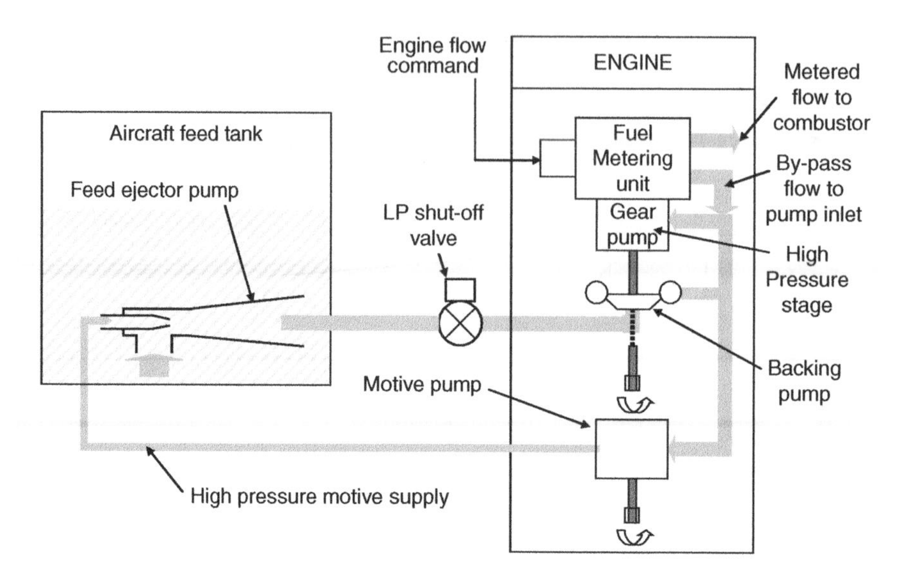

Figure 4.10 Ejector feed schematic.

The output from the feed ejector passes through the LP (Low Pressure) shut-off valve which isolates the engine fuel supply from the airframe when the engine is shut down or in the event of an engine failure situation. The fuel from the aircraft is then fed to the fuel control backing pump which increases fuel pressure from the 20 to 50 lb/in^2 feed pump outlet range to, typically 100 to 150 lb/in^2. This intermediate stage pressure is the input to the engine fuel control high pressure pump which raises the pressure to above combustor pressure levels and can be over 1000 lb/in^2. Excess flow from the fuel control HP pump is by-passed back to the pump inlet. The motive pump provides the feed ejector with sufficient pressure and flow to be compatible with the feed system requirements. Typically the maximum pressure from this device will be limited to below 500 lb/in^2 since very high pressures would result in ejector nozzle sizes that become small enough to be subject to contamination and icing in service.

The motor-driven auxiliary pump used for engine starting or for boost pressure back-up is not shown in the above figure but would be an additional feed pump operating in parallel with the ejector pump. An important point to note here is that the operating time per flight hour for the auxiliary motor-driven pump is greatly reduced (since it is only required to operate during the start phase (or following ejector pump failure and possibly in support of other functions such as fuel transfer) hence the reliability of this device in service will be substantially greater than for a full-time motor-driven feed pump application.

Once a decision is made to use ejector pumps for the engine feed function and an engine source for motive flow has been established, this same high pressure source can be used to power additional ejector pumps for fuel transfer and tank scavenging.

Also, it is good engineering practice to include a flow fuse in the motive flow line so that should a major leak develop in the motive flow line, or if the line fractures, the fuse will sense the higher flow and shut off the motive flow source. Without this feature an undetected motive line fuel leak could lead to an engine fire.

4.2.3 Feed Tank Scavenging

In order to minimize the unusable fuel, scavenge pumps are often employed to suck up fuel from the remote corners of fuel tanks and to discharge this fuel at the inlet to the main feed pump(s). Ejector pumps are used for this purpose and the motive flow for these scavenge devices may be taken from the feed pump outlet if other motive sources are not available. In some cases the main feed pumps may be located in a 'Collector cell' within the feed tank. The scavenge ejector(s) would then discharge into this cell.

One key issue with scavenge systems is that they have access to the lowest points in the tank where water contamination can be significant. Water is condensed into the fuel storage system via the vent system (as described in Chapter 3) and in large, long-range transport applications, the amount of water involved can be significant (several liters per flight). While tank drain valves are usually provided for the purpose of bleeding water from the tank sump, the time available between flights is often insufficient to allow the water to settle out of solution and descend into the tank sump. An important role of the scavenge pump is therefore to break down any water entrained within the fuel into small droplets so that the water can be safely pumped to the engine and consumed.

The established limitations as to how much water contamination can be tolerated by the typical gas turbine engine are of the order of 600 parts per million which is very small and it is the responsibility of the fuel system designers and airline operators to ensure that these limitations are complied with.

It is also important to ensure that a fuel-water mixture when pumped from the fuel tanks to the engine cannot result in the blockage of critical engine components such as fuel filters in the event that the water droplets carried along with the fuel become frozen as a result of the prevailing conditions. Fortunately engine fuel systems generate lots of heat and therefore the operational situations where this can occur are limited to cold start situations.

4.2.4 Negative g Considerations

During normal operation, commercial aircraft spend time almost continuously at or close to one g flight. Transients can occur however and airworthiness regulations mandate that the aircraft must be capable of continued safe operation following a negative g excursion lasting up to 8 seconds.

During negative g the fuel in the tanks will migrate towards the upper surface of the tank and as a result the feed pump inlets can become uncovered with the possibility of feed flow to the engine being interrupted.

The fuel system designer can employ a number of techniques to minimize or eliminate the effects of negative g operation including:

- utilizing a fuel pump design that can automatically separate out air entrained from the inlet and discharge it back into the tank;
- incorporating a collector cell within the feed tank that is kept full and slightly pressurized (by up to 1 lb/in^2) via the scavenge system (see Figure 4.11); baffle check valves between the main feed tank and the collector cell will ensure that fuel will flow into the collector cell if the scavenge ejector fails;
- providing additional means of trapping fuel close to the feed pump inlet so that its migration during negative g operation is prevented or minimized.

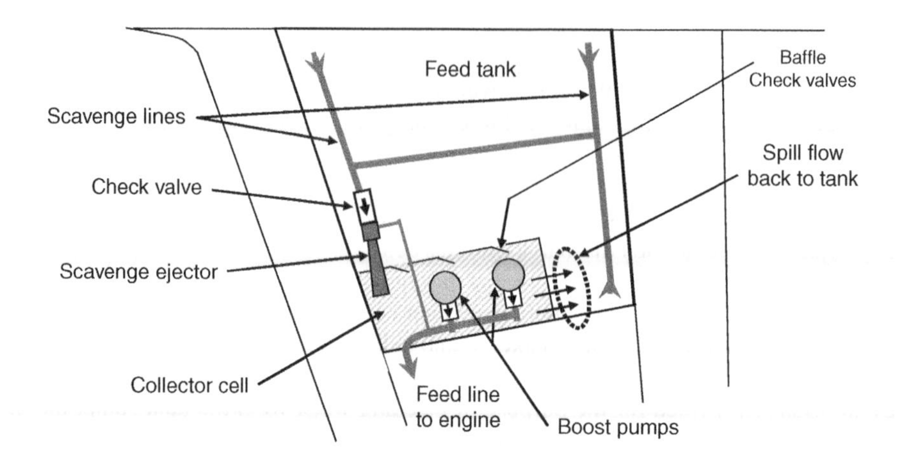

Figure 4.11 Collector cell approach to managing negative g.

Whichever design technique is adopted, physical evidence will have to be generated that demonstrates the system's ability to meet the 8 second requirement.

Pump designs that are capable of separating-out air at the pump inlet and discharging it back to the tank use a feature called a 'liquid ring' which is a separate device attached to the main pumping element. The operating principle of this concept is explained in Chapter 6.

The collector cell solution is perhaps the most common approach used in today's transport designs. The scavenge ejector will continue to pressurize the collector cell thus keeping the cell slightly pressurized and able to continue providing feed pressure and flow. It may be desirable to install a check valve (non-return valve) in the ejector pump low pressure inlet to ensure that motive flow cannot flow back to the feed tank.

The design condition for the scavenge ejector is maximum power take-off when the ejector must be capable of maintaining a full collector tank. This in turn will establish the required motive flow and the additional load on the engine boost pumps.

4.2.5 Crossfeed

In multi-engined aircraft, an engine crossfeed system is required to allow fuel from the feed tank of a failed engine to be consumed by the other engine or engines. Since the crossfeed function is rarely used, operating procedures must be provided that can verify the availability of the crossfeed function. For example, the crossfeed valve may be exercised as part of the start-up procedure (either via the flight crew or automatically) to verify correctness of operation.

In some of today's long-range transport aircraft where ETOPS certification is required a dual crossfeed valve arrangement is employed so that the crossfeed function remains available following a failure of any one of the two crossfeed valves.

An important consideration in the design and installation of crossfeed valves is to minimize the possibility of water collection in or close to the valve assembly since ice formation may cause the valve actuator to stall thus inhibiting the crossfeed function.

4.2.6 Integrated Feed System Solution

This subsection considers the overall engine/APU feed system arrangement illustrating how the various functional requirements come together as an integrated whole.

Figures 4.12 and 4.13 show a feed system example for a twin engine transport application using the collector cell concept. Only the left side of the aircraft is shown in order to simplify the diagrams.

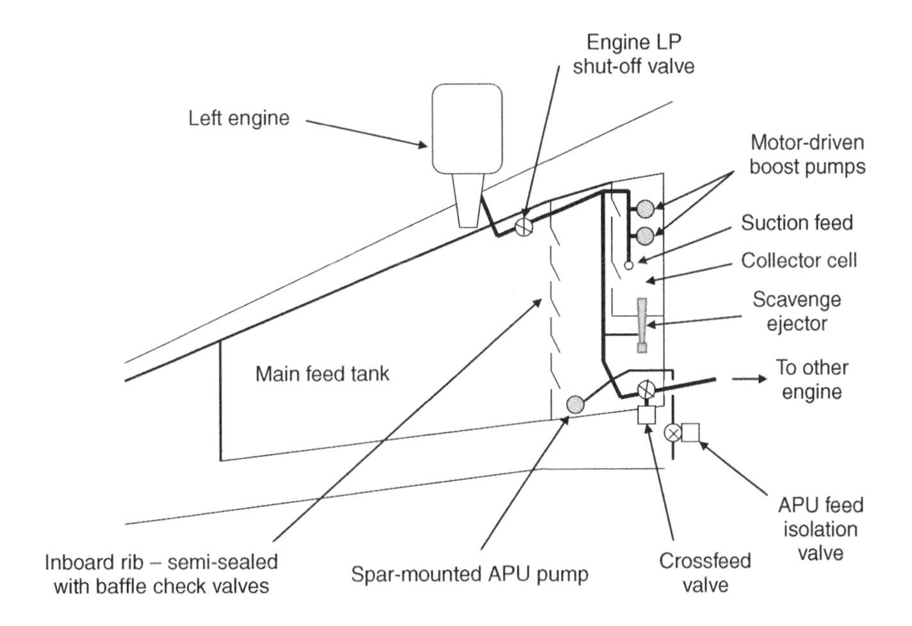

Figure 4.12 Twin engine feed system physical overview.

Figure 4.12 shows a plan of the left wing showing the main feed tank, pumps, lines and control valves in their approximate locations. The inboard section of the tank is where the feed system equipment is located and this section is bounded by a semi- sealed rib with flapper check valves that allow fuel to migrate inboard only. This has the effect of trapping fuel inboard which is desirable. The fuel boost pumps are located together on the lower skin of the collector cell. (Pressure switches are normally connected to the pump discharge to verify correct functioning but these are not shown in the figures for clarity.) Two boost pumps are typically installed to allow dispatch of the aircraft with only one boost pump operative. There is also a suction feed check valve in the collector cell to allow the engine to suck fuel from the tank in the unlikely event of loss of both feed pumps. In this situation the suction capability of the engine fuel system will be limited to altitudes of about 20,000 ft or lower. The actual value of this operational limit will be established during flight testing of the aircraft as part of the certification process.

A scavenge ejector pump is shown in the figure which is used to charge the collector cell. The scavenge lines are not shown for clarity.

The APU pump is shown mounted to the rear spar together with the crossfeed valve. These two items will not be repeated on the right side of the aircraft.

The feed line connects with the engine via the LP shut-off valve shown mounted on the front spar of the wing.

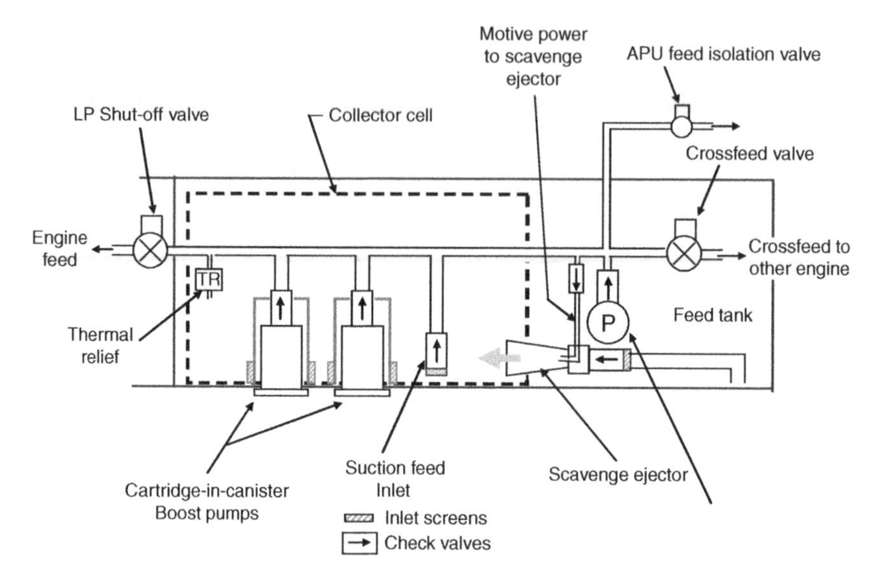

Figure 4.13 Twin engine feed system detailed schematic.

The schematic diagram of Figure 4.13 illustrates the functionality of this feed system example in detail. The discharge of each pumping element is connected to a common feed line that is in turn connected to:

- the left engine via the LP shut-off valve
- the right engine feed line via the crossfeed valve
- the APU via the APU isolation valve
- the suction feed inlet
- the motive flow nozzle of the scavenge ejector
- a thermal relief valve.

Check valves are installed at the outlet of each of the motor-driven boost pumps, the suction feed inlet and at the inlet of the scavenge ejector motive flow line to protect the pressure integrity of the main feed line. A check valve is also installed at the suction inlet of the scavenge ejector to prevent fuel from migrating from the collector cell should both boost pumps lose power. A thermal relief valve prevents feed line pressure build up in the feed line due to temperature changes when the engine is shut down and the LP valve is closed. Thermal relief may also be required downstream of the LP shut-off valve

The feed system example presented is only one of many different approaches used in today's modern aircraft. The principles involved, however are common as are the airworthiness requirements that must be complied with.

The reader is encouraged to study the various examples presented in Chapter 12.

4.2.7 *Feed System Design Practices*

The feed system is perhaps the most critical function of the fuel system since continued safe operation of the engine depends upon it. Presented here therefore are some of the key design issues and engineering practices that should be borne in mind.

4.2.7.1 Hot Fuel Operation

Because of the volatility of jet fuel combined with the high altitude operation of today's modern transport aircraft the functional limitations of the feed system during high fuel temperature operation must be understood and demonstrated to the satisfaction of the regulatory authorities. Most notably the 'hot fuel climb' is the worst case operating condition to be examined.

Fuel type is a major consideration here since the fuel vapor pressure characteristics will determine the vapor evolution process as altitude is increased and the tank ullage pressure declines. The starting temperature for this situation should represent the worst case that could result from the aircraft being parked in direct sunlight in a hot climate. A value of 55 degrees C is commonly used.

Functional verification can be accomplished either by test rigs or via flight test. The cost of a hot fuel climb test is extremely high and the use of a test rig may be considered as a lower cost alternative. Such a rig would be required to:

- provide representative tank geometry and equipment installation
- vary aircraft pitch attitude
- provide representative altitude (tank ullage pressure) variation
- control and monitor fuel temperature, pressure and flow.

It is critical in such a test to use fresh (unweathered) fuel to ensure that vapor evolution during the test will be representative of the worst case for that fuel type.

If the aircraft is to be certified for more than one fuel type then separate tests will be required for each type. Emergency fuels must also be tested to establish if any operating altitude limitations apply.

4.2.7.2 Design Requirements

Presented here are a few established design requirements and practices that should be considered in feed system design:

- When sizing the feed lines the general guide is to keep fuel flow velocity below 10 ft/sec for acceptable pressure loss; however, if line lengths are long (as in the rear engined aircraft designs) an additional margin of 30 % on line cross section is advised to allow for a potential increase in pressure loss due to air and/or vapor evolution since the formation of bubbles effectively reduces the available flow area.
- Feed lines should be smooth with no sharp bends, internal sharp edges or sudden changes in cross section which can encourage vapor or air bubble formation within the feed line when operating close to the vapor pressure of the fuel. This situation can occur in the hot fuel climb scenario, for example.

- Provision for thermal relief is usually required to prevent excessive feed line pressure from developing when all of the feed line valves are closed trapping a fixed fuel volume. If the scavenge ejector motive flow line is used to provide thermal relief care should be taken to ensure that air cannot enter the feed line when operating in the suction feed mode.
- The use of air release valves is favored by some aircraft manufacturers. These devices installed in feed lines utilize a float-actuated mechanism to separate air from the fuel and to discharge it back to the feed tank ullage. One of the failure modes, however, is to allow fuel to leak back to tank thus compromising the integrity of the feed system.

4.3 Fuel Transfer

A fuel transfer system is needed in applications where multiple tanks are used for fuel storage to ensure that fuel is consumed from the various tanks in accordance with a predetermined schedule. This schedule (or fuel burn sequence) takes into account many operational considerations including:

- aircraft CG variation with fuel burn
- wing load alleviation
- feed tank maximum and minimum fuel quantities.

Control of the fuel transfer system can be either under the direct control of the flight crew or via an automated Fuel Management System.

This subsection focuses primarily on the fluid mechanical functional aspects of fuel transfer while the management and control aspects are covered in Section 4.6 below.

4.3.1 Fuel Burn Scheduling

For aircraft with very simple architectures, for example, twin engined aircraft with two feed tanks storing all of the fuel on board; the only fuel transfer issue of significance is to provide the capability to crossfeed fuel from one tank to the other in the event of an engine failure. Without this feature fuel would be trapped and unusable to the remaining good engine. Not only is the remaining fuel unavailable but the aircraft lateral balance would continue to deteriorate as fuel is consumed by the remaining good engine.

Fuel crossfeed is therefore a standard functional capability for all commercial aircraft to accommodate the engine failure situation.

As aircraft become larger and more complex in terms of the fuel storage design, the need for more complex fuel transfer schemes becomes the norm. In each case, however, the basic functional objective remains the same, i.e. to keep the engine feed tanks topped up from the various auxiliary tanks until they have been depleted and to use the feed tanks as the last available source of fuel on board. In four engined aircraft this can become quite complex when taking into account the various failure mode possibilities.

4.3.1.1 Transfer System Architectural Considerations

Transfer system functions include the support of inter and intra fuel system tasks. In addition the fuel transfer system must provide for safe aircraft operation in emergency situations such as loss of an engine or any other serious loss of aircraft functionality.

The basic inter-system transfer task is to sequentially manage the fuel transfer from the auxiliary fuel tanks into the engine feed tanks as fuel is consumed. There are two commonly used architectures used in today's commercial transports, namely:

- the override transfer system
- the quantity sequenced transfer system.

The first of these architectures is in common use on most Boeing transports and has the advantage of requiring no pump or valve cycling throughout the transfer process.

Figure 4.14 shows a schematic of a typical override transfer system in a traditional three tank aircraft. Here center tank fuel is consumed first by employing center tank transfer pumps that produce significantly higher feed line pressures than the main feed boost pumps are capable of. So while the feed tank boost pumps operate continuously their outlet check valves are maintained closed by the override pump pressure so that all of the feed flow to the engine comes from the center tank.

Once the center tank fuel has been depleted, the center tank boost pumps are switched off allowing feed flow to be provided from the main feed tank boost pumps that automatically take over the engine feed task.

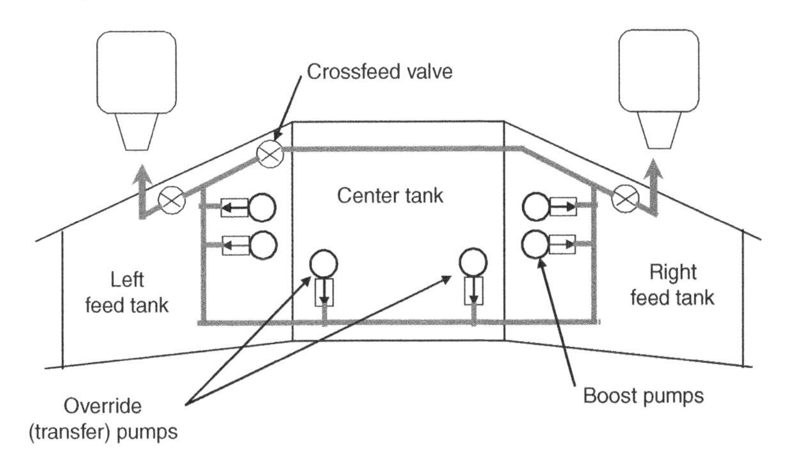

Figure 4.14 Override transfer system schematic.

The second transfer methodology uses sequential transfer of fuel to the feed tanks using selector valves to stop transfer when the feed tanks are full and to resume transfer when the feed tank quantity has fallen below some predetermined value (say, 90 % of maximum). As a result, the feed tank quantities will oscillate between full and the transfer selection value. This transfer system method results in more wear and tear on the associated transfer system equipment while the override system carries a weight and power consumption penalty associated with the additional performance required by the transfer pumps.

So far we have considered only motor-driven pumps in the architecture discussion. Ejector pumps are also used to provide the fuel transfer function, particularly in smaller sized aircraft. In aircraft with variable frequency ac power (which is becoming more common in newer aircraft fuel systems) induction motor-driven pump rotational speeds will vary over a range of about two to one from engine idle to maximum power settings. As a result motor-driven boost pumps may have to be oversized to meet flow requirements at the lowest operational power frequency. The excess flow capacity now available at high engine rotational speeds commensurate with typical climb and cruise power settings can therefore be used as a motive pressure source for transfer ejector pumps.

4.3.2 Wing Load Alleviation

For traditional aluminum wing structures, the fatigue life of the aircraft can be managed through the control and monitoring of the aircraft's operational life in terms of stress cycling due to pressurization cycles or, in the case of the wing structure, the stress cycles imposed on the wing structure each time the aircraft takes off with a full payload. To address this latter issue, it is common practice with large, long-range aircraft, to have an outboard wing tank that is kept full for the majority of the flight thus reducing the wing bending moment and, hence the magnitude of the structural stress cycles. Typically, the outboard wing tank fuel is transferred into the feed tanks during descent or when feed tank fuel reaches some predetermined minimum quantity.

The discussion here addresses the operational flight stress cycling and not the dynamic wing load alleviation that is becoming more common as a high technology feature provided via the flight control actuation system more as a ride quality enhancer than a structural life benefit.

In the case of the A380 Super Jumbo, the fuel system takes on an additional task by keeping fuel inboard whilst on the ground to minimize wing structural stresses due to fuel and engine weight. Immediately after take-off fuel is transferred outboard as fast as possible to minimize wing bending stresses in flight.

In aircraft that have a carbon fiber composite wing structures, the aluminum fatigue issue goes away only to be replaced by a more challenging issue associated with the protection of in-tank equipment from electro-magnetic phenomena. This issue is discussed in Chapter 9.

Fuel transfer is used in some fuel system designs to support the thermal management requirements of the aircraft. Fuel is a convenient heat sink source and maybe used for example to cool the aircraft hydraulic system, aircraft avionics (more commonly in military aircraft applications) or the engine oil lubrication system.

See Section 4.7 entitled 'Ancillary systems' for thermal management examples.

4.3.3 Fuel Transfer System Design Requirements

Flow rates and pressure drop requirements follow essentially the same design rules as for the feed system described earlier except that there is less concern about air and vapor evolution which is a critical factor in engine feed lines. Line sizing to meet the 10 feet per second requirement remains an appropriate goal.

Transfer valve technology has many forms. Perhaps the most common technology in use in today's commercial aircraft is the motor-operated valve. Here the motor and reduction gear is mounted external to the fuel tank and the rotary output which penetrates the tank wall actuates a valve installed in a fuel transfer line within the tank. End-of-travel sensors are usually installed to provide the crew and/or the fuel management system valve position information. Variations in power levels and valve friction can result in significantly different operating times however in fuel transfer system operation this is usually not a problem.

Another form of transfer valve design uses fuel pressure developed by the transfer pump as the motive power source. Valve selection is usually via a small solenoid-operated valve.

A more detailed treatment of valve design technology is presented in Chapter 6.

An important issue regarding transfer system design is the accommodation of failures. It is important that fuel is not trapped and therefore unusable in an auxiliary tank. In considering this issue, the system designer may consider:

- redundancy of function
- design valves to fail to a preferred position
- gravity transfer as a back-up
- transfer system work-arounds.

This section on fuel transfer should be read in conjunction with Section 4.6 which describes fuel management and control.

4.4 Fuel Jettison

The requirement for fuel jettison is brought about by the difference between the Maximum Take-Off Weight (MTOW) and Maximum Landing Weight (MLW) which becomes more and more significant as the size of aircraft increases (see Figure 4.15) which shows a graph of this difference versus MTOW for several commercial transports in service today reference [6]. This difference represents the worst case fuel jettison requirement by assuming a fully loaded aircraft developing an emergency situation immediately after take-off that necessitates a need for a 'Soon as possible' landing.

Such events could be an engine failure, a fire either in the engine or in a critical location of the aircraft or any other failure that threatens the safety of the passengers and crew should

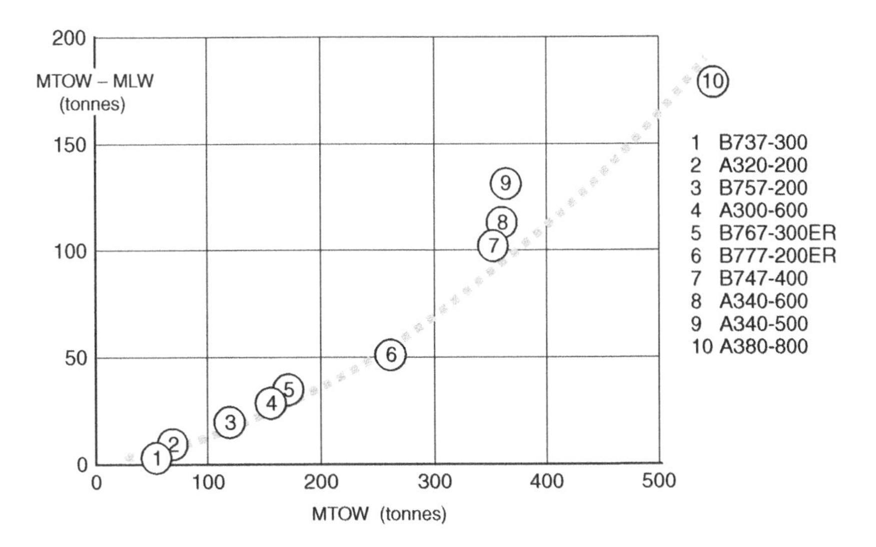

Figure 4.15 Maximum jettison requirements for various aircraft.

immediate action is not be taken to facilitate a landing as quickly as possible. Of course in the worst case situation, the captain has the authority to proceed to land the aircraft with a weight above the MLW but must recognize the risk of landing gear failure with all of the operational issues that follow. Should an aircraft land above the certified MLW, then a formal investigation, analysis and inspection is mandated that must be addressed and closed out before the aircraft can be returned to service.

From the graph of Figure 4.15 it is clear that the jettison problem grows with size of aircraft. The challenge for the large aircraft is how to provide the necessary pumping capacity to dump fuel overboard in a sufficiently short time for the crew to be able to effectively manage the emergency and safely land the aircraft.

Typical worst case dump times are in the range 30 to 45 minutes and even if this may seem to the reader to be a relatively long time, it is nevertheless a challenge to the fuel system designer and for this reason must be considered early in the system design trade studies that examine fuel transfer requirements and solutions.

The design issue here is to minimize the weight, cost and maintainability issues associated with a system where the probability of being used in service is very small. Therefore additional equipment to support the jettison function will be, for the most part, a weight and operational penalty that must be borne by the fleet throughout its operational life.

Consideration should be made therefore to utilize existing equipment (e.g. transfer pumps) that can be used to support the fuel jettison in the unlikely event that it is needed. In fact the flow requirements of the transfer pumps can be adapted to take into account the needs of the fuel jettison flow requirement. In some aircraft applications, the feed pumps have been used to supplement jettison flow needs; however, this approach should be carefully evaluated because of the potential risk to the feed system integrity which must be maintained even during jettison operation.

4.4.1 Jettison System Example

Jettison systems become more challenging as the size of aircraft increases and as the number of fuel storage tanks increases.

Typically the fuel feed and transfer network has to be reconfigured to allow the jettison function to proceed while the engine feed system continues to be supported. During the jettison process it is important that the flight crew workload in managing this function be kept to an absolute minimum thus allowing them to focus on the initial emergency.

To this end most modern aircraft have a 'Dump-to gross weight' capability thus eliminating the need for close monitoring of the jettison process. This control system is addressed from a management and control aspect in Section 4.6.

Figure 4.16 shows a schematic of the fuel jettison system for a typical twin engine aircraft.

This figure shows a jettison system arrangement for a traditional three tank transport aircraft where the center fuel tank provides additional fuel storage which is normally consumed first in the fuel burn sequence. (The APU feed and refuel circuitry are omitted here for clarity.) In normal operation the fuel transfer protocol keeps the feed tanks topped up by sequential transfer from the center tank to each of the two feed tanks via the transfer gallery and the feed tank transfer valves.

Should a fuel jettison command be issued from the flight deck, the fuel transfer valves are closed and the jettison pumps and jettison valves are enabled. The transfer pumps are now

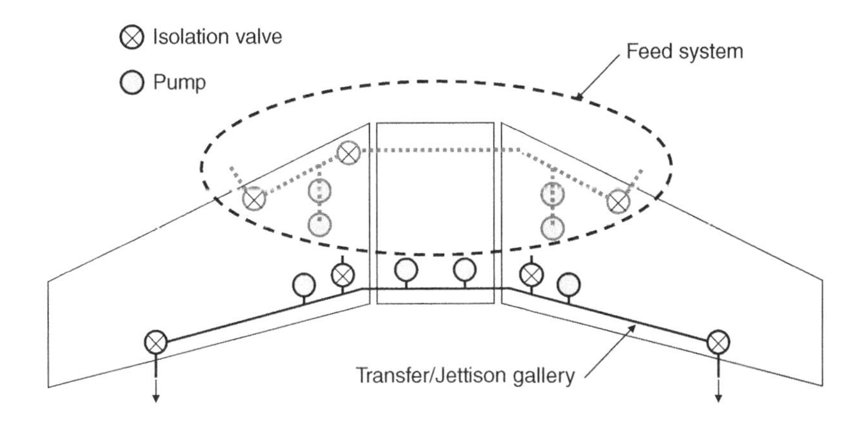

Figure 4.16 Schematic of a simple fuel jettison system.

available to support the jettison flow requirement thus reducing the jettison time. A critical aspect of the jettison system implementation is to ensure that inadvertent selection of the jettison function can only occur with a probability equivalent to catastrophic failure rates i.e. 10^{-9} per flight hour since such an event could result in loss of aircraft.

In very large aircraft, additional pumps, dedicated to the jettison function may be required in order to meet the maximum fuel dump time requirement.

Note that in the example shown, the feed system is completely isolated from the jettison system thus ensuring that the engine feed function is not compromised during the fuel dump process.

The location of the fuel vent outlet must take into account the aircraft configuration so that the vented fuel cannot cause secondary operational problems. For example, the jettisoned fuel must avoid engine inlets under all potential jettison situations (an important issue with rear mounted engines). In the example shown, there are two jettison outlets located on the trailing edge of each wing. This works well with traditional wing mounted engine designs and by providing two symmetrical outlets, good lateral balance is maintained throughout the process.

While the driving performance requirement for the jettison system is the time to complete the worst case dump process, the following requirements related to the functional integrity of the system are critical:

- the probability of an uncommanded jettison;
- the probability of dumping to an unsafe fuel state (i.e. failure of the jettison system to shut-off at the required fuel quantity).

Each of the above failures can result in loss of aircraft and therefore the failure classification is 'Catastrophic'. For each the above failure modes, the probability of occurrence must be less than 10^{-9} per flight hour.

A common practice in jettison system design is to locate the jettison pump inlets some distance above the bottom of the feed tank to ensure that some minimum fuel quantity will remain even if the pumps continue to operate. Also the selection system must be carefully implemented (double pole switches, switch guards etc.) to prevent inadvertent operation of the jettison system.

4.5 Fuel Quantity Gauging

The fuel quantity gauging system measures the fuel contents of all feed and auxiliary tanks and provides this information to both the flight deck and, when on the ground, to the refuel panel display at the refueling station. As mentioned in the introduction, fuel quantity information is provided in mass quantity terms since this is a measure of the fuel's calorific content or 'stored energy'.

The challenge for the in-tank sensing system is to provide accurate, continuous information while coping with substantial variations in the storage environment, for example:

- variations in fuel properties as a result of having to take fuel on board from many different locations around the world. even the same fuel type can vary significantly from batch to batch;
- stratification resulting from loading relatively warm fuel on top of very cold residual fuel from the previous flight;
- variations in fuel surface attitude due to aircraft maneuvers;
- fuel quantity variation from empty to full;
- tank structural distortion (e.g. wing bending and twisting) under the influence of aerodynamic loads.

Unusable or ungaugable fuel, which is often largely dependent upon tank geometry, inter-tank compartments, tank sumps and fuel transfer galleries, is also a major challenge in fuel storage and measurement system design. This can be a significant operational burden for the aircraft and every effort should be made early in the system design phase to minimize the resulting penalties.

In addition to the need for continuous gauging of tank mass quantity, there is also a need for discrete volumetric information. There is a requirement for a low-level warning system that tells the crew that an immediate safe landing is necessary. For integrity reasons, the low level warning system should be functionally independent of the primary gauging system. High level sensing is also used to protect the fuel tanks from overfilling into the expansion space during the refuel process, however, this function may also be provided by hydro-mechanical means.

The most commonly used sensor technology in aircraft fuel quantity gauging systems today is that utilizing capacitance sensors, commonly referred to as probes or tank units. A capacitance probe typically comprises a pair of concentric tubes designed for near vertical mounting at a specific location within a fuel tank to act as an electronic 'dip stick'. The capacitance between each of the two concentric tubes varies with the wetted length due to the permittivity difference between fuel and air.

Fuel quantity gauging system accuracy is typically specified as a percentage of the full tank quantity ± a percentage of the prevailing quantity. The highest accuracy level requirement specified today (primarily for use on long-range aircraft applications) is 1/2 % of full scale ± 1/2 % of point.

To achieve this degree of accuracy requires the use of the following additional sensors:

- *Probe compensators*. These devices are fixed capacitance probes, located near the bottom of the tank in order to remain submerged, that compensate primarily for variations in the permittivity of the fuel due to installation and environmental effects. A fully submerged probe can also act as a compensator and has the advantage of minimizing gauging errors from fuel stratification.

- *Densitometers*. Direct measurement of fuel density is necessary to provide the highest accuracy in converting volume information into mass. Accuracies of the order of ± 0.2 % can be achieved with the latest sensor technology (see Chapter 7).
- *Temperature sensors*. Temperature measurement provides validation of the fuel properties derived from the other sensors and can be used, together with permittivity data, as a back-up for the determination of fuel density in the event of a failed densitometer (although the accuracy achievable via this method is limited). Fuel temperature is also required to advise the crew when the bulk fuel low temperature limits have been reached.

Some recent fuel gauging system designs use a Fuel Properties Measurement Unit (FPMU) comprising one of each of the above sensors in a single assembly. A sample of the uplifted fuel passes through this unit each time the aircraft is refueled thus capturing the key properties required for accurate gauging. The information from the previous refuel process, which applies to the residual fuel from the last flight, is retained in the gauging computer's memory so that the requisite properties of the new fuel combination can be determined.

The number of probes and mounting locations required to gauge a fuel tank depends upon the tank shape and the accuracy requirements for the range of specified fuel surface attitudes. In order to provide independent, contiguous gauging of fuel quantity, the probe array must have a minimum of three probes that cut the fuel surface over the full range of surface attitudes and tank quantities. Three surface penetrations identify the fuel surface plane location within the tank from which fuel volume and hence mass can be determined. Accommodation of failures will add to the probe count unless some loss in gauging accuracy is acceptable.

As a back-up to the main (primary) gauging and level sensing system, transport aircraft incorporate a secondary fuel gauging capability. In the event of an on-ground primary gauging failure, the secondary gauge enables the continued safe dispatch of the aircraft, by providing an independent means to determine the fuel quantity in a tank or tanks. Dissimilar sensing technology to that of the primary gauging is usually employed to offset the risk of a common mode tank sensor failure.

Fuel gauging systems in today's aircraft are extremely sophisticated systems that have to provide the crew with high integrity fuel quantity information that must never be overstated even following equipment failures. Fuel on board must also be continuously reconciled with the fuel consumed by the engines where fuel flowmeters provide an independent source of information.

A major issue for all fuel quantity gauging and level sensing systems that should be mentioned here is the requirement to provide an intrinsically safe implementation. Since the sensors involved are electrical, the possibility of the discharge of electrical energy within the fuel tank at a level sufficient to ignite an explosion must be demonstrated to be adequately remote. This issue has become a major focus of the Regulatory Authorities, who define the certification requirements for new aircraft. Equipment design issues associated with the latest intrinsic safety requirements are covered in Chapter 9.

The following text discusses the functional, system-level issues associated with fuel quantity gauging and level sensing systems.

4.5.1 Architectural Considerations

There are many considerations to be taken into account in arriving at the optimum architecture for a fuel gauging system. These considerations are not just limited to the fuel gauging system itself but its degree of interface with, and its role within, the fuel system and the overall aircraft.

The design of the architecture of a fuel gauging system and its degree of robustness is fundamentally driven by the necessary system safety derived requirements in terms of availability (continuity of function) and integrity (correctness of behavior). A detailed treatment of these issues is presented in Chapter 11 which covers the overall fuel system activities associated with the design, development and certification process.

The following text discusses gauging system architectures that have been applied to commercial transport aircraft over the past thirty to forty years and how operational events have impacted gauging system architecture evolution to the standards that exist in the more recent aircraft to enter service in the past decade.

4.5.1.1 A Brief History of Gauging System Evolution

Historically, fuel gauging systems generally comprised tank sensors and interconnecting harnesses with all the signal conditioning and display being performed in one or more combined amplifier/indicator units. This approach provided an economical and overall reliable system but somewhat limited in accuracy and application. Accuracy improvements were made with the introduction of density measurement. Further improvements to the system were made possible by the introduction of integrated circuit based processor/indicators. But it was the advent of the microprocessor in the mid-to-late 1970s that stimulated a major revolution in gauging systems by enabling, through the application of software, the introduction of:

- enhanced accuracy by the introduction of aircraft continuous attitude correction and precision fuel density measurement;
- individual probe addressing and fault survivability through self-healing;
- Enhanced Built-In-Test (BIT);
- advanced intra and inter system communication.

Once the gauging system became microprocessor-based, it enabled a variety of architectural options to be considered and implemented in order to keep pace with the ever increasing demands of the fuel system and overall aircraft requirements as they grew in complexity.

4.5.1.2 Dual Channel Gauging Architectures

During the 1970's and 1980's dual channel architectures were a common architectural solution. The simplified schematic of Figure 4.17 shows one such solution.

Within each fuel tank are two independent probe/sensor arrays. Probes are interleaved so that either of the two probe arrays can meet the fuel gauging accuracy requirements. With a fully functional system, therefore, the system typically exceeds the specified performance. This design approach allows safe aircraft dispatch in the presence of any single failure of either the avionics or in-tank equipment with accuracy performance compliant with specification requirements. A commonly used alternative to the above approach with simple two-tank systems is to have each channel dedicated to one tank. Loss of one channel would lose the

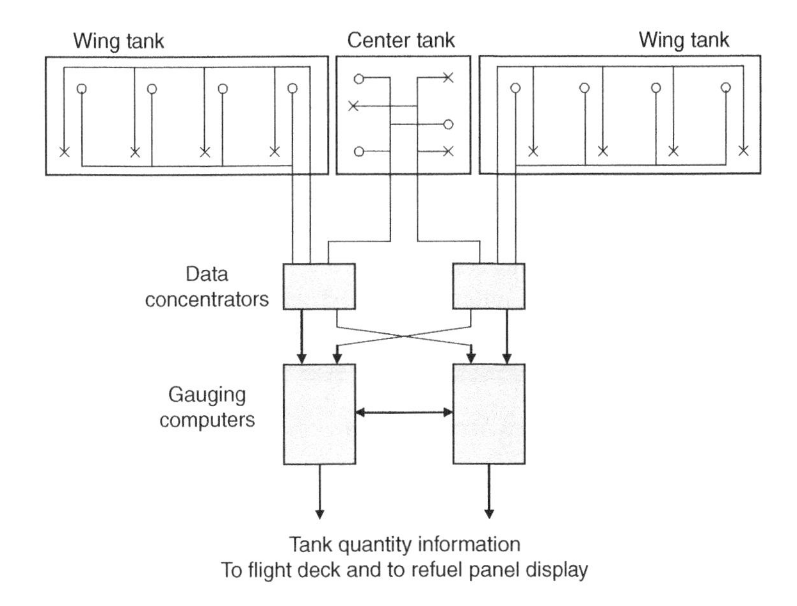

Figure 4.17 Simplified schematic of a dual channel gauging system.

gauging information from its related tank leaving the crew to assume fuel load symmetry to infer the fuel quantity in the failed tank.

Within each of the in-tank arrays are probes, compensators and, depending upon the accuracy requirements, densitometers. Compensators are fully immersed capacitance units that provide a reference capacitance signal which allows probe wetted length to be expressed as a percentage of full immersion rather than as an absolute capacitance value.

The data concentrators are a fairly recent development and provide the functional interface with the in-tank sensor arrays. These units are typically located relatively close to the fuel tank harness penetrations in order to minimize the length of sensor wiring outside of the tanks.

The data concentrators provide excitation for the in-tank sensors and digitize the sensor data for transmission to the gauging computers via digital data busses, typically using the ARINC 429 bus protocol.

The gauging computers usually operate in a primary/standby (or master/slave) arrangement where one computer provides the gauging computation function while the other acts as a standby. In the event of a fault developing in the primary computer, control will be switched over to the standby computer which will then assume the gauging function. The changeover function can be manual or automatic with the latter approach being more common in the more recent system designs.

An interesting operational event occurred in July of 1993 that caused the Boeing Company to reevaluate the dual channel approach to fuel gauging. This event involved an Air Canada 767 scheduled to fly from Ottawa to Edmonton. This aircraft ran out of fuel over Manitoba and was forced to make a powerless but fortunately successful landing at Gimili, Manitoba.

As with many accidents, this incident was attributable to a sequence of events which included a new type of aircraft, lack of familiarity with the Minimum Equipment List, recent adoption of metric units and, most relevantly a faulty dual redundant fuel computer. The aircraft had been dispatched from Ottawa following incorrect calculation of the fuel tank quantities in kilograms by converting drip stick readings using pounds-based density information, this action all being attributable to a faulty gauging system that was providing no fuel quantity indications to the crew.

The subsequent inquiry into the incident revealed a dry solder joint manufacturing flaw in the power supply of one computer channel that was just sufficient to prevent automatic channel changeover but insufficient to provide fuel quantity indication from that channel. The other channel was determined to be working correctly.

Subsequently Boeing chose to offer a 757/767 gauging package that featured a microprocessor-based 'Brick-wall' type of architecture design adapted from that of the 747-400 system. The Brick wall architecture is described in the next section.

4.5.1.3 Brick Wall vs Dual Channel Architectures

There has been much discussion in the industry over the architectural differences of these two fundamentally different gauging approaches. A 'brick wall' system is one in which the gauging of each tank is configured so as to have total independence to the extent that any failures arising from gauging faults associated with one tank can not propagate to effect or disable the gauging of any other tank. Figures 4.18a and 4.18b show how the duel channel architecture of the original 767 aircraft differs from the subsequent brick wall design that was offered as an alternative following the Gimili incident described above.

The brick wall system has been most strongly advocated by Boeing for most of its commercial aircraft with the exception of the 717 and 757/767 families of aircraft. The 717 architecture with its manual changeover dual redundant architecture was inherited from McDonnell-Douglas with the acquisition of that manufacturer's range of aircraft.

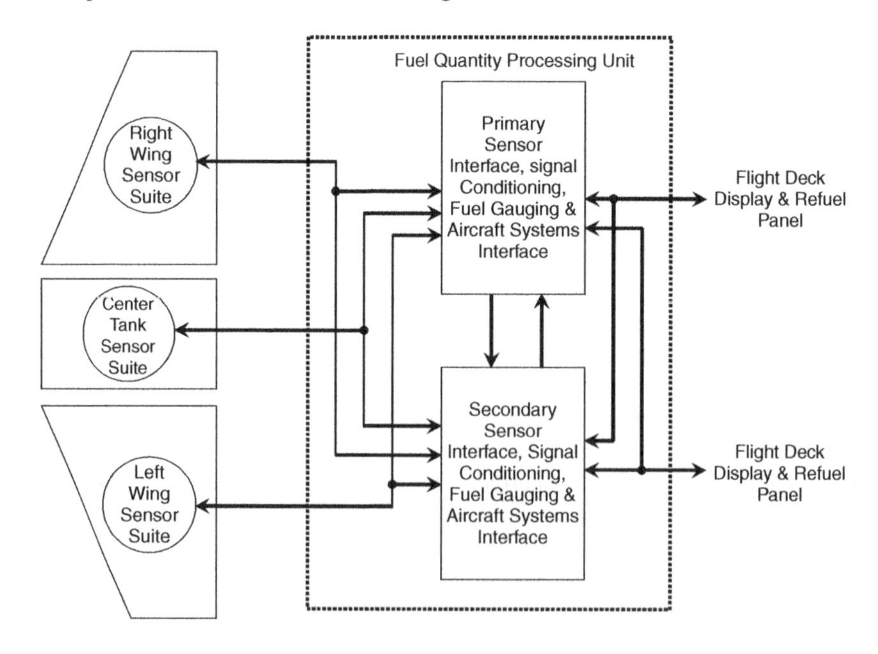

Figure 4.18a Boeing 767 dual channel gauging system schematic.

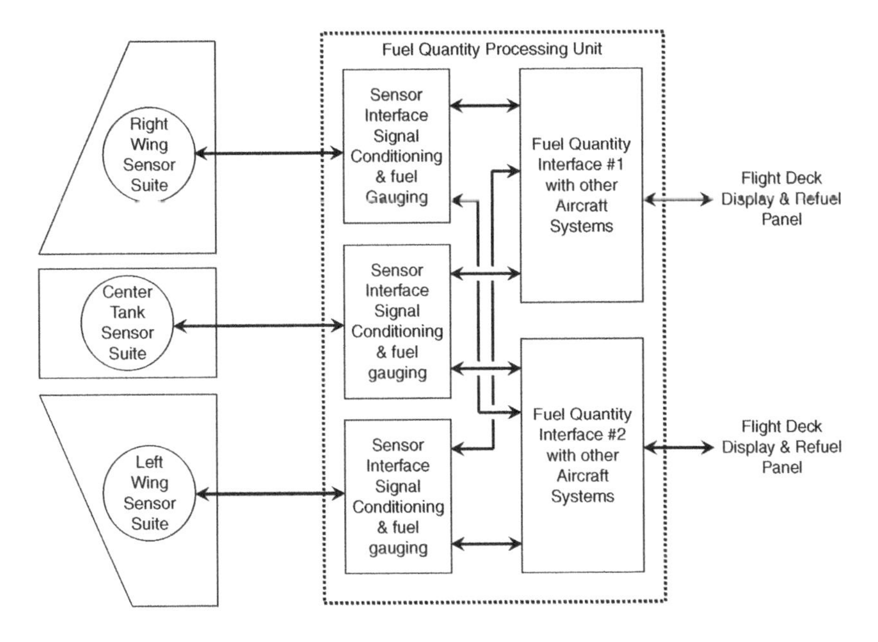

Figure 4.18b Boeing 767 brick wall gauging system.

4.5.1.4 Dual Channel vs. Dual-Dual Channel

The integrity and availability requirements of a gauging system are set by the role of the aircraft. In LROPS (Long Range OPerationS), for example, the operating range and intended routing of the aircraft can have a direct bearing on the architecture of the system since the potential ramifications of an erroneous indication may well be classified as catastrophic in the Functional Hazard Analysis.

The dual channel and dual-dual channel architectures are implementations of dual redundant designs discussed above. Whereas a dual channel architecture features two single processor channels, one operating as prime and the other as standby, a dual-dual channel architecture features two dual processor channels (see Figure 4.19).

In the dual-dual architecture each channel contains two microprocessors; one operating as a command processor and the other acting as a monitor

Again one channel (pair of processors) acts as prime and the other as standby. The key difference provided by the dual-dual channel over the dual channel architecture is the ability to compare processor outputs within each channel thus reducing the probability of an undetected error. The dual-dual architecture can therefore demonstrate compliance with a higher safety requirement.

From the hardware perspective, a single processor channel can be shown to demonstrate compliance with a safety requirement (in terms of probability of loss of function) of better than 1 in 10^5 operating hours. A dual channel architecture using identical hardware in each channel can be shown to demonstrate compliance with a safety requirement of better than 1 failure in 10^7 hours.

Compliance with a safety requirement of better than 1 failure in 10^9 hours can only be achieved by a dual channel implementation, per ARP 4754 reference [1], using a combination

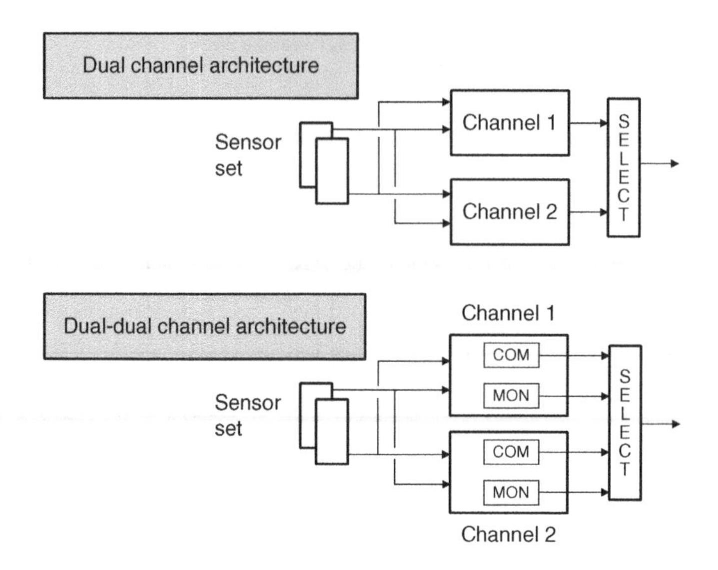

Figure 4.19 Dual channel and dual-dual channel architectures.

of dissimilar hardware and dissimilar software; with software implemented at Level A per RTCA DO-178B reference [7] within a dual-dual architecture.

To achieve this design integrity objective, the dual-dual channel architecture may be configured in one of two ways:

- Each channel is arranged to have the two dissimilar processors executing the system function using dissimilar level A software, failure to compare resulting in channel shutdown.
- Each channel is arranged to have two dissimilar processors, one executing the system function and the other in a monitoring role, failure to satisfy the monitor resulting in channel shutdown.

A word of caution regarding the dual-dual architecture is worth mentioning here. The original premise associated with this architecture is to have two processors within each channel, each executing identical software using the same sensor data. This ability, to do a bit-for-bit comparison of the digital computation process is the cornerstone of the dual-dual integrity concept.

In order to ensure, therefore, that this premise is maintained, both the command and monitor microprocessors must be synchronized in time. If not, the logic executed within the command and monitor machines can become a full clock cycle out of sync leading to inadvertent channel switchover.

This comment is particularly relevant to the fuel management function described in the next section but is mentioned here because it is common practice to integrate both the gauging computation and fuel management functions within the same computer hardware.

4.5.2 Fuel Load Planning

An extremely important aspect of every flight that relates to the fuel quantity gauging system is Fuel Load Planning. This task involves the flight crew and airline dispatcher who together must ensure that the appropriate fuel load is established before each flight and becomes an integral part of the flight plan that is filed under Instrument Flight Rules (IFR). Clearly this function becomes especially critical on long-range, trans-oceanic flights where weather and wind forecasting play a major role in establishing the optimum load. Operating regulations require that, as an absolute minimum, the fuel load must be sufficient to ensure that there is:

- an additional 5 % margin to allow for gauging inaccuracies;
- sufficient additional fuel to allow diversion from 1000 ft above the planned destination airfield to 1000 ft above the planned alternate airfield;
- additional fuel to allow 30 minutes of flying time after reaching the alternate airfield.

The fuel load for the intended flight is normally determined by rotary slide or by computer using the predicted route and the prevailing head or tail winds. The plan may be amended by negotiating a different alternate airfield en route; a process referred to as airborne dispatch.

4.5.3 Leak Detection

Fuel leaks can be a serious threat to aircraft safety particularly in long-range over-water situations where loss of fuel can have potentially catastrophic results as demonstrated by the Air Transat event described in Chapter 1. In this case an engine fuel leak led to loss of both engines while en route over the Atlantic. Luckily the aircraft made a successful emergency landing on the Azores.

Fuel system integrity begins with the refuel process where the accuracy of the total fuel on board is particularly critical to the safety of long-range operations. Here the fuel uplifted from the ground refueling system must be satisfactorily reconciled with the residual fuel and the final fuel load.

Once the engines are started, the fuel flowmeters on each engine monitor the fuel consumed by each engine and by integrating this information, an estimate of the remaining fuel on board can be made.

Engine flowmeters typically measure mass flow directly and are accurate to better than 1 % of point at the cruise condition. Therefore, for cruise flows of, say 5000 pounds per hour, the worst case error per flow meter would be 50 pounds per hour.

For a twin engine transport therefore, it is possible to establish a worst-case boundary against which the fuel quantity remaining, per the fuel quantity gauging system, can be compared with in order to assess the probability of a fuel leak.

Figure 4.20 shows a simplified graph of fuel quantity versus time for a typical long-range mission to illustrate this point.

Following take-off and climb, the fuel consumed stays approximately constant assuming a constant cruise operating condition. The engine flowmeter information is integrated to provide an estimate the fuel consumed. The flowmeter inaccuracy creates an error band around this estimate that grows as the flight progresses. By comparing the worst case fuel remaining with the fuel gauging information the probability of a fuel leak can be determined. Allowance for gauging errors must of course be taken into account.

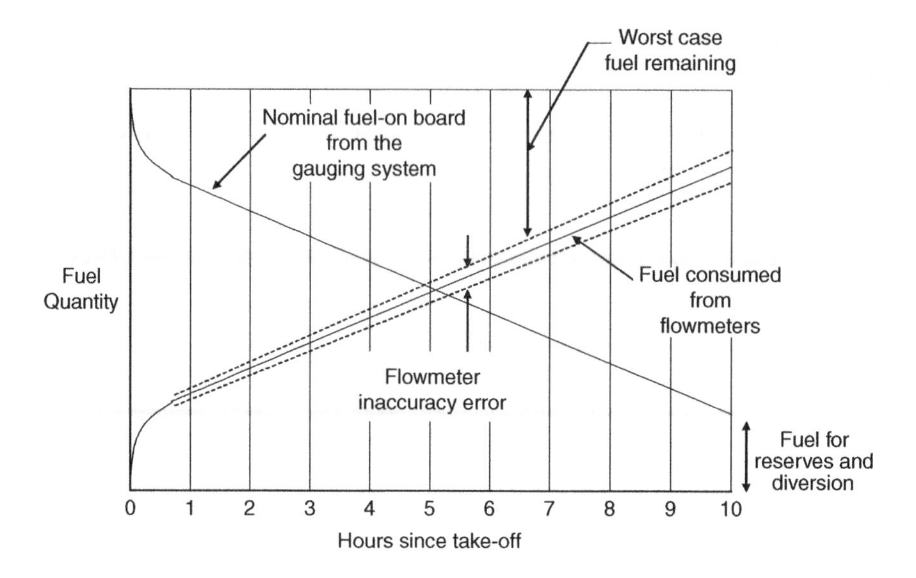

Figure 4.20 Leak detection during a ten hour flight.

4.6 Fuel Management and Control

The fuel management system provides coordination, control and monitoring of all fuel system functions.

In small regional aircraft with simple two tank fuel systems, the fuel management task is limited to the automatic refuel process and accommodation of the engine out situation using the crossfeed system. Here the refuel system is provided with gauging system information to facilitate automatic shut-off of the refuel valves when the preset fuel level has been reached and the crossfeed function is usually controlled directly by the crew.

In large long-range transports with multiple tanks and operational modes, the fuel management task can become much more complex as indicated in Figure 4.21 which illustrates how the fuel management system interfaces within both the fuel system and the overall aircraft.

The level of complexity of this type of system requires that most of the fuel management system is automatic in order to minimize crew workload.

As indicated by the figure, the fuel management and gauging systems may be implemented as an integrated fuel measurement and management system.

The most significant feature shown in the figure is the extent to which many of the aircraft systems provide or receive key information associated with the measurement and management functional process.

On the flight deck and in the avionics suite the interfacing functions and systems include:

- the Flight Management System (FMS) which keeps the fuel management system advised as to where in the overall flight plan the aircraft is and to allocate auxiliary fuel tank contents appropriately;

- the Flight Warning Computer (FWC) receives fuel system status advisories and warnings for display the crew;
- the Display Management Computer (DMC) and Multifunction Displays (MFDs) which receive fuel system status information and present this information together with a system synoptic to the crew on a priority basis;
- the Overhead Fuel Panel (OHP) which presents fuel quantity information to the crew on a tank-by-tank basis as well as the total fuel on board; in addition, the overhead panel typically incorporates switches that allow the crew to select fuel system pumps and valves directly in the event of a failure of the automated management system;
- the Inertial Navigation System (INS) information which is typically made available to the fuel system to provide supplementary aircraft attitude information in support of the fuel gauging system;
- engine master switches (ENG 1and ENG 2) which typically provide signals to the fuel system to open the low pressure feed valves to make fuel boost pressure available to the engine(s);
- finally the landing gear status, 'Weight-On-Wheels' (WOW), which advises the fuel system that the aircraft is either on the ground or in flight; the ground status allows the refuel panel and display to be powered.

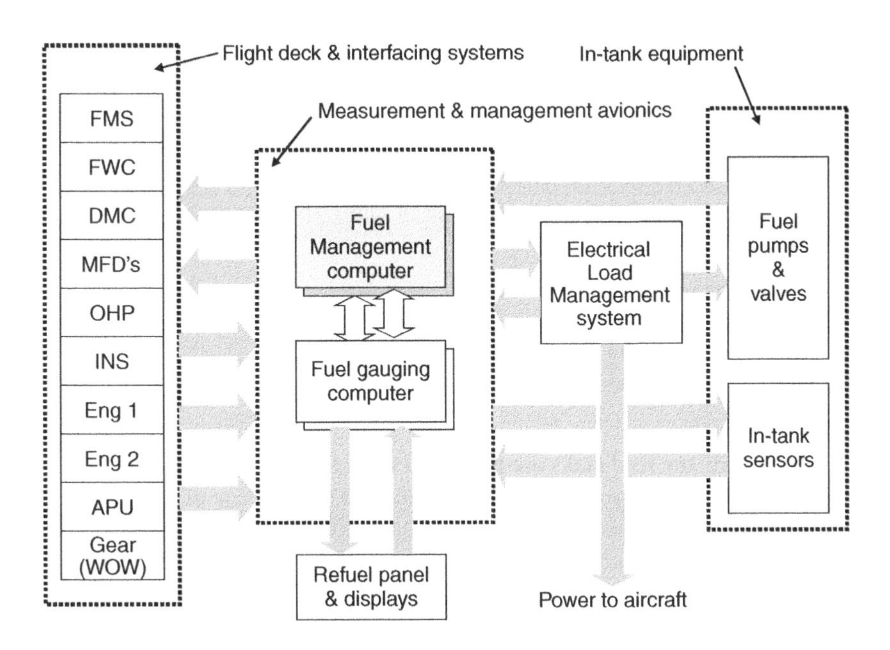

Figure 4.21 The fuel management system and its interfaces.

The electrical power management system also plays a key role in the fuel management system since loss of electrical power to feed and transfer pumps can incur system failures with a 'Serious' functional hazard assessment. For example, loss of boost pump power at high altitudes could result in engine flame-out. This situation can be accommodated by monitoring the status of the electrical power management system regarding power bus availability and when the situation is reduced to the point where one additional failure could result in a loss of boost pressure, the fuel management system would select an auxiliary DC pump on-line as a back-up source of feed line pressure.

The following paragraphs describe fuel management and control modes typical of most commercial transport aircraft including some of the issues and challenges involved.

4.6.1 Refuel Distribution

In large commercial transports, with many auxiliary fuel tanks, it is important to direct the uplifted fuel into the appropriate tank locations to ensure that safe balance is maintained throughout the refuel process. Of particular concern is when the aircraft uses a trim tank well aft of the aircraft's longitudinal center of gravity (CG). The horizontal stabilizer trim tank is a good example of this situation. Such tanks are often intended to provide control of the aircraft's CG during flight in order to optimize range. (A detailed discussion of active CG control is presented later.)

If attention to this issue is neglected, and trim tank fuel is loaded early in the refuel process, it is quite possible, in view of the large moment arm, to cause the aircraft to tip over onto its tail section causing substantial damage. Similarly, lateral balance must be maintained to avoid asymmetric loading of the landing gear.

For this reason, and to minimize the ground crew workload, an automatic refuel distribution function is typically available via the fuel management system.

Figure 4.22 which shows the refuel distribution process for the A340-600 aircraft illustrates this point clearly. This aircraft has a large number of tanks, specifically:

- four feed tanks (one per engine)
- two outboard feed tanks (one in each wing)
- a center tank located between the wings and below the cabin
- a trim tank located within the horizontal stabilizer.

Referring to the figure, the feed tanks are first filled to three tones, each making the total fuel upload 12 tonnes. Outboard wing tanks are then filled to about 95 % of capacity before the feed tank fill process is re-started. When the total fuel on board reaches about 86 tonnes, the center tank and trim tanks begin filling. Also at this point the outer feed tanks (which are about five tonnes smaller than the inner feed tanks) stop filling. At 110 tones, the outer feed tanks stop filling and the center tank and trim tanks continue filling. At 144 tonnes on total fuel load, all tanks begin the final topping-up to a maximum of 171.5 tonnes based on a fuel density of 0.88 kilograms per liter.

For lower fuel densities, volume limits will stop refueling at some lower quantity.

This auto-refuel distribution algorithm greatly simplifies the refuel task from the ground operations perspective, however, there are significant variations in both the ground refuel facilities and in the aircraft on-board control system equipment that provide an additional complexity dimension that is worth some further explanation here.

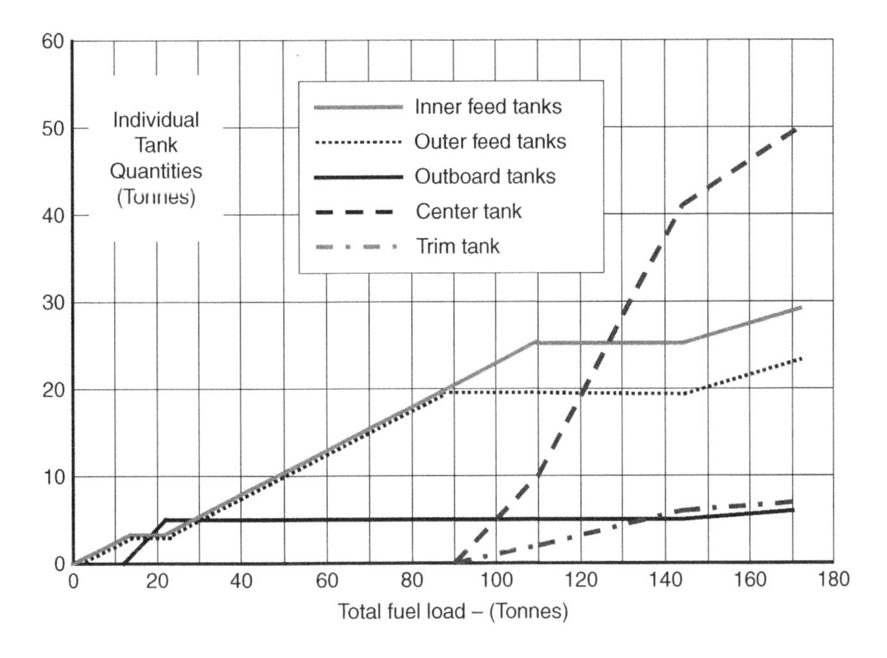

Figure 4.22 A340-600 Refuel up-lift distribution.

4.6.1.1 Auto-refuel Accuracy Issues

In order to load an accurate quantity into each tank, the fuel management system monitors the quantity outputs from the gauging system and commands the refuel valve(s) closed when the set value is reached.

The problem, however, is that there are a number of significant variables involved in this process. For example:

- Refueling pressures from the ground hydrant system can vary significantly from airport to airport and refuel truck to refuel truck. The resulting differences in the fuel flow rate into the tank affect the 'overshoot' (the quantity of fuel passing through the shut-off valve from the time of command to achieving the fully closed position).
- Variations in valve closure rate due to shut-off valve performance variation will also affect the fuel overshoot quantity. For example, motorized shut-off valves can exhibit closure time differences of two to one as a result of power supply voltage variations and other design tolerances.
- There is usually some latency in the gauging information associated with the quantity calculation and filtration algorithms.

The combined effects of the above features can result in some errors between the nominal and achieved tank upload quantities.

Typically, the automatic refuel system is required to maintain tank upload quantities within a predetermined limit which, if exceeded, results in an auto-refuel abort. The refuel process may be continued using the manual refuel mode.

There are a number of upload accuracy improvement techniques that can be employed to minimize the frequency of auto-refuel aborts. One such method is to have the fuel management software 'Learn' the refuel valve characteristics over time and to adjust the shut-off anticipation algorithm accordingly.

4.6.2 In-flight Fuel Management

In-flight fuel management involves ensuring that the fuel on board is located appropriately via the control of pumps and valves in order to ensure continued safe flight and engine operation. In many cases this task is left to the flight crew, particularly where simple fuel system architectures are involved. In today's two-person cockpit crew philosophy however, it becomes important to minimize crew workload associated with utilities systems management so that maximum focus can be maintained on flying the aircraft.

Notwithstanding this observation, there is also a point of view that maximum simplicity in systems implementation results in better function availability and hence dispatchability. Also by minimizing system complexity, aircraft maintenance and direct operating costs will be lower. From a fuel systems perspective this philosophy has been followed in most of Boeing's commercial transport aircraft. For example in the twin engined 737, 757, 767 and 777 aircraft the crew is responsible for correcting any lateral imbalance that develops during flight by opening the crossfeed valve and de-selecting the engine boost pumps on the light side of the aircraft. Thus fuel from the heavy side feeds both engines until lateral balance is achieved. Since the need for lateral balance action occurs only on rare occasions (due to engine burn differences for example) it is not considered to be a significant imposition of workload on the crew.

Airbus fuel system management philosophy is to provide automatic fuel management of all transfer and balance functions except in the event of equipment failures when the crew is required to operate the system manually. On many Airbus aircraft the fuel management system includes active control of longitudinal CG by moving fuel between a trim tank in the horizontal stabilizer and the forward fuel tanks (see 4.6.2.2 below).

4.6.2.1 Control of Fuel Burn Sequencing

As aircraft become larger and more complex in terms of the fuel storage design, the need for more complex fuel burn fuel transfer arrangements become necessary.

To illustrate this point, Figure 4.23 shows the fuel burn schedule for the Airbus A340-600 which has a total of eight storage tanks comprising:

- four feed tanks (one per engine)
- two outboard wing tanks
- a center tank between thee main wing tank structure
- a trim tank located in the tail horizontal stabilizer.

In this application all of the fuel network control valves are motor-operated and in the event of valve failures, a number of 'work-arounds' are available to provide the required transfer functionality. This may require the use of certain valves and galleries not involved in normal (fault-free) flight operation under the control of the automatic fuel management system. In addition to these automatic 'work-arounds' the flight crew has the ability to control fuel transfer manually.

The figure shows the transfer sequencing programmed into the fuel management system. As shown, the center tank is used to keep the four feed tanks topped up via center-to-feed transfer. Transfer is selected when feed tank fuel falls below a predetermined level and deselected when the feed tanks are full. Once the center tank is emptied, the feed tanks are depleted until the trim tank is transferred to the feed tanks after the last way-point in the flight plan is passed.

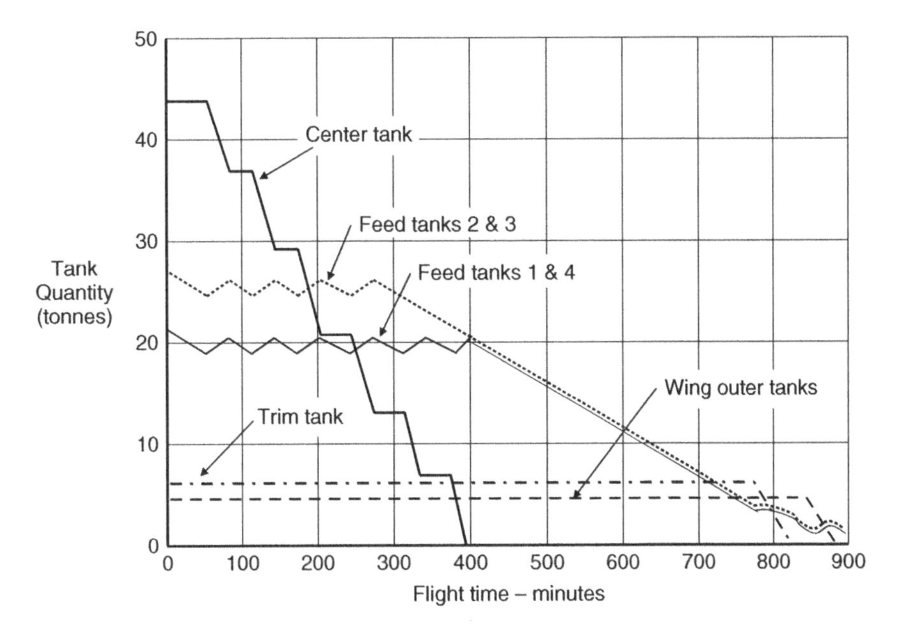

Figure 4.23 A340-600 fuel burn/transfer sequence.

Finally the outer tanks are emptied into the feed tanks as some minimum feed tank quantity is reached. This figure is a simplification of what occurs in practice where, superimposed on the sequence shown, the active CG control function transfers small quantities of fuel between the trim tank and the forward tanks in order to maintain the correct aircraft longitudinal CG. See the next subsection for more detail.

4.6.2.2 Active Longitudinal CG control

Active longitudinal CG control was introduced by Airbus on their A330 and A340 long-range aircraft as a means to minimize trim drag during the cruise phase. This is accomplished by managing the quantity of fuel in a trim tank located within the horizontal stabilizer as fuel is consumed.

It should be noted, therefore, that for the aircraft to be statically stable, the aircraft CG must be located in front of the center of lift. The challenge in active CG control is ensure that the aircraft remains safely controllable even in the presence of equipment failures.

The principle of active CG control is illustrated in Figure 4.24. In this example which shows a swept wing aircraft with a center tank, two wing tanks and an aft trim tank. CG varies as fuel is consumed. The quantity of fuel initially loaded into the trim tank will depend upon the total fuel load. As fuel is consumed the aircraft CG will move aft as center tank fuel is burnt and then forward as wing tank fuel is consumed. The CG control system transfers fuel forward and aft as the cruise continues in order to maintain the aircraft longitudinal CG within predefined limits. The outer wing tanks normally remain full until close to the end of the cruise phase for wing load alleviation. Typically, the active CG control mode is only operative during the end of climb and cruise phases of flight. As the last way point in the flight plan is reached, the Flight Management Systems (FMS) notifies the Fuel Management System that it is time to transfer any fuel remaining in the trim tank forward into the feed tanks.

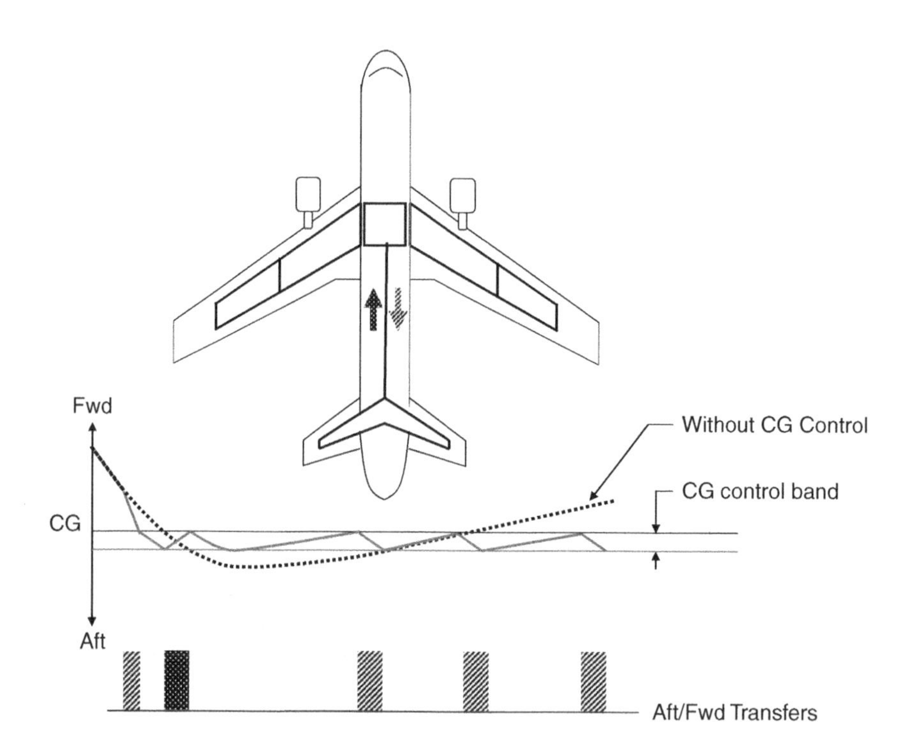

Figure 4.24 Aircraft CG control example.

Incorporation of the necessary tanks and equipment necessary to perform active CG control is significant in terms of both acquisition and operational cost and therefore it is important to demonstrate during the design process that this function makes economic sense from a cost/pay-back perspective. Clearly Boeing and Airbus disagree on this issue based on the fuel system implementations of competing aircraft.

4.6.2.3 Fuel Jettison Control

In view of the criticality of the fuel jettison function initiation of the process can only be done by the flight crew. To prevent inadvertent operation, the system typically uses a two-step selection process whereby the system must first be armed and then selected via a guarded switch. The automatic aspect of the jettison function is to allow the crew to select a fuel quantity at which the jettison is stopped. This is important since while jettison is proceeding the crew may be extremely busy dealing with an emergency. The default fuel remaining quantity will correspond to the maximum landing weight.

Typical jettison system designs arrange for jettison pump inlets to become uncovered at some safe low fuel state thus ensuring that a system fault cannot result in jettison continuing beyond this point.

4.6.3 Fuel Management System Architecture Considerations

Fuel management architecture depends primarily upon the functional integrity objectives that are derived from the operational and safety drivers that are, in turn dependent upon the operational objectives of the aircraft. Many of these issues have already been discussed in Chapter 2 of this book.

While the above (and following) commentary refers primarily to long-range transports with challenging fuel management tasks, the principles discussed herein apply to all aircraft fuel systems no matter what level of complexity is involved.

While the goal of providing automatic fuel management functionality is desirable from the perspective of minimizing flight crew workload, it becomes necessary to consider the implications of 'Loss of function' due to potential failure modes within the fuel management system. This includes the failure probabilities associated with sensor and/or computer failures that may result in loss of critical management functions. Functional integrity requirements that evolve from the design drivers discussed in Chapter 2 demand that specific levels of functional integrity be available and demonstrable to the certification authorities.

The system architecture issues discussed under fuel quantity gauging can be applied directly to the fuel management function. It is common practice today to integrate the gauging and management functions into common avionics and therefore the gauging system architecture and integrity discussion provided in Section 4.5 applies here directly.

Provision of a high level of fuel management functionality integrity is best obtained by providing the crew with a manual back-up capability where transfer control can be selected manually by the crew in the event of an equipment failure.

This approach provides an automatic mode functional integrity of 10^{-7} failures per flight hour for a dual-dual architecture using common software within each channel. The manual back-up capability increases functional integrity to 10^{-9} failures per flight hour which covers the most severe system safety assessment category.

4.6.4 Flight Deck Displays, Warnings and Advisories

Today's modern transport aircraft use Active Matrix LCD flat panel displays to provide all flight attitude, navigation and system status information. Figure 4.25 is a photograph of the Airbus A330 flight deck showing the five flat panels. The primary flight and navigation displays are presented in horizontal formation in front of the Captain and First Officer and the system displays are installed vertically in the center console. These two displays typically show a system synoptic display on one of the panels with status information on the second display.

There is also a fuel system control panel usually installed in the overhead panel. This panel shows pump and valve status. This panel allows the crew to select pumps and valves in the event of a system failure.

The fuel gauging and/or management computers provide fuel system status information via digital data buses with sufficient redundancy to meet the airworthiness design assurance levels required for the display function. The display electronics typically uses triplex redundancy with three separate display management computers in order to support the required display integrity.

Figure 4.25 The A330 flight deck (courtesy of Airbus Industrie).

Figure 4.26 shows the synoptic page for the Airbus A340-600 aircraft. From this display, the crew can see at a glance:

- the total Fuel On Board (FOB)
- the fuel used so far in the flight
- the fuel flow to each of the four engines
- the fuel quantity and fuel temperature in each tank (except for the center tank which does not have a temperature sensor)
- the pump and valve status
- static and total air temperatures
- flight time since departure
- gross weight and CG information.

In the synoptic, it can be seen that all four LP shut-off valves are open while all the crossfeed valves are closed. Also the feed pressure to each engine is provided by the primary boost pump while the back-up pump is shown deselected.

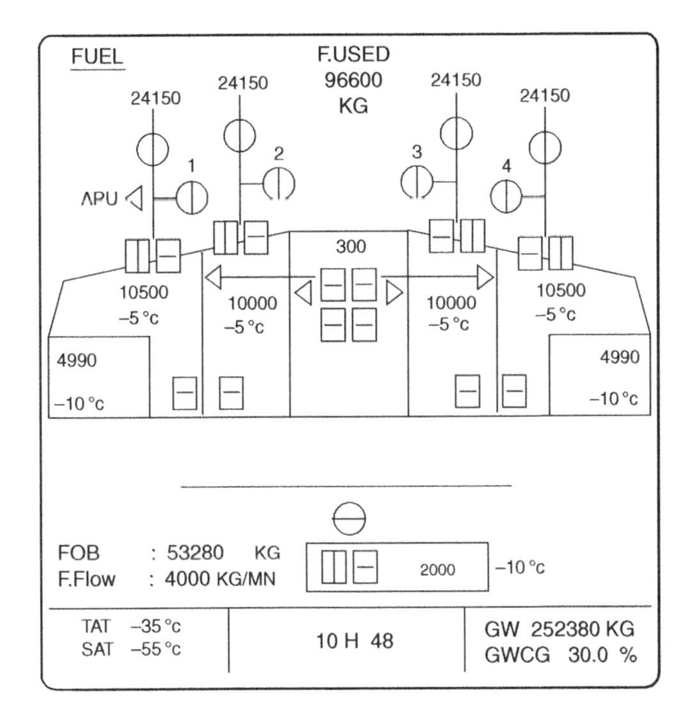

Symbol	Status
	Fuel pump in stand-by mode
	Fuel pump in operation
	Valve closed
	Valve open

Figure 4.26 Airbus A340-600 fuel synoptic display (courtesy of Airbus Industrie).

In the event of a system fault, the status page below the synoptic will provide warning, caution and advisory messages. Warnings are presented in red, cautions in amber and advisories in white. At each abnormal event occurrence, the 'fuel page' is selected automatically to ensure that the crew is immediately aware and to demand action from the crew before the display selection can be changed.

4.7 Ancillary Systems

Aircraft fuel systems are frequently required to provide ancillary functions most commonly related to thermal issues associated with the aircraft and engine systems. Because of the very large quantities of fuel involved particularly with long-range applications, the fuel system becomes very attractive for use as a sink for excessive heat that can be generated in many of the aircraft systems.

The aircraft hydraulic system's biggest challenge is heat dissipation due to the fact that the many services requiring hydraulic pressure have quiescent leakage through servo valves even not in use. This leakage manifests itself as waste heat and without a convenient source to dump this heat, excessively high hydraulic fluid temperatures can occur. Air-oil coolers (heat exchangers) provide one option; however, they do introduce drag with obvious operational penalties. The natural alternative is to use fuel-oil heat exchangers to transfer fuel in to the fuel. When large quantities of cold fuel are available, this approach appears to be attractive; however, consideration must be made of the full range of operational conditions that can occur and provisions made to ensure that there are no safety risks involved, for example:

- Location of the heat exchanger in the fuel storage system. Under what circumstances can the heat exchanger become uncovered and if so can fuel or oil over-temperature situations occur?
- During ground maintenance operations with low fuel quantities on board, can unsafe fuel or oil temperatures occur and if so what provisions can be made to protect against this situation?

Another common area that is often used to take advantage of the fuel system as a heat sink is the engine oil lubrication and scavenging system. This system provides oil under pressure to the engine bearings and to the reduction gearbox. The fuel oil cooling system is shown schematically in Figure 4.27. Here the excess fuel from the high pressure fuel pump (the spill flow back to pump inlet) is used to cool the engine lubrication oil.

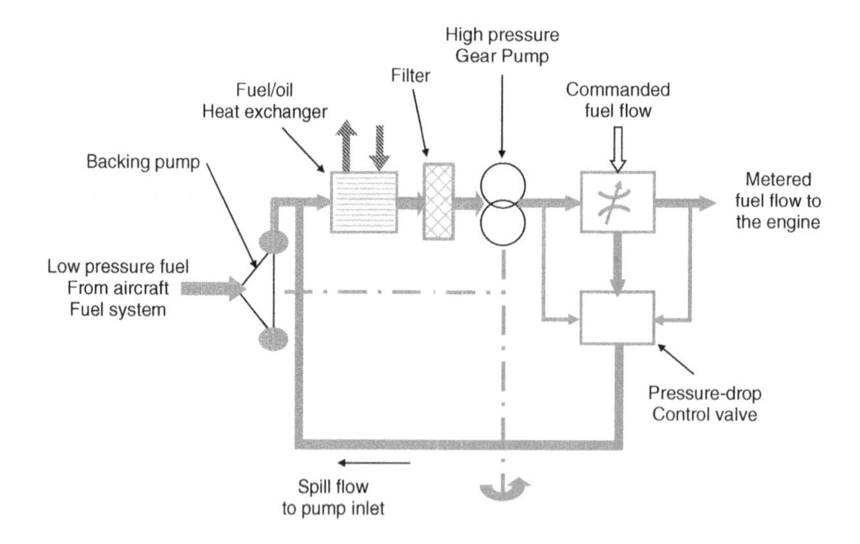

Figure 4.27 Fuel-to-engine oil cooling schematic.

There are constraints to this approach; however, since the maximum fuel temperature at the fuel combustion nozzles must not exceed about 350 degrees F otherwise 'Coking' (carbonization) of the nozzles will occur.

This becomes a challenge when operating at high altitudes, when the fuel flow from the cold fuel tanks is low and the engine mechanical speed is high and therefore spill fuel temperatures can become so high as to be ineffective as a cooling medium.

To alleviate this problem, fuel may be recirculated into the engine feed tanks; however, this in turn can introduce problems within the fuel system, particularly with the fuel quantity gauging system. This system is highly dependent upon the properties of the fuel as it affects density and permittivity (for capacitance gauging). Since the premise for accurate gauging is that the fuel stored within the tanks is of a homogeneous and consistent form, introducing pockets of fuel within the tanks that have significantly different characteristics due to local heating, can significantly impact the accuracy of the gauging system.

This situation also applies to recirculated fuel for the purpose of maintaining tank fuel temperatures above the fuel freeze point with adequate margins during high altitude, long-range operations.

Using the engine as a source of motive flow for ejector pumps is not as critical, since the motive flow is typically low and mixes with the tank fuel within the discharge piping.

5

Fuel System Functions of Military Aircraft and Helicopters

This chapter describes the fuel system functions associated with modern military aircraft from the fixed wing combat fighter up to the much larger transport and tanker aircraft. Fuel system issues associated with rotary wing vehicles are also included in this chapter. The topics covered focus on the system-level aspects of these aircraft where they differ from the methodologies employed by commercial aircraft which were covered in the preceding chapter.

The functions described herein cover refueling, fuel transfer, engine feed and fuel management. Two areas of particular relevance to military applications are concerned with fuel tank pressurization (coming from the need to operate at extremely high altitudes) and aerial refueling, which is an essential feature of almost all military aircraft. These functional requirements bring with them unique fuel system design features in order to ensure safe and secure execution of critical military missions between take-off and landing from either land or carrier bases of operations.

The schematic diagram of Figure 5.1 is an overview of the fuel system design process which emphasizes the system-level functions that are associated with the contents of this chapter and how they relate sequentially to the overall design and development process.

Where the military function is identical or very similar to the commercial function, the reader is directed to the applicable commercial section.

In addition to the system functions outlined in Figure 5.1, ancillary functions may also be provided by the fuel system; as with commercial aircraft these functions depend upon the fuel system as a heat sink which can lead to significant design and operational issues that must be addressed early in the design of the fuel system and the aircraft.

The following sections address each of the major fuel system functions separately.

Aircraft Fuel Systems R. Langton, C. Clark, M. Hewitt, L. Richards

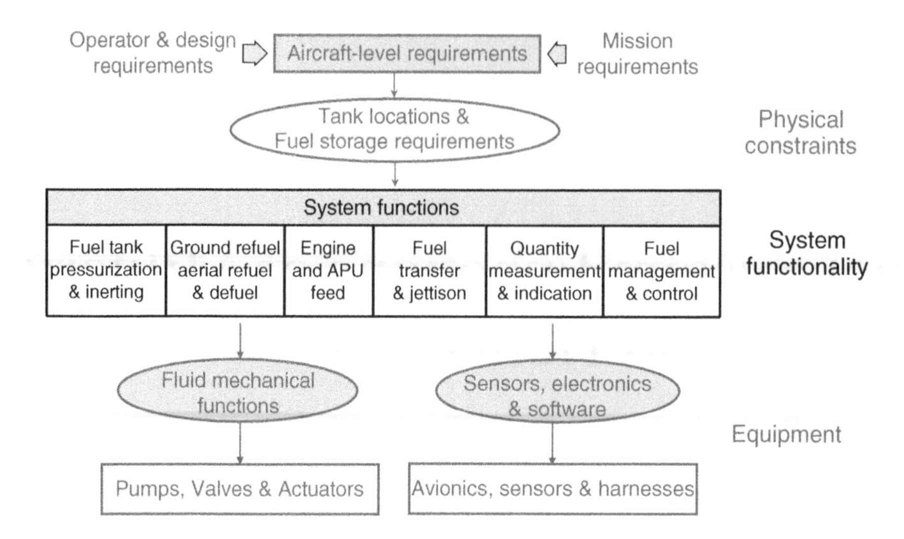

Figure 5.1 Military aircraft design process overview – system functionality.

5.1 Refueling and Defueling

Ground refueling and defueling of military aircraft is very similar to that for commercial aircraft. Although military aircraft are typically refueled at military bases, they utilize the same type of refueling equipment as used at commercial airports. The most significant issues leading to additional complexities associated with refueling of military aircraft are related to aerial refueling which is described in detail later in this chapter. Carrier-based aircraft operation also imposes special refueling procedures based on the demanding safety requirements associated with such applications. Military aircraft typically do not utilize gravity, or over-the-wing refueling.

All current day military jet aircraft use pressure refueling with the possible exception of small unmanned military aircraft. Because military combat aircraft tend to put fuel everywhere possible, there tends to be a large number of small fuel tanks in wing and fuselage areas. Although today's combat air vehicles carry considerably less fuel than a commercial transport aircraft, the refuel/defuel system of a military aircraft is typically much more complex than a commercial aircraft. The large number of small fuel tanks also greatly complicates fuel transfer, engine feed, fuel tank venting and the fuel management subsystems. Figure 5.2 illustrates the large number of fuel tanks potentially required in combat aircraft.

5.1.1 Pressure Refueling

As with commercial aircraft the fundamental need for pressure refueling of a military aircraft is to provide safe, quick aircraft deployment. In designing the refuel system, there is typically a contracted refueling time. The required refueling time will always result in the need for a pressure refueling system. Military aircraft present additional safety issues that need to be considered in the refuel system when carrier based operations and/or aerial refueling is required.

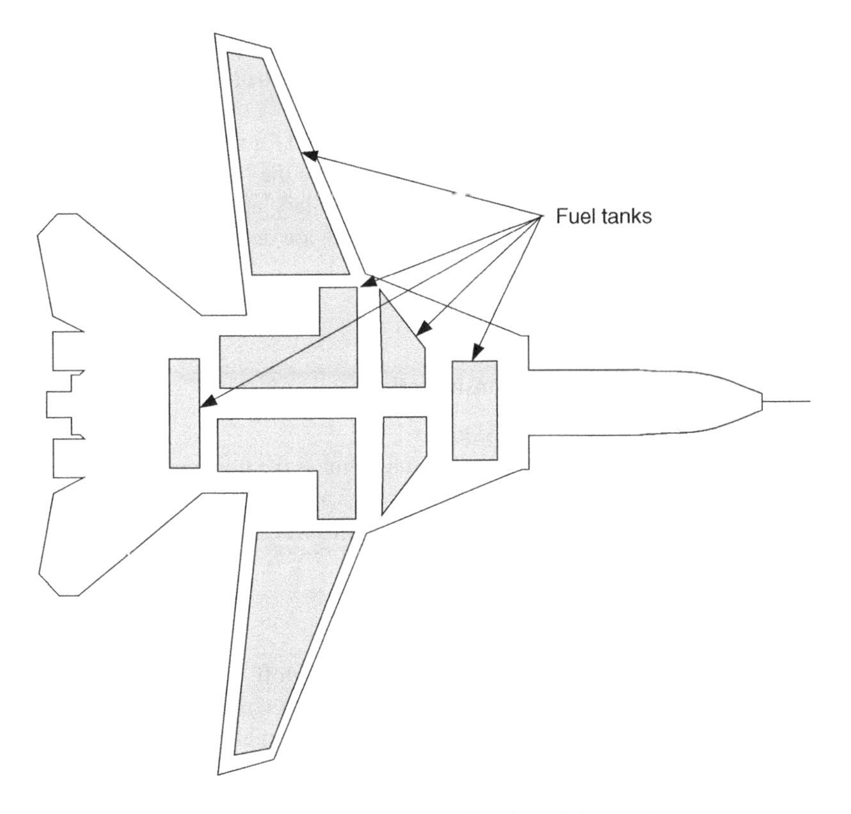

Figure 5.2 Representative combat aircraft fuel tankage.

Failure of a refuel system to shutoff on an aircraft carrier can result in a hazardous fuel spill while failure to shutoff during aerial refueling could be equally hazardous.

Typical military aircraft refueling system requirements include:

- fast refuel times for combat aircraft, not as critical for large transport aircraft;
- for most aircraft platforms, aerial refueling capability;
- ground and aerial refueling of aircraft external stores (when equipped);
- accurate loading of the required fuel quantity, often via an automated system on board the aircraft that allows the refuel operator to preset the total fuel load required at the refuel station. (similar to commercial aircraft);
- accurate location of the fuel on board to ensure that aircraft CG limits are complied with;
- protection against overboard spillage of fuel out of the tank vent system (especially critical for aircraft carrier based operations where a resulting fire could have catastrophic implications);
- protection against over-pressurizing of the tank structure with failure of the refuel system to shut-off similar to commercial aircraft);
- the refueling system aboard the aircraft must be compatible with facilities at commercial airports and military bases throughout the world.

The military ground refueling equipment has the same physical interfaces as on commercial aircraft. The commercial ground refueling standard originated from the United States Military Standard MS24484 and is described in detail in Chapter 4.

There are two different aerial refueling systems being used for military aircraft. The 'Flying Boom' system is used primarily by the US Air Force while the 'probe and drogue' system is used by the US Navy, US Marines and international military operations. Aerial refueling is not used on any commercial aircraft. Both these systems are described in greater detail later in this chapter.

5.1.1.1 Refuel System Architecture Considerations

Single Point and Multipoint Refueling Stations

Most military aircraft only require a single refueling point as the flow rate required to meet the refueling time is typically within that achievable through a single ground refueling connection. Larger transport and tactical aircraft may require multiple refueling points to meet the required refueling times.

Number, Size and Configuration of Tanks

The number of tanks determines the number of refuel shutoff valves and the size of each tank determines the flow rate required into that tank which determines line and valve sizing. The configuration of the tanks (sealed/baffled ribs, low/high points, vent space location – see Chapters 3 and 4) determines the location of the fuel entry points and location of high level sensors. The arrangement of the tanks determines if 'cascade refueling' can be used to properly refuel the aircraft. Cascading allows the refuel flow to enter one tank and when full, provides cascade flow to a second tank connected by a fuel line. The cascading system can fill a series of interconnected tanks. The high level sensor is located at the top of the last tank to fill. See Figure 5.3 for an illustration of a cascading refuel system.

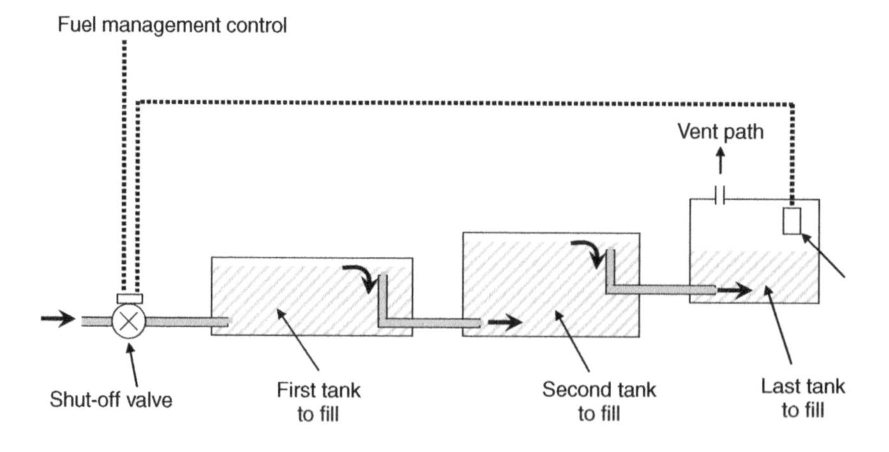

Figure 5.3 An example of the use of a cascading refuel system.

An additional benefit of a cascading system is that it simplifies the vent and fuel transfer systems since it only requires a single vent path in the last tank to fill. Also it provides inherent control of fuel transfer such that the last tank to fill is the first tank to empty.

Dedicated Gallery vs Combined Refuel/Transfer Systems

Both commercial and military aircraft utilize both stand alone refueling systems and refueling systems that are combined with the transfer system however the potential weight benefit of the combined refuel/transfer system approach may prove to be particularly attractive in military applications. Combining these system functions introduces significant design considerations as discussed in the commercial section covering this issue, specifically:

- the serious consequences of a failure of the system to shut-off during an in-flight transfer;
- the high number of operating cycles of the shut-off valves compared with valves used for ground refueling only (which, in turn aggravates the shut-off valve failure rates of combined refuel/transfer valves);
- oversizing of the refuel/transfer valves to accommodate refuel system requirements;
- the need to drain the refuel/transfer gallery to minimize unusable fuel and the subsequent refueling of a completely empty gallery.

Figure 5.4 is an illustration of a combined refuel and transfer system.

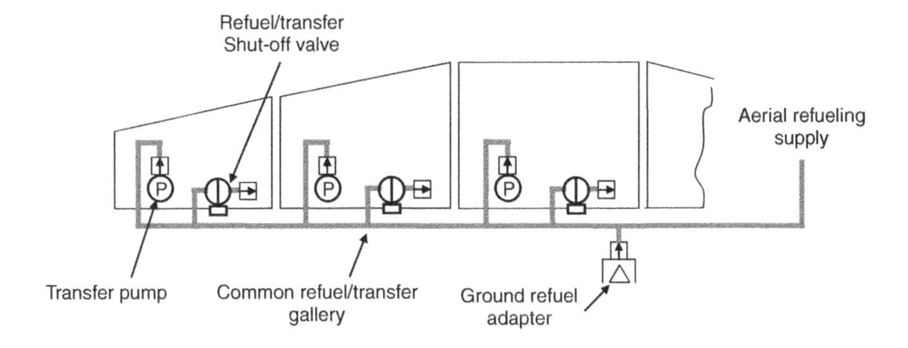

Figure 5.4 Combined refuel and transfer system.

Failure Modes

As with commercial aircraft, failure modes require an in-depth evaluation during the design of the refuel system. Aircraft with aerial refueling capability require special considerations regarding integrity of function since a failure to abort, or to continue the refueling process, could be extremely hazardous for both the receiver and tanker aircraft. As a minimum, a means of protecting the receiver aircraft against a failure to stop the refuel process during aerial refueling must be provided. Specifically the aircraft vent system should be designed to accommodate the maximum aerial refueling flow without exceeding aircraft tank structural limits. In cases where aerial refueling capability has been added to an existing aircraft, such as the C-141, the vent system sizing is unlikely to be capable of handling a failed open refuel valve flow rate and therefore extensive system redundancy must be provided. Dual refuel shutoff valves are frequently used to provide redundancy of the refuel shutoff function. Typically dual refuel valves have dual shutoff features in the shutoff valve and dual high level sensors with dual pre-check capability.

5.1.1.2 Refueling System Design Requirements

The design process for a new military aircraft fuel system is essentially identical to that for a commercial program. Technical requirement documents are established for each of the major subsystems from top level aircraft configuration/operational requirements. These requirements typically establish system design requirements which influence the following:

- line and component sizing – derived from required refuel time, and available ground refuel 'hydrant' and aerial refueling tanker characteristics;
- control of surge pressure and overshoot – derived from allowable maximum system pressures and final shutoff level;
- allowable line and exit velocities;
- use of electrical or fluid-mechanical equipment – derived from availability of electrical power and preferred equipment failure state; additionally, US Navy aircraft have historically prohibited the use electrical equipment inside fuel tanks (except for quantity gauging components) to provide added safety for carrier based operations;
- fail safe provisions;
- prevention of ignition sources;
- tank over-pressure protection – reference Chapters 3 and 4;
- precheck function – electrical and fluid mechanical methods;
- material selections based on higher fuel system operating temperatures of military aircraft.

5.1.2 Defueling

Defueling of military aircraft is typically more frequent than that of commercial aircraft due to the much harsher operating conditions requiring more frequent maintenance actions and checks. Carrier-based aircraft are typically defueled when not in the operational mode. Defueling of a military aircraft is identical to that of commercial aircraft.

5.1.2.1 Defuel System Architecture Considerations

Several factors influence the defuel system architecture:

- use of suction or pressure or both
- method to shutoff tank defuel when empty (suction)
- allowable remaining fuel
- availability of engine feed and transfer pumps for pressure defuel
- final defuel through in-tank drain valves.

5.2 Engine and APU Feed

Generally, engine and APU feed systems of military aircraft are similar to commercial aircraft. The following summarizes the major differences:

- High performance combat aircraft and long-range strategic strike aircraft typically require much higher engine feed flows than required for similar size commercial aircraft. This is to support super sonic flight and high speed dashes.
- Because military aircraft contain a significant amount of on board avionics and electronic combat system equipment, the fuel is typically used as a heat sink for cooling of this equipment. As a result, feed tank fuel can be very hot compared to that of commercial aircraft. The engine feed pumps need to operate with this high temperature fuel and in some cases this high temperature operation may be with high vapor pressure fuels such as JP-4 at high altitudes. Fortunately, most high performance aircraft have pressurized fuel tanks which makes it possible to pump fuel at high altitude, high temperature conditions.

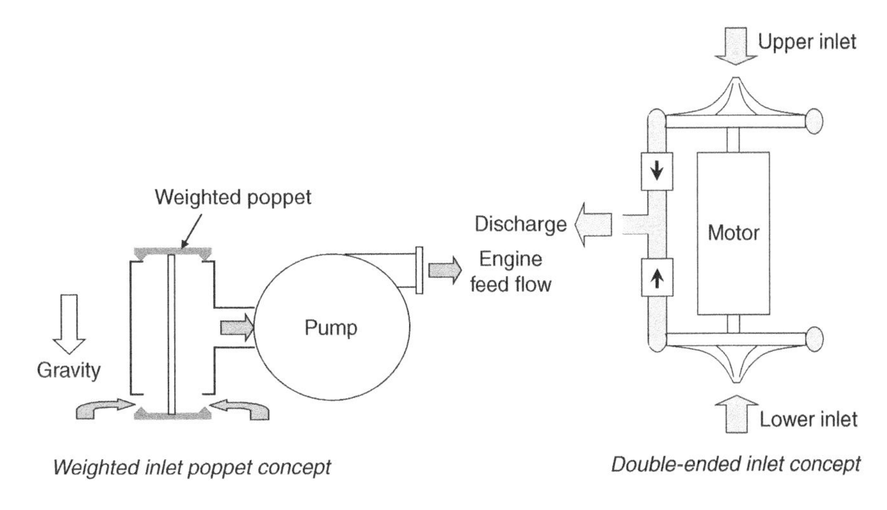

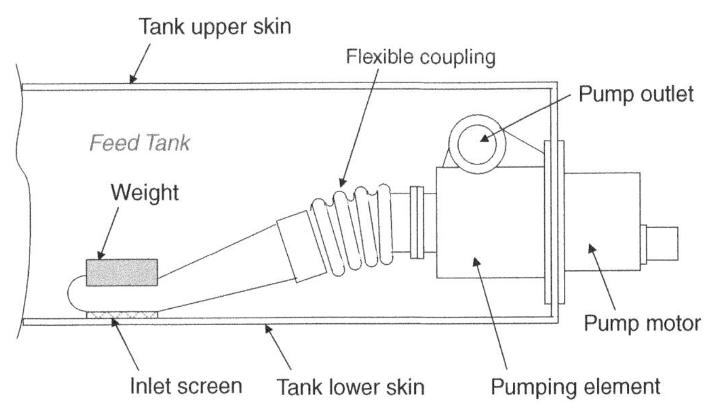

Figure 5.5 Feed pump design concepts for negative g operation.

- Due to the nature of combat aircraft operational requirements, engine feed must be maintained during extreme negative 'g' conditions. There have been various means used to provide this capability including the following:
  - provision of a containment box around the fuel boost pump inlet (this is similar in concept to the collector cell in commercial aircraft described in Chapter 4);
  - the use of double inlet fuel boost pumps within a containment box;
  - the use of a weighted fuel pump inlet shuttle valve;
  - the use of a pendulum-type attachment to the inlet of the fuel pump that moves with changes in the direction of gravity and/or acceleration forces.

Figure 5.5 shows three of the above negative g management concepts used in military aircraft.

All of the negative 'g' schemes have limitations (such as zero 'g') and typically require flight crew operational considerations.

5.3 Fuel Transfer

Fuel transfer systems on military aircraft tend to be more complex than on commercial aircraft. This is especially true of for the combat aircraft with the large number of distributed fuel tanks as illustrated in Figure 5.2. There have been many different transfer systems used on modern-day military aircraft. Basically, the types of transfer systems fall into three categories:

1. gravity
2. transfer fuel pump – electric or motive flow powered
3. fuel tank pressure differential.

Although gravity transfer offers the simplest systems, its use on military aircraft is very limited due to the large number and location (relative heights) of fuel tanks and the limited transfer rates achievable with low differential fuel heads.

Dedicated transfer pumps are probably the most widely used method of fuel transfer in today's military aircraft. Both electric motor driven pumps and motive flow powered ejector pumps are often used to power fuel transfer. Motor driven pumps tend to be used where higher transfer flows and pressures are required or if motive flow is not readily available. Motor driven pumps tend to have better high altitude and high temperature performance than motive flow jet pumps. The use of jet pumps offer the advantages of lower weight, higher reliability, easier installation, lower cost, and greater safety by elimination of the need for electrical power. For this reason, Navy aircraft have used jet pumps extensively on their carrier based aircraft.

Fuel tank pressure differential is also widely used for military fuel transfer systems. This method is very reliable, low weight and cost provided that fuel tank pressurization is available. Figure 5.6 illustrates a fuel tank pressure differential transfer system with several tanks in series. Some military aircraft have used both fuel transfer pumps and fuel tank pressure differential.

Military aircraft equipped with external fuel stores typically use air pressure to transfer fuel from the store to the internal fuel tanks. The shape, size, and method of construction of the external fuel stores allow use of higher air pressurization than normal aircraft fuel tank

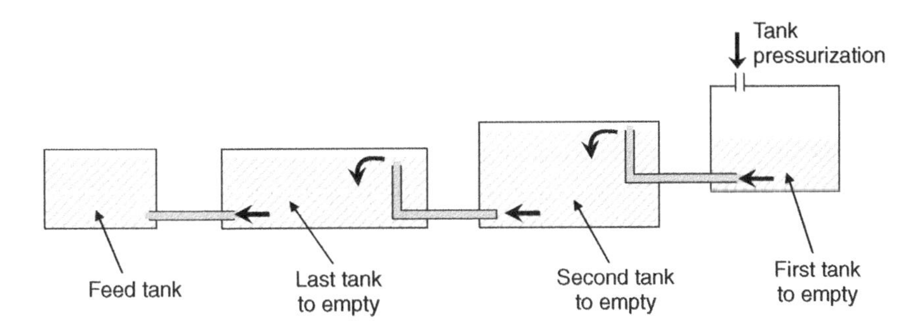

Figure 5.6 Pressure differential transfer system.

pressurization. A valve located inside the external fuel store controls the refueling and transfer functions. In the transfer flow direction, the valve closes at low (empty) fuel levels to prevent the higher pressure air from entering the aircraft fuel tanks using a 'Fuel-no air' feature. This function is typically accomplished using a float actuated valve that initiates valve closure when the fuel level in the store falls below some minimum level. The external fuel store is a good example of a cascading refuel and transfer system.

For some aircraft, it is important to control the center of gravity of the fuel entering (refuel) and exiting (transfer) the external store. The variation of the center of gravity of the fuel in the external store is typically controlled by compartmentalizing the fuel storage area of the tank. The compartmented fuel store lends itself well to the cascading refuel/transfer concepts described above.

Figure 5.7 shows a functional schematic of a typical external fuel tank illustrating how the refuel and transfer functions are accomplished. Note that store attachments to the mother aircraft are via quick disconnects to allow store jettison in combat situations.

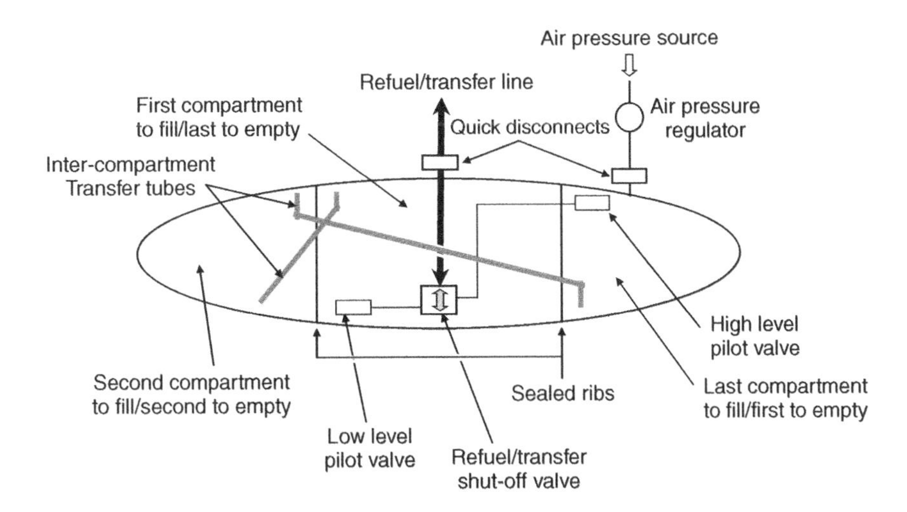

Figure 5.7 External fuel tank store refuel/transfer system schematic.

5.4 Aerial Refueling

Aerial (or in-flight) refueling has become a major force multiplier in the execution of modern day air warfare. A typical scenario in military air combat operations is to have attack aircraft take-off with a full fuel and weapons load, usually requiring the use of afterburners to ensure adequate take-off margins, followed by a fast climb to operating altitude. At this point a substantial portion of the on-board fuel has already been consumed and the ability to rendezvous with a refueling tanker and top up the fuel tanks allows the aggressor aircraft to accomplish mission objectives, in terms of operating range, that would otherwise be impossible. In addition to providing a substantial extension to the mission operating envelope, aerial refueling allows both fighter and ground support aircraft to remain in a combat area for extended periods of time. Also by careful pre-planning the location of aerial refueling tanker aircraft, extremely long-range missions can be accomplished using relatively short-range aircraft. An important example of this point involved the use of the SR-71 Mach 3 reconnaissance aircraft to fly missions to the Middle East during the Arab-Israeli Six Day War in the late 1960s. These aircraft took off from and returned to the USA with invaluable surveillance information that became a key factor in bringing the conflict to a halt.

While the operational benefits of the aerial refueling function are without question a major enabler in the execution of modern air warfare the provision of this function from both the receiver and supplier aircraft perspective adds a number of unique and complex requirements to the design of the fuel system, for example:

- An in-flight hook-up system is required which has fluid-tight connections together with an appropriate safe disconnect capability in case of unforeseen emergencies.
- Compatibility between the tanker fuel off load system and the receiver aircraft. This compatibility must address the combined tanker and receiver flow and pressure ratings, aerial refueling equipment structural issues and physical interfaces.
- Safe accommodation of potential failure modes during aerial refueling operations. This requires a detailed evaluation of the potential impact on both the tanker and receiver aircraft. In some cases, especially where aerial refueling is being added to existing aircraft, the provision of additional fuel system functional redundancies may be necessary.

There are two different aerial refueling systems in operation today. The earliest system, which was originated in 1934 by Alan Cobham is the 'probe and drogue' system. With this system the tanker aircraft holds station and deploys a tension-controlled fuel hose with a drogue coupling attached (see the photograph in Figure 5.8). The receiver pilot must fly his aircraft so that his probe mates with the drogue coupling in order to commence the refuel process. There is a simple fuel pressure assisted spring loaded mechanical latching mechanism in the drogue that holds the probe in place during refueling (see Chapter 6 for a detailed description of the probe and drogue equipment). On completion of the aerial refueling process, the receiver aircraft simply drops back to pull the probe from the drogue. The receiver aircraft probes may be either fixed or deployable depending upon the mission requirements of the aircraft.

In the late 1940s, General Curtis LeMay, commander of the Strategic Air Command (SAC), asked Boeing to develop a refueling system that could transfer fuel at a higher rate than was available with the probe and drogue system. Boeing engineers came up with the concept of the 'Flying Boom' which was first deployed for use with the US Air Force in 1948.

Figure 5.8 Probe and drogue aerial refuelling.

The flying boom aerial refueling system (see Figure 5.9) is fundamentally different from the probe and drogue system both in hardware and operating method. In the flying boom case it is the receiver aircraft which has the female coupling or 'Receptacle' and must hold station while the tanker boom operator 'Flies' the boom with its interfacing nozzle into the receiver receptacle.

Figure 5.9 A C-17 being refueled via the flying boom (courtesy of US Air Force/Master Sgt. Rick Starza).

The receiver receptacle latches onto the boom nozzle using hydraulic pressure supplied by the receiver aircraft. Under normal operating conditions, either the tanker or receiver can initiate a disconnect. Should the normal disconnect function fail, a 'Brute Force' disconnect method can be used. This requires the receiver pilot to maneuver his aircraft to assert a tension force at the boom/receptacle coupling that is sufficient to overcome the holding force of the latching mechanism thus disengaging the boom nozzle from the receptacle.

Since the initial deployment of aerial refueling systems there has been significant advancement in equipment and procedures however the two original system concepts described above are still in common use today.

There is a great deal of public information available on the general topic of aerial refueling systems, application, interoperability, and success stories. A dedicated community and an excellent source for additional information is the Aerial Refueling Systems Advisory Group (ARSAG). ARSAG is dedicated to the improvement of aerial refueling systems and its interoperability throughout the international community.

5.4.1 Design and Operational Issues Associated with Aerial Refueling

Although there are two different types of aerial refueling systems, there are issues and requirements common to both systems:

- The tankers need to provide flows and pressures compatible with the flow and pressure capability of the receiver aircraft. This means some sort of pressure control must be provided by the tanker aircraft. The nominal pressure at the entry point of the receiver aircraft for both probe and drogue and flying boom systems is 55 ± 5 psi. This pressure control limitation must be provided over a wide range of flows. For probe and drogue systems the flow range is from 600 US Gallons /Minute (USGPM) down to 10 cc/min. For the flying boom system the flow range is 1000 USGPM down to 10 cc/min.
- Surge pressure control must be provided by both the tanker aircraft and the receiver aircraft. Surge pressure control is especially important for aerial refueling operations as the potential surge pressure created by aerial refueling tankers is much greater than ground refueling operations due to the required high flow/pressure output of the tanker aerial refueling pumps. Uncontrolled surge pressures could damage the receiver aircraft and potentially result in loss of aircraft.
- Analysis of the refuel system should include both the receiver and tanker delivery fuel systems as one integrated system. Verification of system performance should be verified by extensive ground testing before any flight testing takes place.
- During aerial refueling, in-flight pre-check capability of the refuel system must be provided. This can be complex were dual refuel shutoff valves are installed to meet system redundancy requirements.
- One of the difficulties associated with the design of a probe-equipped receiver aircraft fuel system that is compatible with all drogue equipment aerial refueling systems is the lack of data on the performance of all the various tanker/fuel delivery systems throughout the world. This is not a big problem on the Flying boom system although there are some differences between the current day in-service tankers such as the KC-135 and KC-10.
- There are a large number of probe and drogue fuel delivery systems in service throughout the world. Some of these systems are integrated into the tanker aircraft while other aerial refueling systems are add-on features such as refueling pods that can be attached to a wing or underside of an existing aircraft.
- Some receiver aircraft can also be a tanker aircraft. An example of this is the A-6 aircraft which can be equipped with a 'Buddy store' that allows the aircraft to operate temporarily as an aerial refueling tanker. Many of these delivery systems where designed years ago when

standardization of this important function was not adequately emphasized and, as a result specific requirements were controlled and passed along by the then system experts.

- Some of these systems are hydraulically powered, some electrically powered, some powered by fuel pressure. The mandate of the ARSAG organization is to support the drive for interoperability of aerial refueling systems and equipment and to document the performance characteristics of these varied systems that are in use throughout the world and to provide standard guidelines for future systems.
- Although the tanker systems are specified to provide 55 ± 5psi at the entry to the receiver aircraft throughout a wide flow range, in actuality many (if not most) of the current day systems are not able to provide 55 ± 5 psi at the higher flow rates. The achievable tanker / receiver flow rate is determined by the tanker's actual delivery flow/pressure performance and the flow/pressure drop of the receiver's fuel system. In practice maximum flow occurs when the receiver aircraft is near empty and all tanks are being filled. Flow rate decreases as the tanks become full and the associated tank refuel valves close. Figure 5.10 illustrates the achievable system flow characteristics as a function of tanker delivery capability and the receiver aircraft pressure drop characteristics.

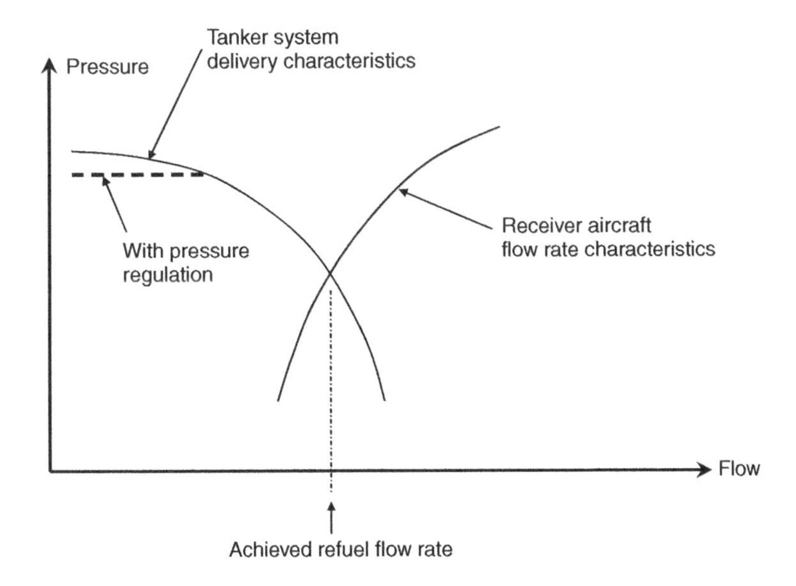

Figure 5.10 Achievable tanker/receiver refueling system flow rates.

The following subsections describe the two established aerial refueling systems in more detail covering some of the key operational aspects involved in this critical strategic function.

5.4.2 Flying Boom System

In the early 1950s, the B-29 (designated as the KB-29P) was the first aircraft to use the Boeing designed 'Flying Boom' system. Subsequently, Boeing developed the world's first production aerial tanker, the KC-97 Stratotanker. Later Boeing received contracts from the USAF to build Jet Tankers based on the Boeing 707 airframe. This modified B707 became the KC135 Stratotanker. Of the 732 that were built, over 300 are still in operation as of early 2008.

The 'Flying Boom' installed on the Boeing KC-135 consists of a long, rigid, telescoping tube attached to the rear of the aircraft. The telescoping fuel tube has a nozzle (boom nozzle) at the end that mates with a 'receptacle' in the receiver aircraft. A poppet valve in the boom nozzle prevents fuel from exiting the boom until the nozzle is inserted into the receiver aircraft. The engagement phase of the refueling process is known as 'the contact phase'. Once the boom nozzle is fully engaged and latched into the receiver receptacle, the system advances to the 'Contact' or 'Latched' mode. Once the refueling of the receiver aircraft is completed, the system is triggered to the 'Disconnect' state. Disconnect can be initiated by ether the tanker or receiver aircraft.

The operation of the flying boom is controlled by the 'boom operator' or 'boomer', who, on the KC-135 tanker, is stationed at the rear of the tanker aircraft with his line of sight close to alignment with the flying boom tube. Mounted at the far end of the fixed portion of the boom are small wings, or 'Ruddevators'. These control surfaces are manipulated by the boomer in order to 'fly' the boom into alignment with the receiver receptacle. Once aligned, the fuel tube is hydraulically extended to effect contact. Toggles in the receptacle engage recesses or pockets in the boom nozzle, holding the boom nozzle firmly in place during fuel transfer.

Once the boom nozzle is latched into the receptacle, the boom follows the movement of the receiver aircraft. The boom has prescribed allowable up and down (elevation), side to side (azimuth) and inward limits of motion. Should the receiver aircraft exceed any one of these limits, an automatic disconnect is initiated. Should the receiver aircraft drop back too far, reaching the fully extended length of the boom, the disconnect process will be initiated by means of a pressure relief feature integral with the receptacle latching device. This is known as a 'Brute force' disconnect.

Brute force disconnects are also used should the receiver aircraft fail to properly release the boom nozzle on conclusion of fuel transfer. The allowable brute force loads are 4800 lbs minimum, 9000 lbs maximum at disconnect velocities of up to 10 feet per second. Brute force disconnects are not desirable as it results in high structural loading of the boom and frequently requires ground checkout and maintenance to ensure that equipment damage has not occurred. Automatic disconnect is also initiated by a receiver aircraft fuel manifold over-pressure switch should the tanker provide excessive fuel pressure to the receiver aircraft.

A rarely used, but potential alternative mode of aerial refueling is termed 'Stiff Boom' refueling. This method is only used in extreme cases where the normal boom nozzle latching system has failed and the receiver aircraft is in a 'must have fuel' state. Stiff boom, as the name implies is simply the insertion of the boom nozzle into the receiver receptacle and holding it there by the tanker keeping a compression load on the boom. Receiver positioning is aided with either voice or visual commands from the boom operator. Current day tankers and receiver aircraft have the capability to communicate through mating induction coils located within the boom nozzle and the receptacle. This allows radio silence communications when required for operational security. Visual communication is provided by signal lights located on the bottom of the tanker aircraft.

In addition to the KC-135 tanker, which was developed from the Boeing 707 airframe, there are a limited number of other Boeing 707 tankers with flying booms in operation. Currently, the only other USAF flying boom tanker in service is the KC-10. There were approximately 50 of these built from existing DC-10 airframes. The KC-10 has significantly more fuel load capacity than the KC-135 and utilizes a flying boom with several improvements over the KC-135 system. These improvements include a fly by wire 'Ruddevator' system, automatic load alleviation, and a boom nozzle with independent disconnect capability. The boomer sits in an upright position in front of a heads-up display with a real time image of the refueling process. All of the KC-10 tankers are equipped with a receptacle which allows in-flight transfer of fuel from any other tanker to the KC-10. The KC-10 tanker also has probe and drogue tanker capability with both centerline and wing mounted hose reel systems.

5.4.3 Probe and Drogue Systems

In probe and drogue aerial refueling system applications the tanker carriers a hose reel system consisting of a Hose Drum Unit (HDU), a long, re-enforced flexible hose, a drogue (sometimes referred to as a para-drogue) and a coupler that mates with the receiver probe. The interface between the probe and coupler is closely controlled by NATO standard STANAG 3447 reference [8].

The HDU is a hydraulic powered reel assembly that controls the release and retraction of the flexible hose. In addition to the HDU providing control for extension and retraction of the hose/drogue, it must sense the engagement of the probe into the coupler and instantly take up any slack that is produced by the receiver aircraft pushing the drogue forward. The hose has markings visible to the receiver pilot that tells him when he in the correct position (distance) relative to the tanker aircraft. When he reaches this position, the tanker pumping system is automatically turned on. The receiver aircraft is equipped with a probe at the end of a relatively long probe mast. Some aircraft use a fixed position mast while others use retractable probe masts. The probe mast must be long enough to ensure that the probe is located outside of aircraft boundary layer so that the approaching drogue will not be easily disturbed. The mast should also locate the probe in a good position for pilot observation during engagement into the drogue and provide best possible centerline alignment with the trailing drogue.

The key interface equipment between the tanker and receiver aircraft is the coupler which is an integral part of the drogue assembly. The coupler mates with the nozzle at the end of the refueling probe. Originally, these devices were referred to as the MA-2 Coupler and MA-2 nozzle. These probe and drogue systems performed adequately with most aircraft systems however, operational problems developed with certain receiver and tanker delivery systems with the main concern being the occurrence of high surge pressures when topping-off the receiver tanks at the end of the refuel process. As a result, it was decided that there was a need to control or limit the fuel pressure entering the receiver aircraft. This led to the introduction of the MA-3 pressure regulated coupler in the late 1970s. Later, it was determined that although the MA-3 pressure regulated coupler, when working properly was doing what was intended, there was the potential for an in-service undetected failure. This lead to a dual pressure regulated coupler, which was designated as the MA-4.

Equipment descriptions of the coupler and nozzle designs are presented in Chapter 6.

5.4.3.1 Boom to Drogue Adapter Units

In order to provide the capability to refuel a probe equipped aircraft with a flying boom tanker system the 'Boom to Drogue Adapter' (BDA) kit was developed. The BDA kit consists of a short, nine-foot hose, a coupler, a rigid drogue, and a swivel fitting. A portion of the boom nozzle is removed to provide a flange that the BDA is attached to. In this system, the tanker boom operator (the 'boomer') 'flies' the boom into a position optimal for the receiver aircraft. From this point the pilot of the receiver aircraft flies the probe into the rigid drogue/coupler. Once the probe nozzle is engaged into the coupler, the receiver pilot calls 'contact' and the boom operator starts the refueling process. The receiver maintains his position during refueling, keeping an eye on the hose to make sure he remains in a suitable position. When fueling is complete, he slowly backs off until the probe disconnects from the coupler. Although the BDA equipped tanker system has been used extensively for critical needs, it is not recommended for inexperienced boom operators and receiver pilots. The short nine-foot hose is very rigid and can impose higher loads into the probe equipped receiver aircraft than it was originally designed for.

5.5 Fuel Measurement and Management Systems in Military Applications

5.5.1 KC-135 Aerial Refueling Tanker Fuel Measurement and Management System

In the early 1980s the USAF embarked on a KC-135 Fuel Savings Advisory System (FSAS) avionics installation upgrade program. Figure 5.11 shows a simplified schematic of this system.

A significant portion of this upgrade focused on reducing crew workload during the aerial refueling process through the introduction of a new Integrated Fuel Management & Center of Gravity System designated the IFMCGS.

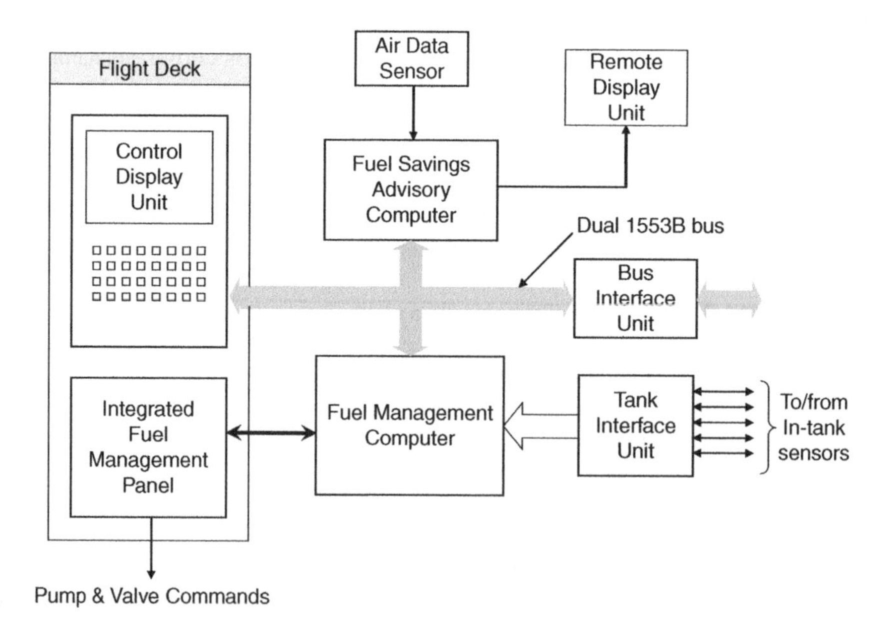

Figure 5.11 KC-135 Fuel Savings and Advisory System schematic.

This system replaced most of the older technology fuel gauging indicators allowing the installation of a new enhanced radar display and Data Entry Unit (DEU) in the vacated space. This new system communicated with the aircraft avionics via the new military Digital Information Transmission System format MIL STD 1553 reference [9] and as such broke new ground in fuel management system technology. The electronics equipment at that time utilized first generation digital microprocessor technology using an 8-bit bus protocol. Nevertheless this was a major step forward from the world of analog electronics that were dominant at that time. IFMCGS components consisted of a Tank Interface Unit (TIU), Fuel Management Computer (FMC) and an Integrated Fuel Management Panel (IFMP).

Key features of this new IFMCGS were as follows:

- The IFMCGS was required to maintain the existing 'Brick-wall' fuel quantity gauging system architecture wherein each tank's probes and indicators remain isolated from the other fuel tanks thus any single failure could only affect a single tank.
- The IFMCGS was required to interface with the existing fuel quantity gauging system tank located capacitance fuel probes.
- Provision was made to enter cargo weight and aircraft navigation data via the DEU. From these inputs, the IFMCGS calculates aircraft CG and CG limits from fuel quantity information. This replaces prior manual entry procedures and reduces crew workload significantly.
- The system calculates and displays fuel transferred via the aerial refueling boom.
- The system provides comprehensive Built-In-Test (BIT) information associated with the complete IFMCGS including annunciation and storage in non-volatile memory of all detected faults

The older technology analog fuel gauging indicators that the IFMCGS replaced had 'Empty' and 'Full' calibration adjustment potentiometers at the rear of the units. These adjustments had to be reset whenever aircraft maintenance was necessary to replace a malfunctioning fuel probe or indicator. Their purpose was to harmonize the manufacturing and installation tolerances of the system. With the new IFMCGS it was required that the Fuel Management Computer (FMC) or Integrated Fuel Management Panel (IFMP) could be replaced without the need for any calibration adjustments, thus reducing the mean time to repair the system.

The purpose of the Tank Interface Unit (TIU) was to normalize the fuel probe capacitance data and the affects of aircraft wiring stray capacitance for input to the FMC. The TIU contained a set of high reliability trimmer capacitors that were electrically in parallel with the fuel probes. These trimmer capacitors were set at initial installation of the IFMCGS and reset when a malfunctioning fuel probe was replaced. This facilitated rapid replacement of the relatively lower reliability FMC and IFMP equipment without the need for calibration adjustments.

The IFMCGS functions were divided into 'mission critical' and 'non-mission critical' categories in order to ensure the appropriate integrity of the various system functions. As such mission critical functional requirements included:

- Gauging system integrity shall not be compromised by the IFMCGS and therefore no single failures of the IFMCGS should affect the fuel quantity gauging of multiple tanks.
- The electrical circuitry and wiring independence of the existing pump and valve switches shall not be compromised.

Non-mission critical functional requirements were identified as follows:

- display of aircraft CG
- calculation and display of total fuel transferred
- Built-In-Test.

Figure 5.12 is a simplified block diagram of the Fuel Management Computer (FMC). The FMC contains two independent microprocessor channels consisting of the Primary Microprocessor (PUP) and the Redundant Microprocessor (RUP). The PUP is responsible for all mission critical functions and the RUP for non-mission critical functions.

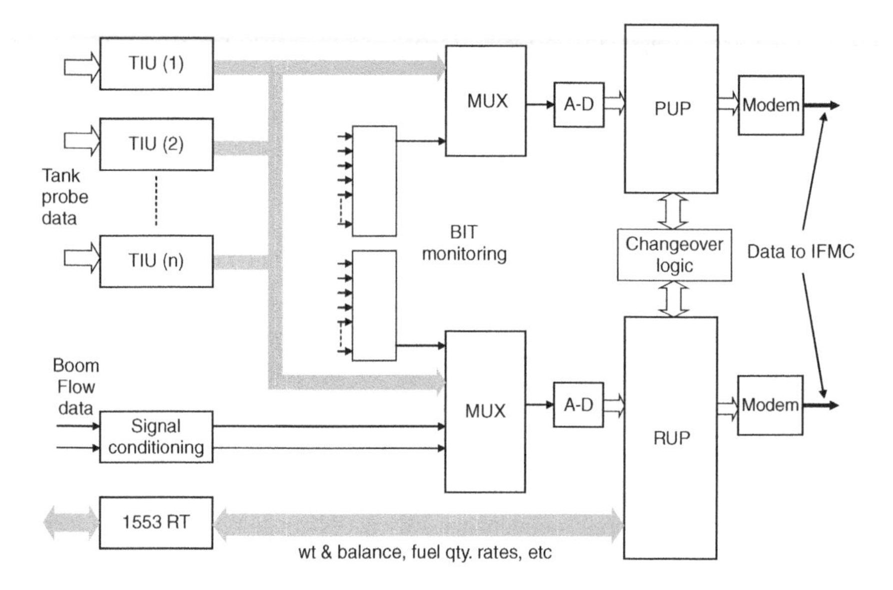

Figure 5.12 Simplified Fuel Management Computer (FMC) schematic.

In the event of a PUP failure, the RUP stops processing its non-mission critical functions and switches to a PUP mode. Thus, no single failure renders the loss of mission critical functions.

The fuel tank capacitance data enters the FMC through isolated Tank Processor Unit (TPU) channels, one for each tank, thus preserving the original 'Brick-wall' architecture of the aircraft. The TPUs feed both PUP and RUP.

The Integrated Fuel Management Panel (IFMP) which interfaces with the FMC is shown schematically on Figure 5.13. As in the FMC, it also has two independent microprocessor channels consisting of the Display Primary Microprocessor (DPUP) and the Display Redundant Microprocessor (DRUP).

The signal bus interconnection of the FMC and IFMP has the PUP feeding the DPUP and the RUP feeding the DRUP to provide redundant operation of the FMC/IFMP set.

Key features of the Integrated Fuel management Panel include seven-segment, direct view incandescent displays, a graphic display of aircraft CG and illuminated valve switches showing direction of flow. A photograph of this unit is shown in Figure 5.14.

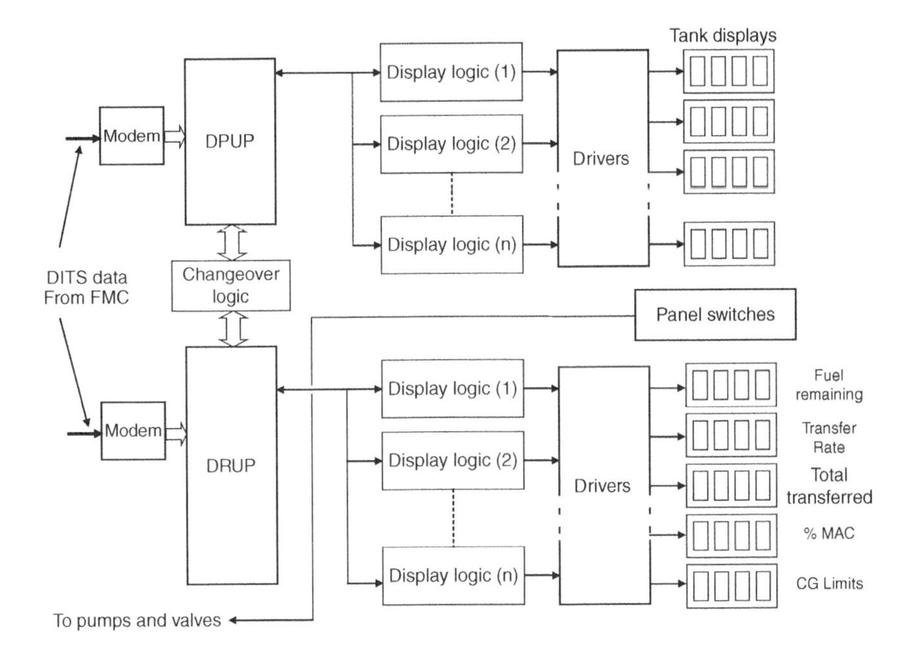

Figure 5.13 Integrated Fuel Management Panel (IFMC) schematic.

Figure 5.14 Integrated Fuel Management Panel (courtesy of Parker Aerospace).

When the C/KC-135 was originally designed in the late 1950s, intrinsically safe short current limiting to the in-tank fuel quantity probes was limited to 200 milliamps per MIL-G-26988. In the mid 1970s, Boeing was a leader in reducing the short circuit limit to 10 milliamps for its B757/767 aircraft. This became an industry standard and became a requirement for the IFMCGS.

At the time of IFMCGS installation, the then aging in-tank capacitive fuel quantity probes were subject to silver and copper sulfide electrically conductive surface contamination of their concentric tubes. The lower exciting currents of the IFMCGS, in conjunction with probe degradation caused by this contamination, resulted in numerous gauging anomalies. As a result, the new system was subject to many nuisance faults until a probe replacement campaign was instituted.

It is interesting to note that the most likely cause of the loss of TWA flight 800 on July 17, 1996 was an ignition source in the center fuel tank causing an explosion of fuel vapors in that tank. Extensive post-crash investigations have concluded that a potential cause of the explosion *may have been* conductive probe contamination in conjunction with an external tank electrical harness short that coupled electrical energy into the tank to produce a spark.

The issue of SFAR 88, reference [10] in 2001 by the United States Federal Aviation Agency (FAA) took these findings into account by requiring modifications to in-service aging aircraft and subsequently amended FAR 25 to impose enhanced intrinsic safety standards for new aircraft.

In summary the KC-135 IFMCGS was a major step forward in the application of digital avionics technology to fuel management in the critical area of aerial refueling. Here it is important to minimize the crew workload in order to maximize safety in a critical operational situation. Even though there were many shortcomings of this system, which was introduced as an aircraft modification/upgrade the IFMCGS played a key role in improving the effectiveness of the USAF aerial refueling capabilities.

5.6 Helicopter Fuel Systems

Helicopter fuel systems can be both very simple and extremely complex, depending on the number of tanks and mission requirements. Obviously the most complex helicopter fuel systems are on military vehicles. These military helicopters typically have many fuel tanks, along with significant survivability requirements.

Commercial helicopter fuel systems tend to be very simple. So the following commentary will focus primarily on the military helicopter fuel system considerations. The basic functions of fuel system are the same as commercial and military aircraft. Many have pressure refueling and defueling, engine feed, fuel transfer, and vent systems. Additionally some military helicopters require aerial refueling (see Figure 5.15 which shows MH-53 helicopters being refueled by a C-130 tanker). Military helicopters frequently have the need to change configuration to carry troops or to carry more fuel. This may result in the fuel system having to be very adaptable for different configurations.

Survivability is a major player in fuel system design for helicopters. This added complexity includes provision of the following features:

- fuel spillage prevention following roll-over maneuvers
- self-sealing break-away fittings where tubes penetrate the fuel tank walls
- bladder tanks that will not rupture during a crash landing.

Typically military helicopters require the use of pressure ground refueling. Due to the complexity of the tank arrangements and the fact that the vent system normally cannot handle a failed opened refuel valve, dual refuel valves are frequently used. Helicopter fuel systems typically require the ability to be defueled for maintenance or in some cases to use

Figure 5.15 Aerial Refueling of MH-53 Helicopters (courtesy of US Air Force/Senior Airman Emily Moore).

the fuel in other vehicles. This requires the capability to defuel each tank by means of suction.

The engine feed system for helicopters can be complex due to the fact that there are several tanks that need to provide fuel to the engines. In contrast to fixed wing aircraft, helicopters typically do not have dedicated feed tanks to serve as a collection point for fuel to be fed to the engines. Engine feed is provided by suction created by the engine mounted fuel pumps and the tanks supplying fuel to the engines is typically selected by the flight crew. Being able to selectively feed the engines from any set of fuel tanks greatly enhances survivability of vehicles operating in hazardous combat areas.

The suction feed arrangement in military helicopters is important since the engines are usually located above the flight deck and battle damage to fuel feed lines will not result in loss of fuel into the fuselage with significant fire potential if positive fuel boost pressure is used.

For engine start and ground APU operation, a small DC powered fuel pump is normally provided. Figure 5.16 illustrates the general complexity of a military helicopter fuel system.

Components used in military helicopter fuel system are typically identical to those used in fixed wing aircraft however there are a few exceptions.

The requirement mentioned above, that the vent system not spill fuel should the helicopter rollover, has been met in several different ways including:

- configuration of a vent line that prevents fuel flow overboard (frequently limited by fuel tank locations and configuration);
- positive and negative pressure relief valves with positive relief cracking pressures higher than the potential fuel head when the helicopter has rolled over;
- weight operated vent line shutoff valves;
- float operated vent valves with negative 'g' overrides.

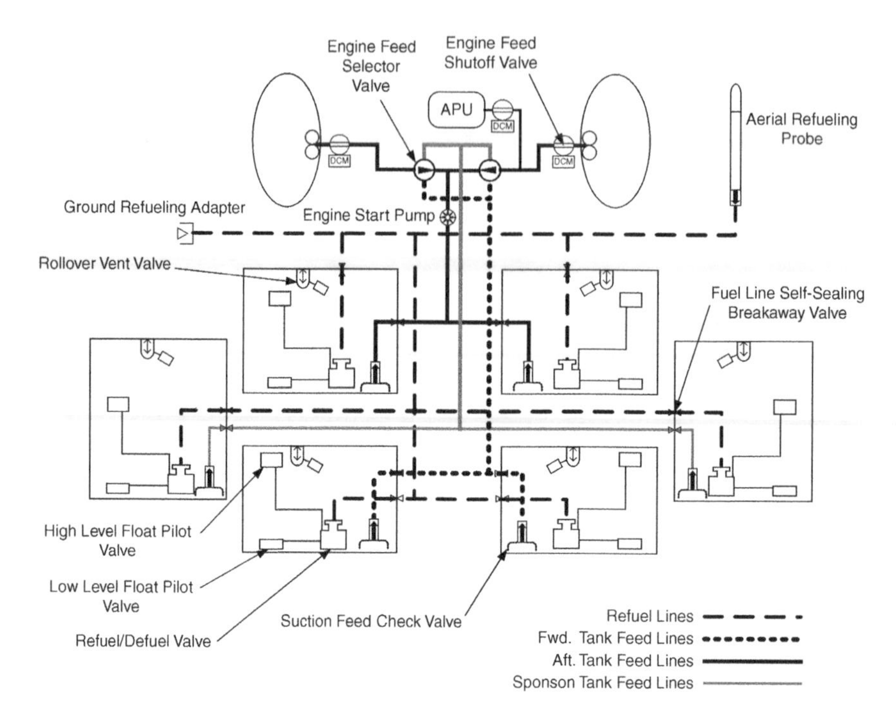

Figure 5.16 Representative military helicopter fuel system.

Another somewhat unique requirement for helicopter fuel systems is that leakage shall not occur from a fuel line or tank should the fuel lines separate from the tanks. This requirement has been met by use of a very specialized fitting generally referred to as a breakaway fitting. The breakaway fitting has a controlled break location with internal valves that close both the separated fuel line and opening at the tank wall.

6

Fluid Mechanical Equipment

This chapter provides a detailed description of the major fluid mechanical equipment typically used in aircraft fuel systems. The schematic diagram of Figure 6.1 shows again the overall design and development process highlighting where, within this process, the design, fabrication and qualification of these fluid mechanical products takes place. With the wide range of different fuel system architectures used today, many disparate design concepts have evolved over the years using many different technologies and as a result there are very few products in this area of fuel systems that fall into a standard design category. Almost every new program has new equipment designs or at the very least, modifications of designs used on prior programs.

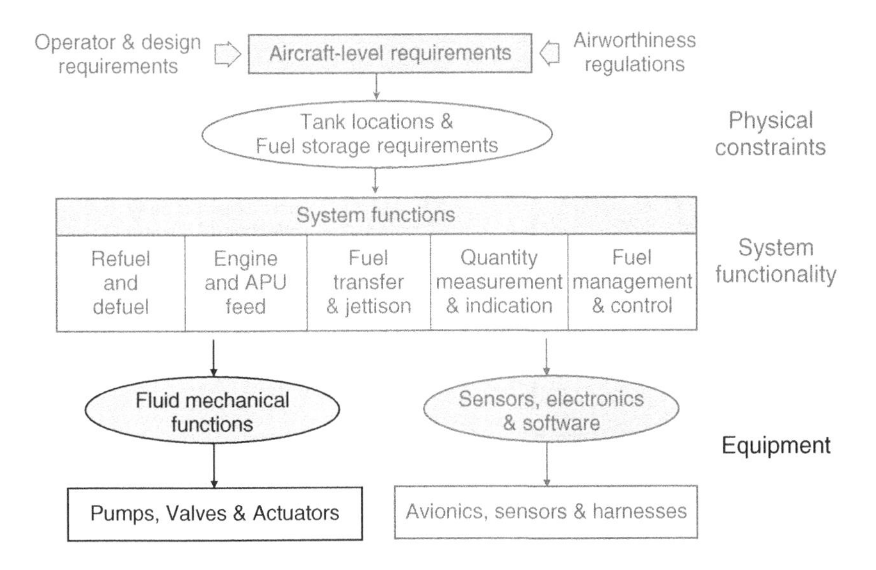

Figure 6.1 Fuel system design process overview – fluid mechanical equipment.

Aircraft Fuel Systems R. Langton, C. Clark, M. Hewitt, L. Richards

Fluid mechanical components fall into the following four main categories:

1. shut-off valves, adaptors and related equipment associated with enabling or stopping fuel flow during refuel, defuel, transfer and engine/APU feed situations;
2. fuel pressurization equipment including motor-driven pumps and ejector pumps;
3. fuel tank venting and ullage pressure control equipment;
4. fuel distribution including in-tank piping and connectors.

Fuel tank inerting equipment may also be considered as an additional equipment category; however this is covered in Chapter 10 which describes both the system and component issues associated with Fuel Tank Inerting.

Each of the above listed equipment categories is described in detail in the following sections of this chapter.

6.1 Ground Refueling and Defueling Equipment

6.1.1 Refueling and Defueling Adaptors

Ground refueling equipment begins with the refuel/defuel adapter that facilitates the refuel (and defuel) process. This device which mates with the ground refueling nozzle has a three-lug bayonet attachment (see Figure 6.2) that conforms to international standards. A poppet valve within the adapter is mechanically opened by a poppet in the ground refueling nozzle following correct installation of the refueling nozzle with the aircraft adaptor.

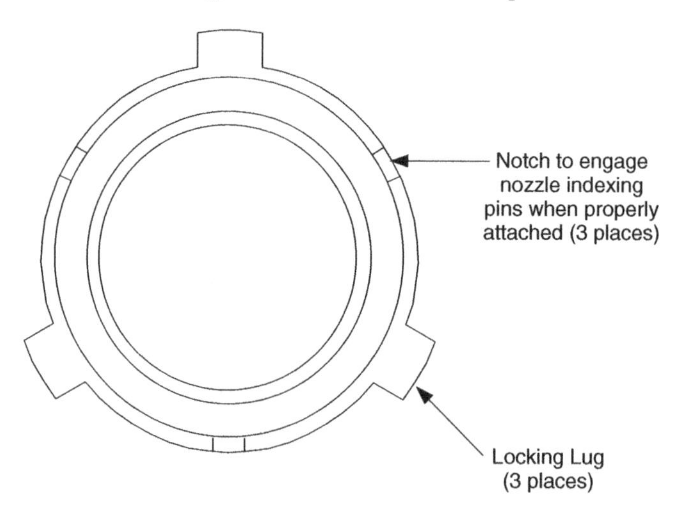

Figure 6.2 Refuel adapter locking lugs and nozzle indexing slots.

Proper alignment during refuel hook-up is facilitated by slots in the adapter that mate with spring-loaded pins in the refueling nozzle. The adapter is typically installed into a housing that provides a fluid coupling at its outlet. On most installations there is a refueling cap that attaches to the adapter when not in use. This cap provides a secondary seal that helps keep the refuel adapter interface clean.

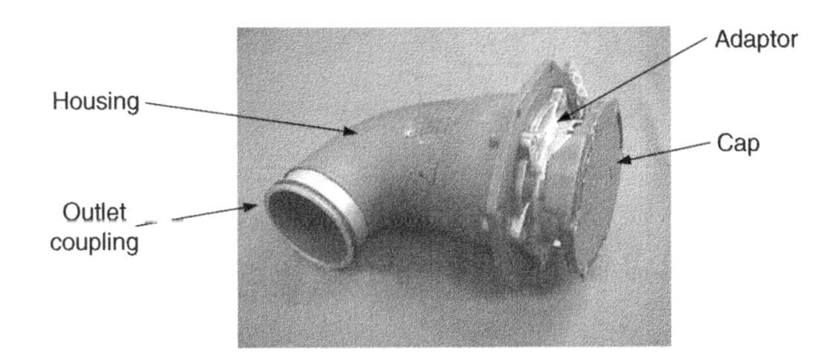

Figure 6.3 Refuel adaptor with housing (courtesy of Parker Aerospace).

Figure 6.3 shows a refuel adaptor installed in a housing that connects to the refuel gallery within the aircraft fuel system.

Typically the ground refueling adapter is installed lower than the fuel level in the fuel tanks and the refuel distribution lines. Therefore to prevent accidental spillage of a large amount of fuel should the refuel adapter connection fail, a check valve (non-return valve) is installed in or just downstream of the adapter. Because the refuel adapter is also typically used for defueling, a means to open this check valve is required to facilitate defueling operations. This can be accomplished in several different ways including mechanical override of the check valve, a fuel line that bypasses the check valve along with a valve that can be opened for defueling (mechanically or electrically), or the use of a check valve that has two-way flow capability with the reverse flow opening pressure set higher than the maximum fuel head. These three options are illustrated in Figure 6.4.

Basically, all refueling/defueling adapters fall into one of the above schemes although there have been some unique implementations of some of these concepts. One particularly clever implementation of the manually overridden check valve is shown in Figure 6.5. Here the refuel adapter outlet check valve is integrated into the adaptor such that the check valve can be mechanically held open for defueling simply by rotating a screw slot on the face of the adapter poppet.

6.1.2 Refuel Shut-off Valves

The function of the refuel shutoff valve is to control the flow of fuel into the fuel tanks. These valves can be selected open or closed either by the ground support or maintenance people at the refuel station or by the fuel management system during the auto-refuel process

6.1.2.1 Motor-Operated Shut-off Valves

One commonly used shut-off valve concept is the motor-operated valve where a small dc motor driving through a reduction gear which rotates a ball valve through 90 degrees as indicated by the concept diagram of Figure 6.6.

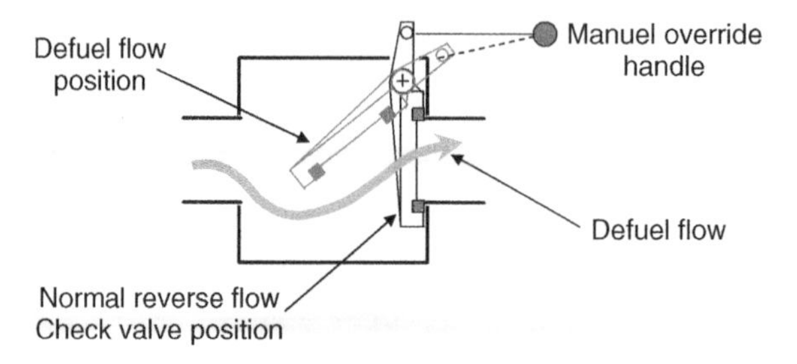

(a) Manually overridden check valve

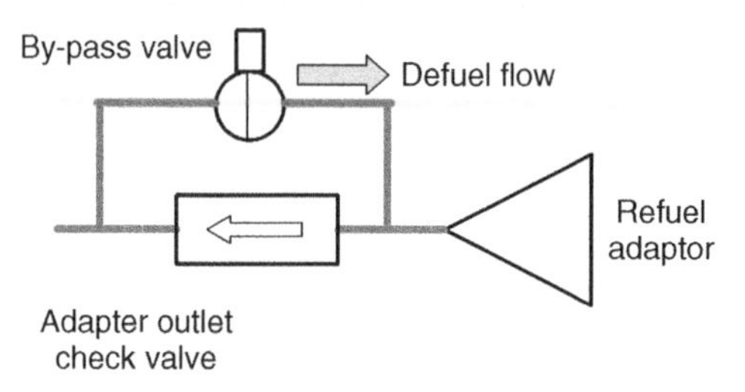

(b) Refuel adaptor outlet check valve by-pass

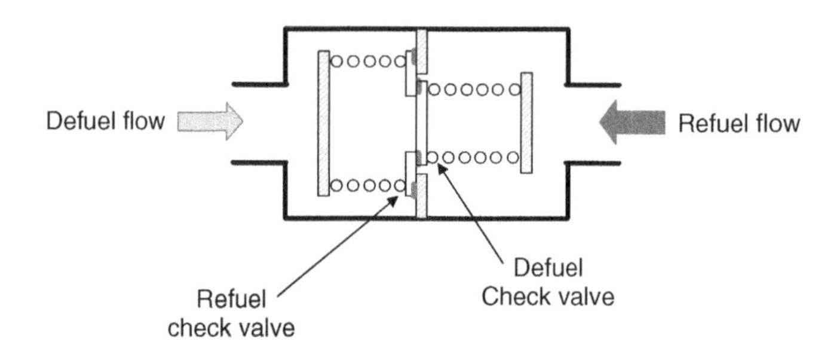

(c) Refuel/defuel check valve

Figure 6.4 Defuel accommodation options.

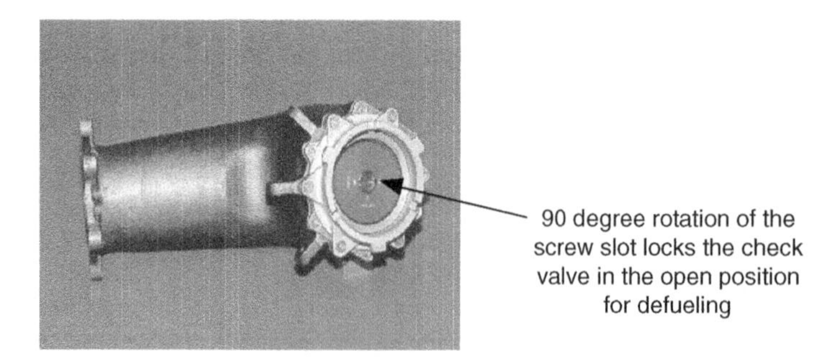

Figure 6.5 Ground refueling adapter with manual override of refuel check valve (courtesy of Parker Aerospace).

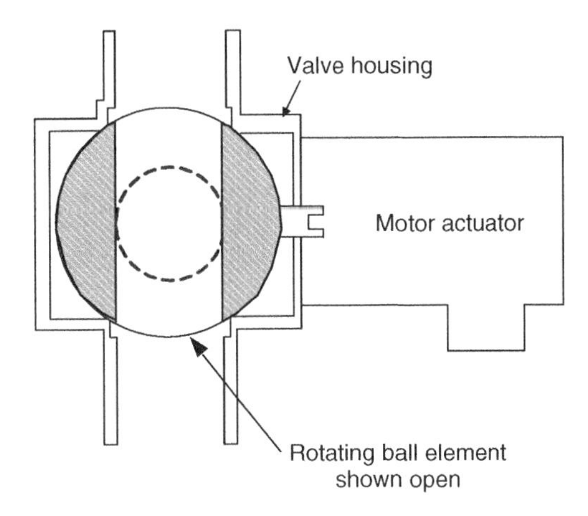

Figure 6.6 Motor-operated valve concept.

There are several unique characteristics of a motor operated valve that must be taken into account in the fuel system design process:

- The valve stays in the last energized position; therefore both open and closed position failure modes need to be considered in system design. This can result in the need for component redundancies such as a dual motor actuator or redundant valves in the fluid circuit. Depending on the criticality of a specific failure mode it may require a redundant valve in series (redundant valve shutoff) or a redundant valve in parallel (redundant valve opening).
- There can be a wide range in operating characteristics due to variation in aircraft system electrical power, temperature, and fluid pressure. Typically, available aircraft dc power ranges from 18 to 32 v dc. At the higher voltages, the valve will close at its fastest rate. At the lowest voltage it will take longer for the valve to close. This variation in operating time is made even greater with variations in the fuel pressure and temperature. This variation in operating time of a motor operated valve can result in high system surge pressures (fast shutoff) or large fluid volume overshoot (slow shutoff).
- Motor operated valves only require electrical power when commanded to change position. When the motor actuator reaches the required position, internal switches turn off the power to the motor. Typically, these switches are actuated by the rotation of the motor actuator output shaft and are precisely adjusted for the required open or closed position of the valve element. Two dual element switches are typically used in the motor actuator. One of the switch elements controls actuator travel and the other provides position indication to the aircraft. When an aircraft fuel system requires position feedback, the motor operated valve provides the most reliable, robust implementation of position switches when compared to implementation of position switches in fluid mechanical shutoff valves.
- Motor actuated valves lend themselves well to mounting the shutoff valve element internal to the tank with the motor actuator mounted external to the tank. This is desirable because it keeps electrical power outside of the fuel tank and it allows easy removal/replacement of the actuator without entering the fuel tank.

- A pre-check of a motor operated valve system is very simple. At any time after starting the refuel process, the motor-actuator is selected closed. After reaching the closed position and verifying (via the end of travel position switches) that the closed position has been reached the refuel process can be resumed. Figure 6.7 shows an electrical schematic of a typical motor-operated valve.

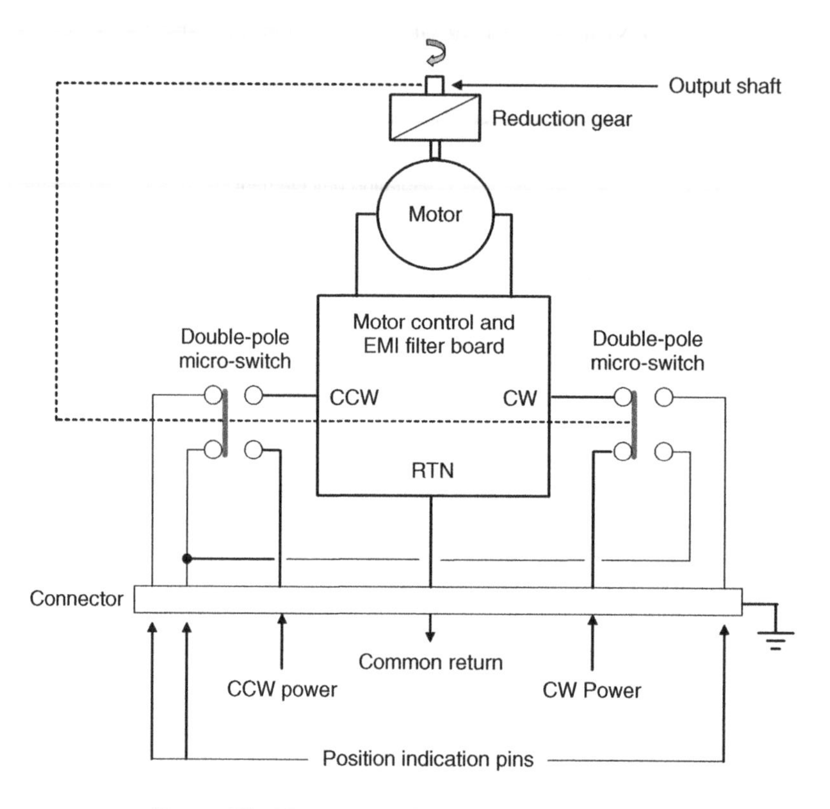

Figure 6.7 Motor operated valve electrical schematic.

In some applications, where functional redundancy is considered critical, the motor operated valve can utilize a dual motor actuator. This approach is adopted when critical failures (e.g. an uncontained engine rotor burst) could result in loss of electrical power to a single motor actuator. For critical functions such as fuel jettison and engine feed shut-off, this may be considered unacceptable. By providing independent routing of power to two separate motors in a dual motor actuator arrangement, power to either of the two motors will ensure availability of function.

The physical designs of the traditional motor-actuator come in two main versions:

1. The spur gear or epicyclic reduction gear design, sometimes referred to as the 'Tower (or chimney) actuator'
2. The worm and wheel reduction gear design, sometimes referred to as the 'Pancake actuator'

As implied by the names, the tower/chimney design is taller requiring a larger installation envelope whereas the pancake design is relatively flat and easier to accommodate physically. From a functional perspective, however, the 'Tower' approach offers the possibility of manual reversion following a failure provided that there is a sufficiently high reduction gear efficiency to provide 'Back-drive-ability'. Such an arrangement typically uses an external lever that also serves as a valve position indicator. The worm and wheel approach is, by definition, irreversible and therefore manual reversion would require a clutch to facilitate disconnection of the output drive shaft from the reduction gear. Alternatively manual reversion can be achieved by removal of the actuator and rotating the valve directly using a special tool.

Figure 6.8 shows the worm and wheel actuator concept for single and dual motor arrangements. As indicated, the functional redundancy is limited to the motor and its power source only. The reduction gear and the valve itself remain simplex.

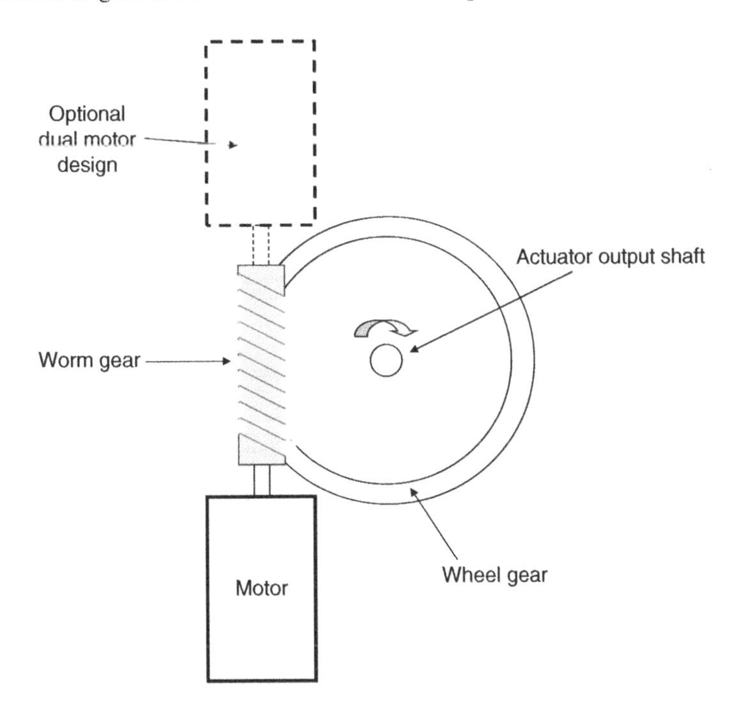

Figure 6.8 Pancake motor actuator concept.

Both the 'Tower' and 'Pancake' motor operated valves are in common use on many of today's commercial aircraft. Figure 6.9 shows conceptual drawings of each type of motor-actuator.

6.1.2.2 Hydro-Mechanical Shut-off Valves

The alternative to the motor-operated shut-off valve is the 'hydro-mechanical' or 'pressure operated' valve.

There are many different configurations/designs of hydro-mechanical valves in service today including both piston and diaphragm operated valves controlled by fluid mechanical

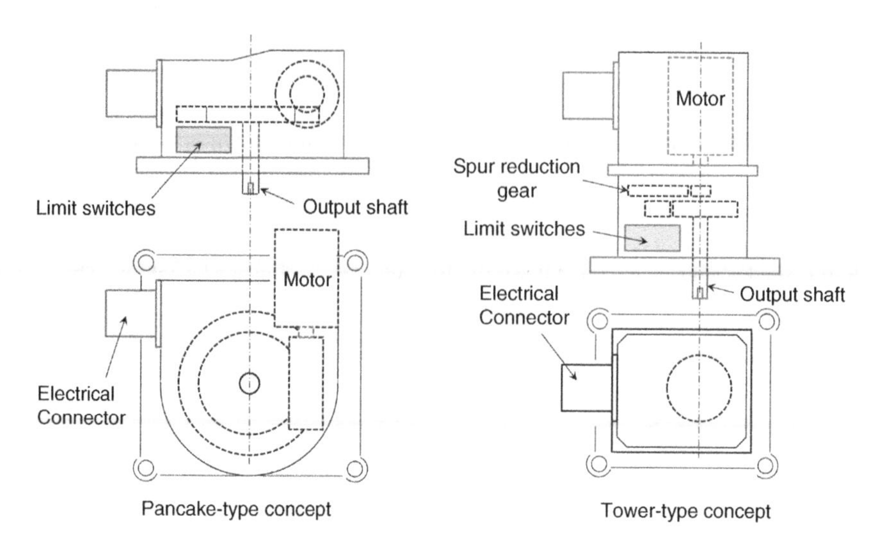

Figure 6.9 Conceptual drawings of Pancake and Tower type motor-actuators.

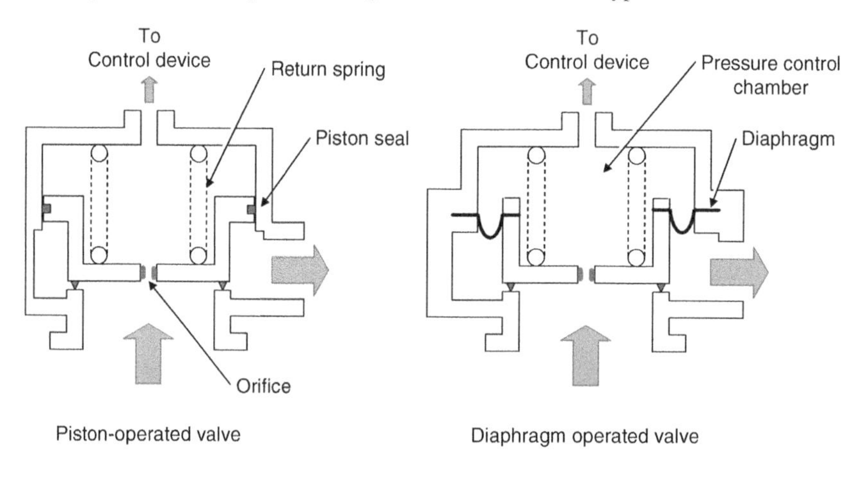

Figure 6.10 Shut-off valve design alternatives.

signals, electric signal or a combination of both. Figure 6.10 shows examples of the piston and diaphragm alternatives.

These types of hydro-mechanical shut-off valves operate in conjunction with a control device. For example, a remotely mounted pilot valve may be positioned to sense the maximum fuel level in the tank as shown in the schematic diagram of Figure 6.11 which shows two pilot valve concepts operating in conjunction with a piston type shut-off valve in the closed position.

Referring to Figure 6.11(a) it can be seen that as refuel pressure is applied to the valve inlet, the piston will be forced open allowing fuel to flow into the tank. Fuel will also flow through the pilot line and out into the float bowl where it drains back into the tank. The piston orifice ensures that the pressure behind the piston is not sufficient to overcome the refuel inlet

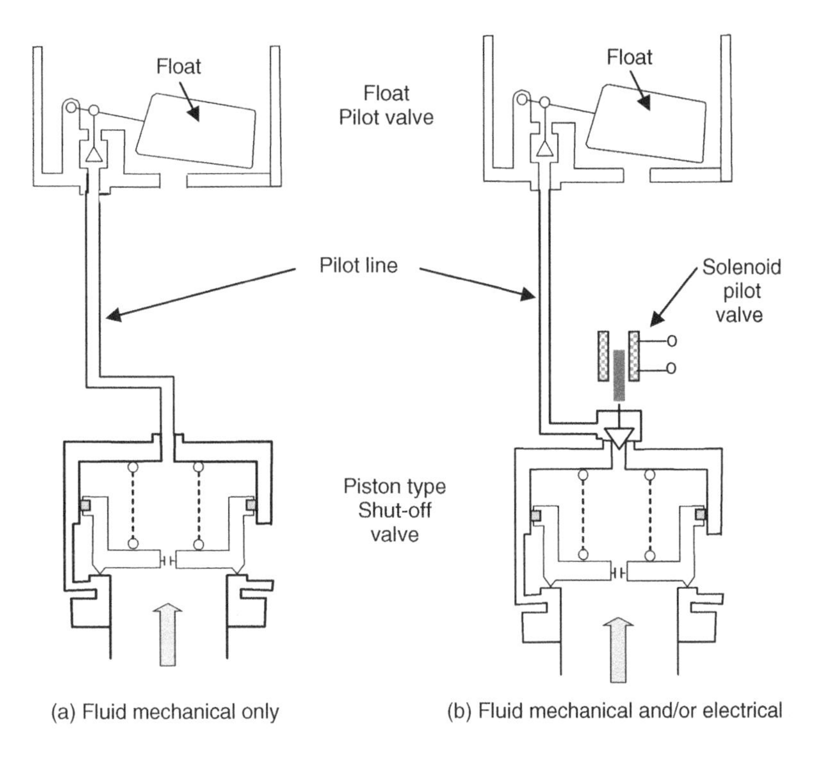

Figure 6.11 Pilot-operated shut-off valve concepts.

pressure. When the fuel level in the tank rises above the float drain, the bowl will begin to accumulate fuel until the float lifts shutting off the pilot line flow. This causes the pressure behind the piston to approach the refuel pressure causing the valve to close.

Figure 6.11(b) has an additional solenoid valve in series with the float pilot. In the figure the solenoid is shown de-energized. Once the solenoid is energized the pilot line is opened and the shut-off valve operates in the same way as the fluid mechanical arrangement of Figure 6.11(a), however, in this design the shut-off valve can be selected to close via the solenoid valve under the control of the fuel gauging system when the desired (pre-set) fuel quantity has been reached.

The above examples represent a simplified illustration of the typical hydro-mechanical fuel shutoff valves.

Diaphragm operated shutoff valves tend to offer lower internal leakage than piston operated valves since the diaphragm is clamped between flanges on the valve body and poppet. Diaphragm operated valves also tend to have lower friction than piston operated valves due to the rolling nature of the fabric/elastomeric diaphragm material. The effective area of the piston valve is much more controllable and remains constant through the valve stroke as compared to a diaphragm operated valve that typically has a change in effective area as the valve strokes. Packaging of diaphragm and piston operated valves tend to result in diaphragm valves being shorter and larger in diameter whereas the piston valves are small in diameter but longer. Generally, piston valves offer the lowest weight designs.

There are some unique characteristics of hydro-mechanical valves that are specifically related to fuel system and equipment design, for example:

- Because the force to operate hydro-mechanical valves is derived from low pressure differentials, the available operating forces are typically very low compared with high pressure aircraft hydraulic systems. These low force margins create a challenge to the design engineer where design requirements include low leakage, minimum size and weight, high reliability and capable of working in an environment that can involve contamination and icing conditions, as well as low and high temperatures.
- Hydro-mechanical valves typically have a discrete failure mode which is normally closed. They can also be configured as a fail open valve if required. When a fluid mechanical valve includes a solenoid operated pilot valve, see Figure 6.11(b), the de-energized position can be arranged to be either normally open or normally closed so that if electrical power is lost the shut-off valve will go to either the fully open or fully closed position respectively. This is not the case for the motor operated valve which stays in the last commanded position if power is lost. (A motor operated valve can be designed to have a preferred failure position; however, this adds significant complexity to the design.)

One particular hydro-mechanical valve design worth describing here is the co-axial, in-line shut-off valve. While the principle of operation of this valve is very similar to the tank-mounted piston valve shown above, its in-line implementation offers substantial functional benefits. The conceptual drawing of Figure 6.12 shows this valve with a solenoid valve actuation device. This valve can be configured with a float pilot or combined electrical/hydro-mechanical actuation mechanism as before. In the configuration shown, loss of electrical power will close off the pilot line resulting in closure of the valve.

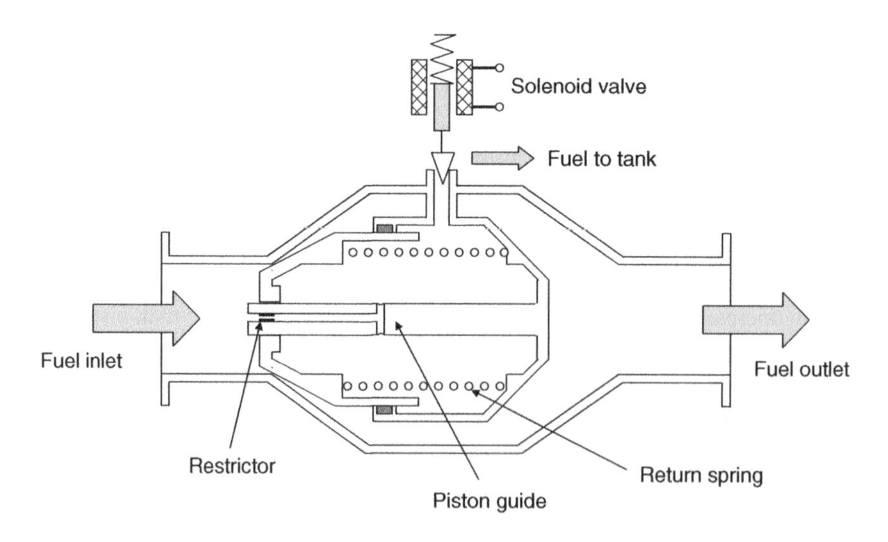

Figure 6.12 Co-axial fluid-mechanical shut-off valve.

Pre-check Considerations

Since failure of a pressure refueling valve to close when the tank is full could result in fuel spillage or in some cases structural tank damage, some form of 'pre-check' feature is required that can be exercised before the 'tank full' state is reached. For the motor or solenoid operated shut-off valves this is relatively easy since these valves can be signaled closed via an electrical command. For the fluid-mechanical shut-off valves the pre-check function becomes more complex because the float in the float pilot valve must be made to rise in order to verify the functional integrity of the complete shut-off process. Figure 6.13 shows four different pre-check techniques that have been employed for refuel systems employing hydro-mechanical shut-off valves with float pilots.

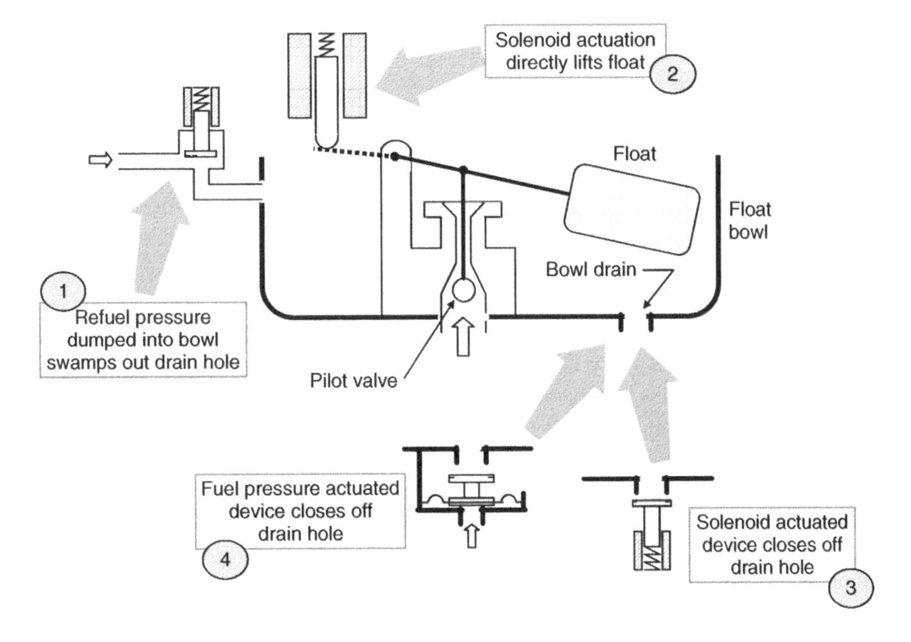

Figure 6.13 Pre-check flotation methods.

Float Valve Design

Since many hydro-mechanical valves operate in conjunction with a float valve, a brief description of float valve design is considered appropriate here.

In addition to the high level pilot actuated refuel shut-off function, equipment using a float for level sensing also includes low level pilot valves, float-actuated vent valves and drain valves.

With only a few exceptions, floats are attached to a metallic float arm which translates relatively large float movement to shorter, higher force actuation of a small poppet valve. The actuation force derived from a float operated device is simply the result of the displaced volume of fuel when the float is submerged.

The basic principles of a float operated device are illustrated in Figure 6.14.

Prior to 1960, floats were almost exclusively made of polyurethane-coated cork which had limited life in the fuel tank environment and buoyancy characteristics were not consistent.

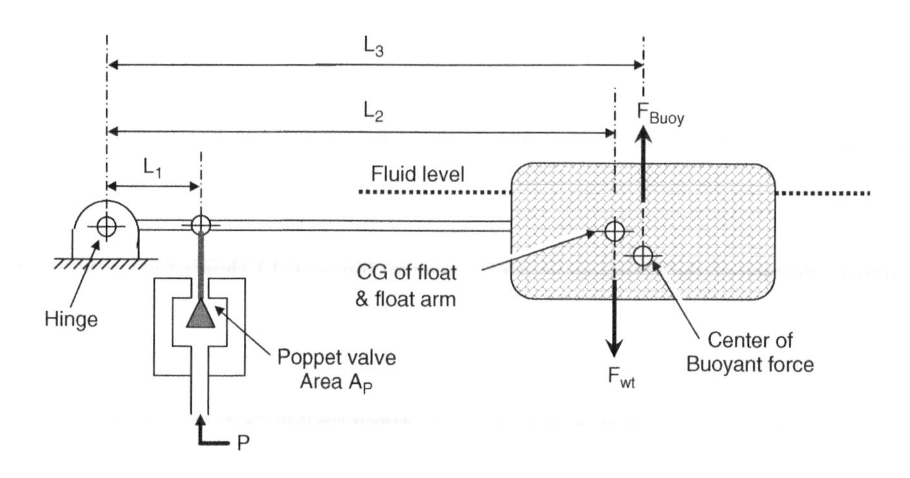

Figure 6.14 Float pilot force balance diagram.

The use of different float materials started in the early 1960s when two new float materials were developed for use in aircraft fuel systems. One of the new materials was a unicellular rubber that was formed in a mold and the other material was a foam urethane. Both materials provided a truly unicellular structure that did not need to be protected with an external coating. It was also found that the density could be precisely controlled as the float is formed in a mold which can be injected with a closely controlled weight of float material. Not only could the density be controlled, but it could also be varied for different design applications. The unicellular and urethane foam constructed floats also had the benefit that the float could be molded around the float arm thus eliminating the need for secondary mechanical attachment. It was also found, in most applications, that these new float materials could withstand the higher temperature and pressure conditions associated with military aircraft thus eliminating the need for the high cost metal floats.

The urethane foam material is the primary choice for floats today as it is the easiest to work with and provides the greatest flexibility in design. The urethane material also has the benefit of being able to be molded with a specific color. US Military fuel system components require identification with a 'Red' mark and therefore dyeing floats red satisfies this requirement.

Redundancy

Some aircraft require shutoff valve redundancy where failure of the system to shutoff could result in an unsafe condition. This is primarily the case for military aircraft with aerial refueling capability. This is particularly critical for aircraft that were initially designed for lower flow rates on the ground and later in life converted to provide aerial refueling capabilities. An example of this is the USAF C-141B. Some of the USAF military helicopter programs also require this type of shutoff redundancy.

To provide this redundancy, dual shutoff valves have been used. A dual shutoff valve consists of two shutoff elements packaged into a single valve assembly. Each shutoff element has its own high level sensing device that controls the shutoff function. Dual shutoff valves can contain two totally independent valves or utilize a single shutoff poppet operated by two independent shutoff operators. The latter approach is illustrated in Figure 6.15.

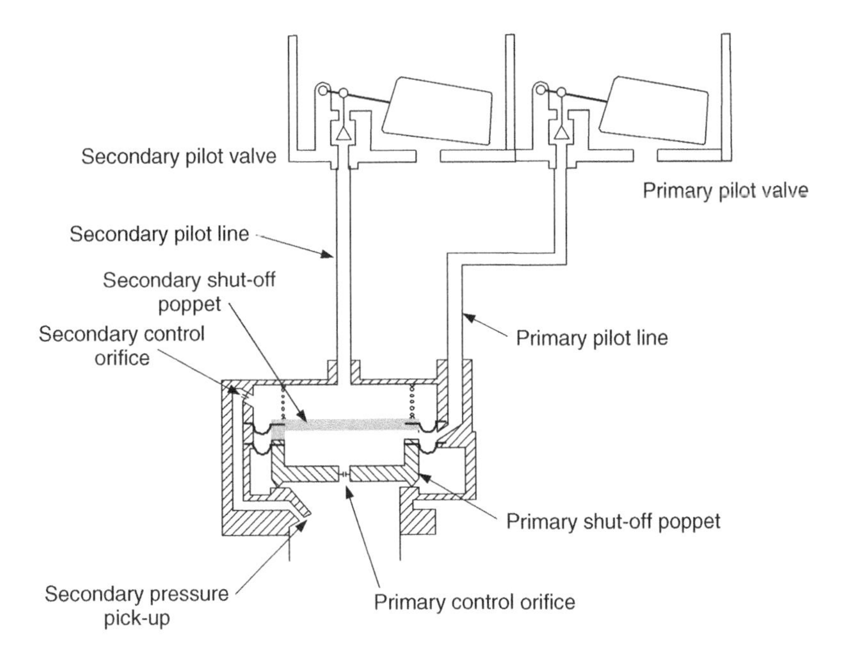

Figure 6.15 Dual redundant shut-off valve concept.

6.1.2.3 The Refuel Distribution Manifold

There are a wide range of packaging and installation arrangements for refuel shut-off valve equipment that have been used to satisfy specific customer and/or program requirements but one particularly novel approach worth describing here is the refuel distribution manifold.

All of the above illustrations show single shutoff valve arrangements whereas the distribution manifold approach uses multiple refueling shutoff valves within a single assembly along with the ground refueling adapter and defuel provisions to provide one integrated unit design.

An example of this approach is the Boeing 737 series ground refueling manifold. This manifold has three solenoid controlled shutoff valves (for each of the aircraft's three fuel tanks), three outlet check valves, the ground refueling adapter, and a defuel port. With this design the refuel shutoff valves are readily accessible for manual override or maintenance. Also, and most importantly, any surge pressures that can occur following valve closure are kept outside of the aircraft plumbing. A photograph of this unit is shown in Figure 6.16.

6.1.3 Fuel Transfer Valves

The applications described so far have been for valves that control the flow of fuel into a fuel tank. These valves can also be used for transferring fuel out of a tank. Typically in these applications, a low level pilot valve is used to shut-off the valve when the tank in near empty. For these reverse flow applications, the pressure drop across the valve is reversed from the re-fuel situation. Since the control chamber pressure is connected to what is now the outlet (which must be lower pressure than the tank pressure) via the restrictor across the piston or

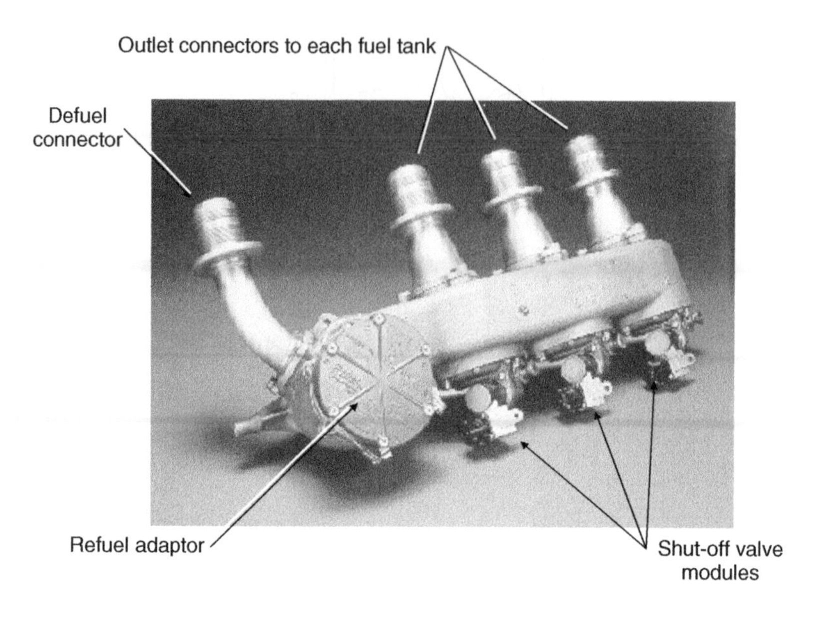

Figure 6.16 Refuel-defuel manifold (courtesy of Parker Aerospace).

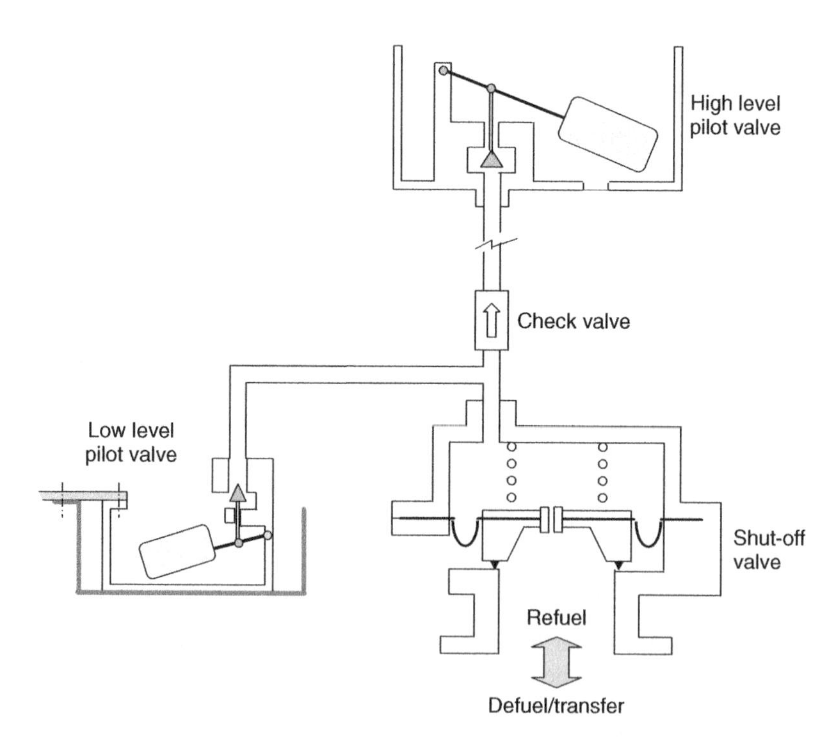

Figure 6.17 Refuel/transfer shutoff valve with low level pilot for fuel/no air.

diaphragm, the tank pressure acts over the differential area between the valve seat and the piston (or diaphragm) area causing the valve to open.

When the level in the tank drops to a near empty state, the low level pilot valve opens to allow tank pressure into the control chamber causing the valve to close. To allow this type of valve to operate, a check valve (non-return valve) must be added to the high level pilot valve lines to prevent tank pressure from prematurely entering the control chamber.

The pressure differential required for this reverse flow condition can be from tank pressurization or suction as would be the case for defueling. A common name used for the low level shutoff function is 'Fuel/No Air'. A very common application for this type of valve is in military aircraft external fuel tanks. This valve configuration is illustrated in Figure 6.17.

For military applications, the low level pilot valve often includes a 'negative g' override device so that the valve will close or remain closed under negative g conditions. Without this override, the fuel and the low-level pilot float would be forced upwards during negative g situations and in the transfer mode, the main shutoff valve would remain open, thus allowing a large volume of air to pass through the valve.

6.2 Fuel Tank Venting and Pressurization Equipment

Fuel tank venting and pressurization equipment covers a very wide range of products with a substantial range of functional complexities.

At the low end of functional complexity are float operated vent valves. These valves are relatively simple devices used to allow air to enter the vent lines and to close when exposed to fuel to prevent fuel from entering the vent system and, ultimately spillage overboard. Most float vent valves are direct acting (not pilot operated or pressure assisted) devices.

They rely on the float buoyancy to close the valve and the float weight to cause the valve to re-open. A float vent valve in one of the simplest forms is illustrated in Figure 6.18.

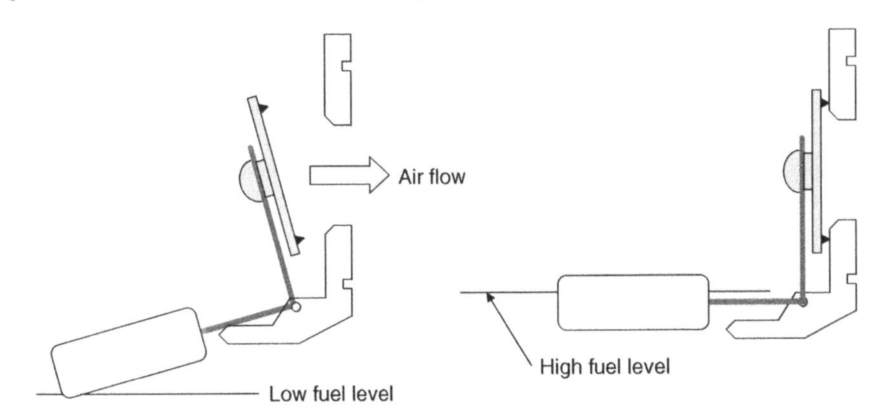

Figure 6.18 Direct acting float vent valve.

Although mechanically this is a very simple device, there are some characteristics that must be considered when designing this type of equipment. Unless the float is made very large, the force margins associated with a direct acting float vent valve are typically very small. The buoyant force created by the displaced fuel volume must be enough to over come the weight of the float valve mechanism with enough margin to provide an adequate seal load for very low (if not zero) leakage. On the other side of the force/moment equation, there must be enough weight moment when the float is not submerged (or even partially submerged) such that the valve does not blow closed prematurely due to air flow past the poppet.

A key limitation of a direct acting float operated vent valve is that it has a very limited reopening capability in the presence of a pressure differential across the closed poppet. For most designs, the limiting reopening pressure differential is less than 1.0 psid. For commercial aircraft, this is generally not a problem since the vent space is always open through at least one open vent passage to outside ambient. For military aircraft with a closed vent systems, however, this issue may be a problem.

In some aircraft applications float actuated vent valves can be used to drain any fuel from the vent lines back into the fuel tank. The design and construction of these valves is essentially the same as described above.

Open vent systems provide a small amount of pressurization during flight by a ram air or NACA scoop. Also in the vent line there is usually a flame arrestor to prevent propagation of an external fire or lightning strike into the fuel tank. The flame arrestor consists of many very small long parallel passages that prevent propagation of a flame. As a safeguard against possible restriction of the flame arrestor due to icing or blockage within other portions of the vent system, there is normally a separate fuel tank pressure relief device. This can be a burst disk or an actual pressure relief valve that will open when normal tank pressure limits (either positive or negative) are exceeded. Airbus typically uses a burst disk while Boeing prefers to use a pressure relief valve. Both are effective means of providing redundant fuel tank over pressure protection and both require ground maintenance after exposure to an over pressure condition causing the valve or burst disk to open. In the Boeing application, the relief valve is mounted to the wing upper surface and following operation it is clearly visible and must be manually reset to the closed position. In the Airbus application, more significant maintenance is required to replace the frangible portion of the burst disk. Figure 6.19 shows a photograph of the Boeing style relief valve.

For some of the smaller regional and business aircraft, it has been found advantageous to use a small vent line with the addition of a fuel tank vent valve that is opened during the ground refueling operation. For some system architectures, it may be mandatory that this vent valve be open before refueling can take place. When this is required, there is a position indication switch in the valve that must be actuated.

This valve is opened by sensing refueling pressure at the ground refueling adapter and locks out the opening of the refuel shutoff valves until the valve is full open. This type of valve is shown in Figure 6.20.

For military aircraft vent and pressurization systems are typically much more complex than seen on commercial aircraft. Most of the military aircraft have closed vent systems requiring pressurization typically from engine bleed air. The pressurization required is just sufficient to prevent the fuel from boiling at high altitudes with the pressure above outside ambient in the range 1 to 5 psi. In some applications the pressurization device (pressure regulator) and vent valve are integrated into a single unit. By combining the two functions into one unit, it provides a link between the pressurization function and the vent function such that there is no overlap of pressure set points that could allow pressurization flow and vent flow at the same time. This can result in a significant waste of bleed air. Some of these devices control tank ullage pressure

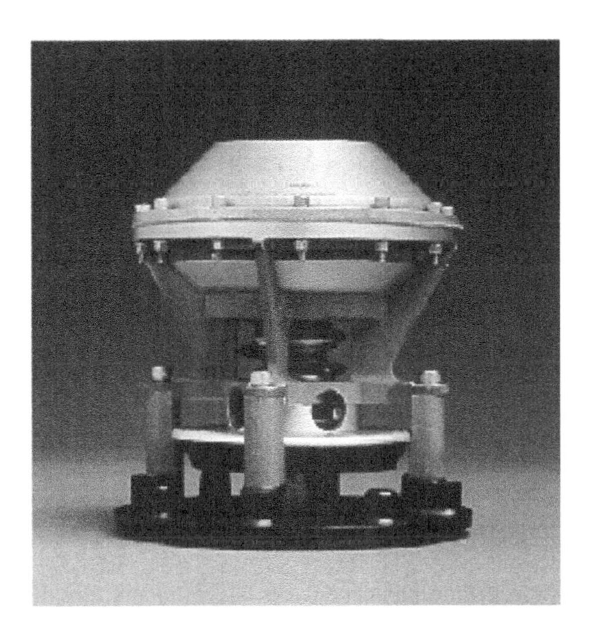

Figure 6.19 Fuel tank pressure relief valve (courtesy of Parker Aerospace).

Figure 6.20 Pressure operated vent valve (courtesy of Parker Aerospace).

with respect to local ambient pressure conditions. This is known as 'gauge pressure control'. In other applications gauge pressure control is provided up to some specified altitude. Above this altitude the ullage pressure control device reverts to an absolute pressure control mode. In many military aircraft applications pressure and vent control devices are completely hydro-mechanical (with possible exception of a solenoid) often resulting in very complex mechanical

Figure 6.21 Mechanical pressurization and vent valves (courtesy of Parker Aerospace).

designs. More recently electro-mechanical devices have been used for this function; however failure modes may dictate the need for functional redundancy to ensure that the probability of an unsafe condition cannot occur as a result of an equipment failure.

Figure 6.21 shows a purely mechanical pressure and vent control valve. In the example shown the pressurization and vent pressure control functions are integrated into a single unit.

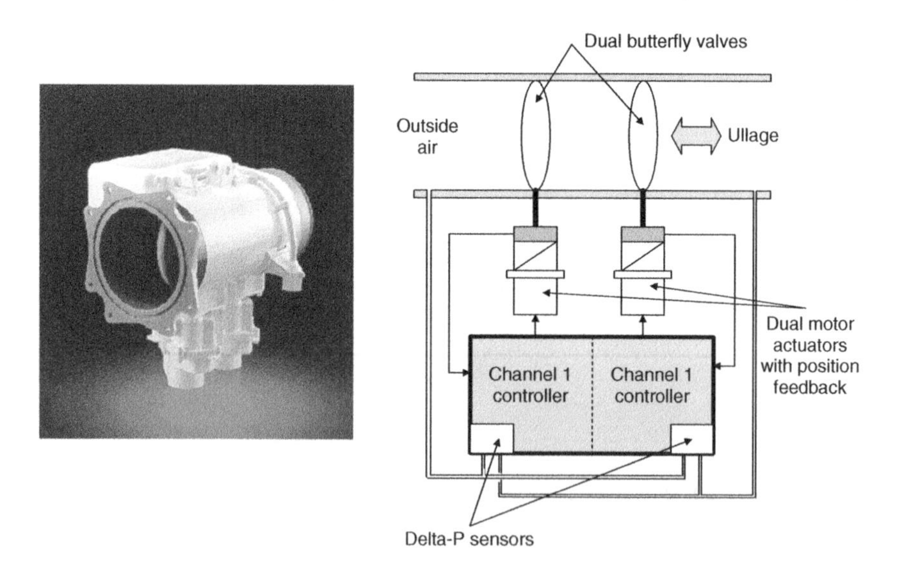

Figure 6.22 Electro-mechanical vent valve (courtesy of Parker Aerospace).

An example of an electro-mechanical vent valve is shown in Figure 6.22. In this application the valve controls ullage pressure relative the outside ambient according to a predetermined schedule. A separate pressurization valve controls the bleed air input to the ullage.

To ensure adequate functional integrity, the valve is fully redundant having dual channel electronics, sensors and valve drive. Thus no single failure can result in an open tank vent situation. In addition to the control function, an independent open command is provided to each valve actuator for the refuel mode.

The above represents jus a brief insight into the many different types of vent valves currently used on commercial and military aircraft.

6.3 Aerial Refueling Equipment

As described in Chapter 5, there are two different aerial refueling systems in operation today; the flying boom system used by the US Air Force and the probe and drogue system used by the US Navy, the US Marines and NATO countries.

The interface between the tanker and receiver aircraft for the USAF flying boom system is the boom nozzle located on the tanker and receptacle located on the receiver aircraft.

For the rest of the world, the probe and drogue system has the coupler which is deployed by the tanker and mates with the probe mounted on the receiver aircraft. Photographs of this equipment are shown in Figure 6.23.

The following paragraphs describe the above equipment in more detail.

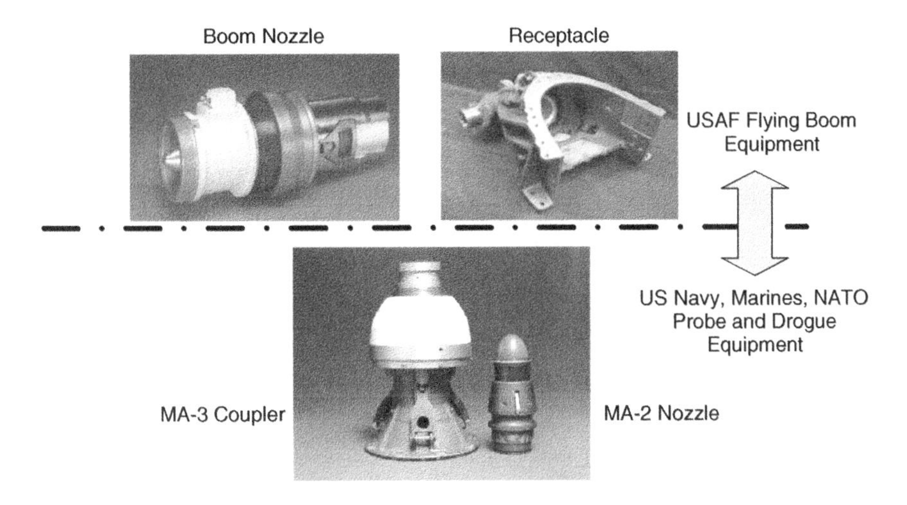

Figure 6.23 Aerial refueling equipment overview (courtesy of Parker Aerospace).

6.3.1 The Flying Boom System Equipment

6.3.1.1 The Boom Nozzle

The primary fuel system interface equipment on the refueling tanker aircraft is the boom nozzle which attaches to the tanker flying boom by means of a large threaded coupling. The attachment comprises a nozzle rotation centering device that allows rotation of the nozzle when engaged with the receptacle. Following a disconnect of the receiver from the nozzle, the centering device automatically rotates the nozzle back to its correct.

The key elements of the boom nozzle design include:

- a large swivel joint that allows a 60 degree included angle movement of the receptacle with respect to the centerline of the boom nozzle;
- a large spring loaded poppet valve that is opened when inserted into the receptacle;
- an embedded induction coil that aligns with a mating coil in the receptacle when fully inserted. This induction coil allows transmission of receptacle status to the tanker aircraft. it also provides voice (intercom) communication capability between the tanker and receiver aircraft to facilitate communication between the tanker and receiver in radio silence;
- two recessed pockets that allow mechanical latching of the nozzle into the receptacle;
- the capability for the tanker to initiate a disconnect from the receiver aircraft.

6.3.1.2 The Aerial Refueling Receptacle

The receptacle is installed in the receiver aircraft. In nearly all installations, the receptacle is in a fixed position in the receiver. There have only been a few installations where the receptacle is actuated into a refueling position. In most cases, there is a door that covers the receptacle when not in use. The entrance area leading into the receptacle is called the 'Slipway' as it helps to guide the nozzle into the receptacle. In many cases, the receptacle installation is designed by the airframe manufacturer. In these cases, the receptacle, the door, and system interconnections

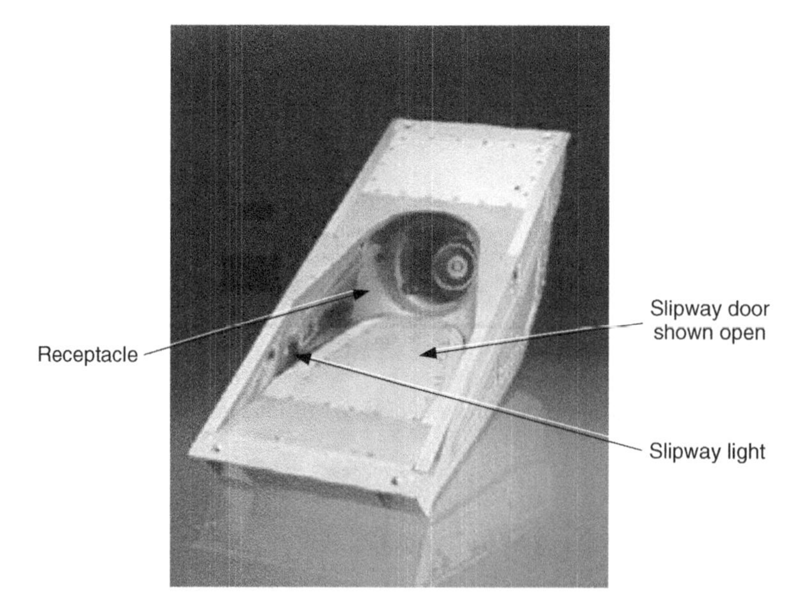

Figure 6.24 The Universal Aerial Refueling Receptacle Slipway Installation (UARRSI) (courtesy of Parker Aerospace).

are installed as discrete components. In the early 1970s, the USAF initiated a program to develop a standard aerial refueling receptacle installation termed UARRSI (Universal Aerial Refueling Receptacle Slipway Installation). This unit contains the receptacle, access door and door mechanism, slipway lighting, interconnection harnesses, and control valves all installed in a structural box. The UARRSI with the use of different installation kits can be installed in a wide range of receiver aircraft models. A photograph of the UARRSI is provided in Figure 6.24.

The key elements of the refueling receptacle include:

- a tapered entrance to guide nozzle into the bore of the receptacle;
- a fixed position pedestal that pushes the poppet in the nozzle open when properly aligned;
- a sliding sleeve valve that seals against the nozzle when fully engaged;
- a position indication switch that signals 'contact' and initiates the actuation of the latch actuator to latch the nozzle firmly into the receptacle;
- a position indication switch that sends a fully latched-in signal to the tanker aircraft; this is required before fuel transfer from the tanker to the receiver can begin;
- a hydraulically actuated latch actuator that holds the nozzle in the receptacle.

Under normal operating conditions, the latch actuator unlatches the nozzle when a disconnect signal is sent by either the tanker or receiver. The latch actuator also includes a relief valve that will cause the latch actuator to release the nozzle should the normal unlatching function fail and there is a high tension load. This is called a 'Brute Force' or 'Tension' disconnect. These features are illustrated in Figure 6.25.

6.3.2 The Probe and Drogue System Equipment

6.3.2.1 The Probe and Nozzle

The probe is attached to the receiver aircraft and typically consists of a relatively long tube section that terminates with the attachment of an MA-2 nozzle. In some cases, the probe is in a fixed position while in other applications, where the aerodynamic and/or stealth penalties associated with a fixed probe are unacceptable, the probe is stored within the receiver aircraft when not in use and is deployed for refueling.

The probe is normally installed such that the MA-2 nozzle can easily be seen by the receiver aircraft pilot so that he can fly the aircraft to engage the MA-2 nozzle into the drogue being trailed by the tanker. A photograph of an extendable aerial refueling probe with MA-2 nozzle attached is provided in Figure 6.26.

Key elements of the MA-2 Nozzle includes:

- a sliding sleeve valve that opens when installed into the coupling;
- a fixed position nose that pushes the poppet in the coupler open when engaged; the nose, although in a fixed position, is flexible to accommodate off center disconnects without damaging the nozzle;
- three nozzle latches that are disengaged from the sliding sleeve valve when inserted into the coupler.

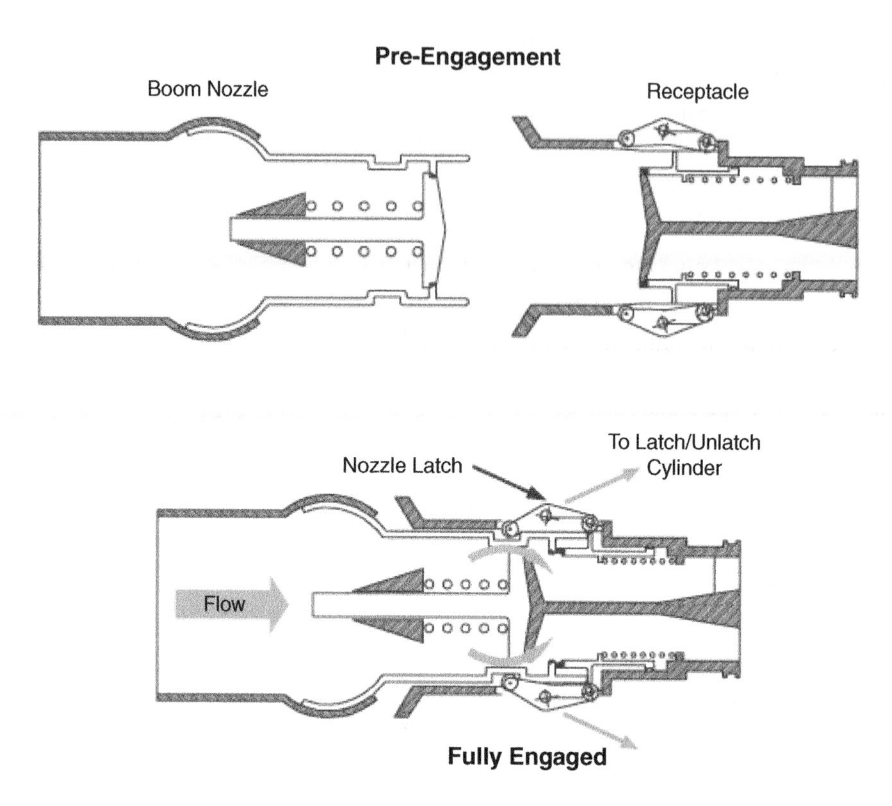

Figure 6.25 Illustration of engagement features of nozzle and receptacle.

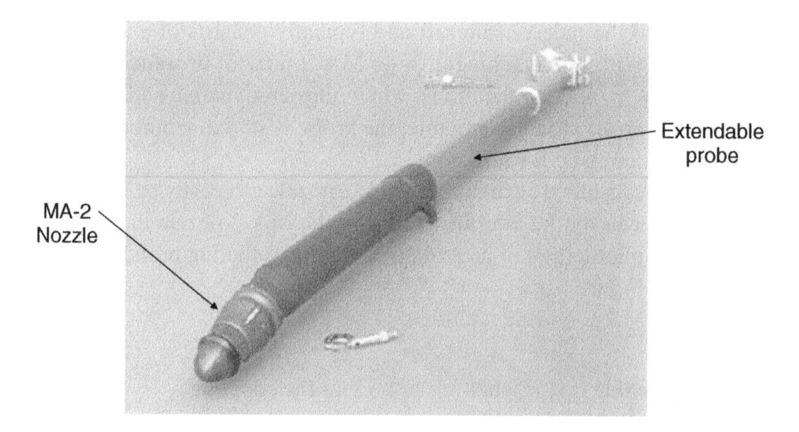

Figure 6.26 Extendable refueling probe with MA-2 nozzle (courtesy of Parker Aerospace).

6.3.2.2 The Drogue Coupler

The drogue is attached to the tanker aircraft by means of a long extendable/retractable hose. Within the drogue is a refueling coupler that is currently available in three different standards; the MA-2, MA-3 and MA-4 couplers with the MA-2 being the original, basic unit.

In the 1980s a new model was introduced that contained a pressure regulator / surge protector in an envelope only slightly longer (approx 3/8 inch) than required for the MA-2 unit. This is identified as the MA-3 coupler. Although the MA-3 coupler was very effective in controlling pressure and surge, there was concern that the unit could fail with resulting unacceptable performance. A second pressure regulator with a slightly different set point was integrated into the unit at the expense of a significantly longer length. This unit is identified as the MA-4 coupler.

Key elements of the coupler include:

- a conical entrance to guide the nozzle into the coupler;
- a poppet that is opened by the MA-2 nozzle;
- an internal pressure regulation and receiver aircraft surge protection device (for MA-3 and MA-4 couplers only); the MA-3 has one such device while the MA-4 has two for system redundancy;
- roller latches (3) that engage a groove in the MA-2 nozzle body;
- internal fuel pressure actuated pistons that assist holding of the nozzle when engaged.

These features are illustrated in Figure 6.27.

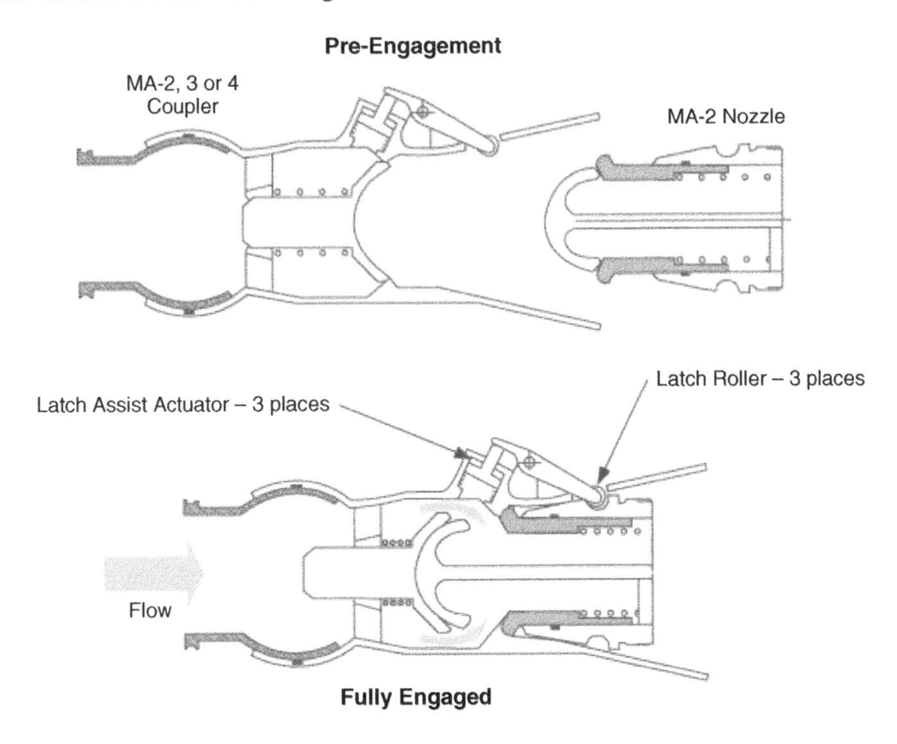

Figure 6.27 Probe and drogue nozzle and coupler.

6.4 Equipment Sizing

6.4.1 Valve Configuration and Pressure Drop Estimation

There are many system analysis tools available today that can be used to accurately determine the required size of fluid control equipment. Some of these tools are available to the general public for purchase and some have been developed for specific programs by companies for their own private and/or customer use. Most of these proprietary tools are very sophisticated requiring very detailed input data. For preliminary sizing, however, there is a very easy method that can be used to estimate the required equipment size based on the flow requirements and the allowable component pressure loss. This method is based upon the fact that for a given valve type, the pressure loss through the valve will have a constant pressure loss factor called the 'K' factor. For many applications, the selection of an appropriate K-factor relies heavily on the experience of the design engineer doing the preliminary analysis. Figure 6.28 shows how K-factor can vary with valve configuration with the in-line shut-off valve (number 3 in the figure) having a K-factor as high as 6.0 while the gate valve (number 5 in the figure) can achieve K-factor as low as 0.3.

Once a K factor has been established for a valve, the pressure drop across the valve can be determined from the following equation:

$$\Delta P = K_\rho V^2/2g$$

where: ΔP is the pressure drop

K is the K factor for the valve

ρ is the fuel density (nominally 50 lb/ft^3)

V is the fuel velocity which varies with line size.

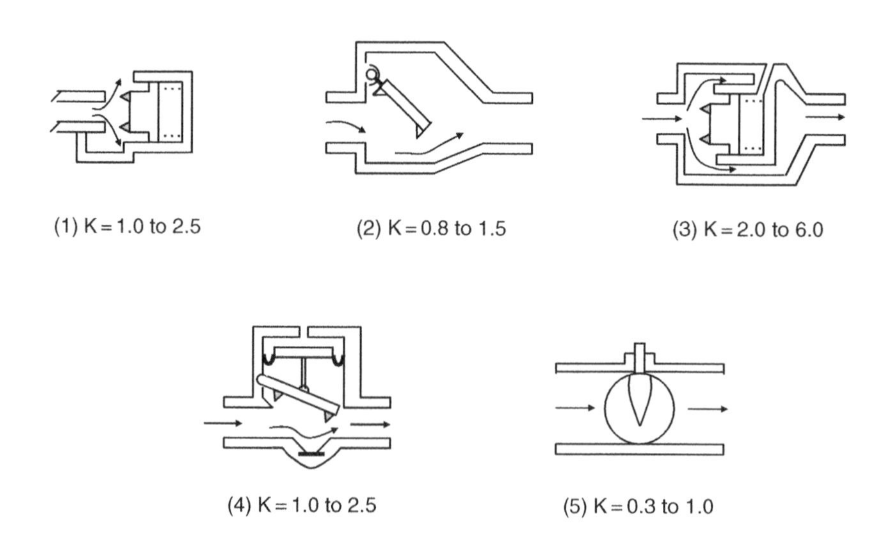

Figure 6.28 K factor versus valve configuration.

6.5 Fuel Pumps

6.5.1 Ejector Pumps

In their simplest form, ejector pumps (also known as jet pumps or eductor pumps) have no moving parts and therefore represent a highly reliable, low cost means to move fuel in an aircraft. All jet pumps contain the physical characteristics illustrated in Figure 6.29.

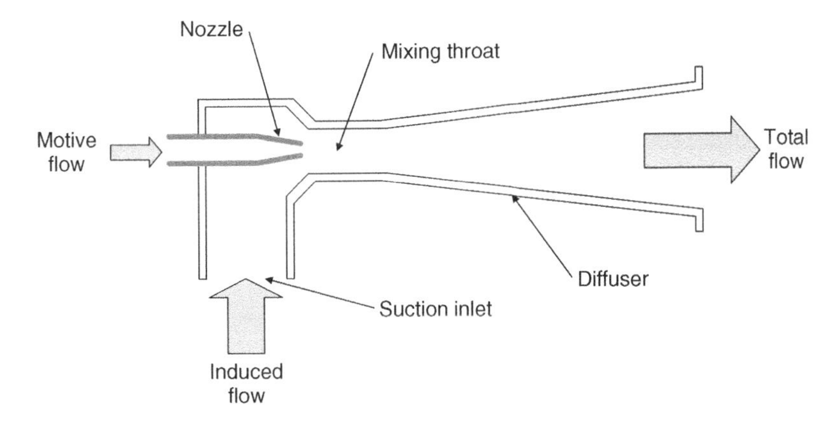

Figure 6.29 Basic ejector pump physical characteristics.

The motive flow power source for the ejector pump can be derived from several sources as follows:

- Feed line pressure provided by the motor-driven fuel pump. This source has limited pressure capability since traditional fuel boost pumps cannot deliver pressure higher than about 50 psig and this pressure upper limit varies with feed pressure in accordance with the boost pump pressure/flow characteristic.
- A dedicated engine-driven fuel pump. With this arrangement, much higher motive pressures are available, typically up to 500 psig, and while there is now no burden on the feed system, the added cost, weight and reliability penalties associated with this additional hardware must be considered in the design trade studies.
- A third source of motive flow is from the engine fuel control system. Since the engine high pressure pump is sized for the engine start condition, there is usually a significant excess capacity that can be used to support the fuel system ejector pumps. If not used this excess flow is spilled back to the high pressure pump inlet with resultant local heat generation.

When using the second of the above options minimum nozzle diameter considerations from a contamination and/or icing perspective may require some pressure limiting device to be installed with the motive pump.

The third option can be challenging since it requires close coordination between the engine fuel system and aircraft fuel system design teams.

In all of the above cases, motive flow is provided at relatively high pressures at relatively low flow rates. This motive flow induces flow into the suction inlet. The total flow then consists of the induced flow plus the motive flow. The resulting total flow rate is high and the resulting pressure is low when compared to the motive flow source.

The overall efficiency of an ejector pump is low when compared to electric motor-powered pumps but they are frequently used due to their extremely long life, high reliability and low cost. A discussion of jet pump use from a system perspective is provided in Chapter 4.

Ejector pump performance usually defined in terms of the ratio of delivered head rise, as defined in the pressure ratio equation below; head rise being the pressure increase from the suction inlet to the pump discharge.

$$N = \frac{(P_d - P_i)}{(P_m - P_d)}$$

where: **N** is the pressure ratio
P_d is the discharge port pressure
P_i is the induced port pressure
P_m is the motive pressure.

The efficiency of the ejector pump is the product of the induced flow rate and the pressure ratio divided by the motive flow rate as defined by the following equation:

$$\eta = \frac{(W_S N)}{W_m}$$

where: η is the pump efficiency
W_S is the induced (suction) flow rate
W_m is the motive flow rate.

The efficiency of typical ejector pump designs ranges from 20 % to 35 % with performance limits depending upon the vapor pressure of the fuel being used. For wide cut fuels, with high vapor pressures (fuel starts to boil at relatively low altitudes), this limitation can preclude the use of ejector pumps. For most commercial applications this is not a problem as they do not typically require the system to be designed for wide cut fuels. If wide cut fuels are used, there will be operational limits (altitude) imposed.

This limitation is much more of an issue on high performance military aircraft because they frequently operate at high altitude / high temperature conditions. Fortunately, most high performance military aircraft have pressurized fuel tanks which allow the ejector pumps to operate at these high altitude/high temperature ambient conditions.

For some applications, the ejector pumps contain additional elements. These can include inlet screens, inlet and outlet check valves, motive flow shutoff capability, or induced port shutoff capability.

Ejector pumps are used in a wide range of applications with discharge flow rates varying from a few hundred pounds per hour to more than ten tons per hour as illustrated by the photograph of Figure 6.30 which shows a large feed transfer ejector for a large transport aircraft and a small ejector used to scavenge fuel from the tank sump to the feed pump inlet.

The large ejector provides very high flow rates following take-off in order to fill quickly and maintain full a sub-compartment within a large feed tank. The small ejector is used to mix any

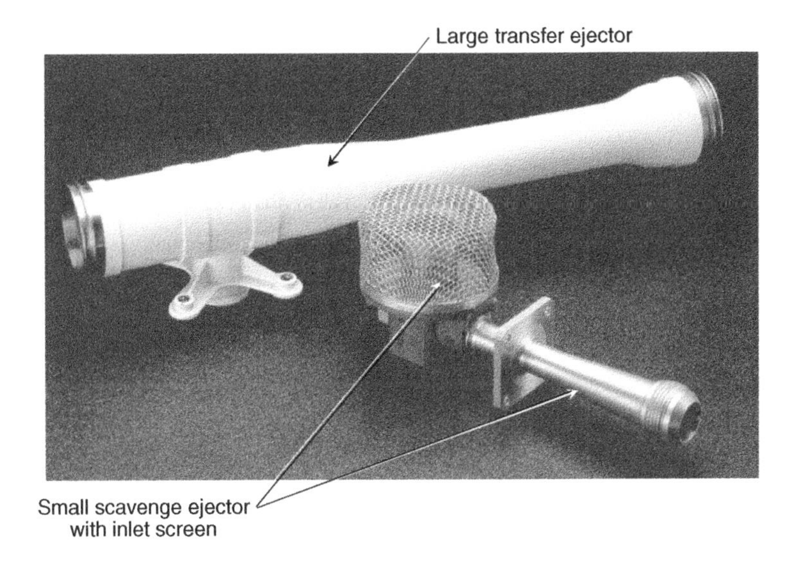

Figure 6.30 Fuel ejector pump examples (courtesy of Parker Aerospace).

free water from the tank sump with the local fuel and to discharge this water in small droplet form directly at the feed pump inlet. This removes small amounts of water from the feed tank by having it pumped to the engine and burnt. This technique is one of the techniques used today in the management of water in fuel tanks which can be a major problem for long-range aircraft.

6.5.2 Motor-driven pumps

The most common type of fuel pump for fuel boost and transfer uses a centrifugal pumping element driven directly by a fuel flooded electric motor. Alternative drive devices include hydraulic motors and air-motors but these types of fuel pumps are relatively rare and are usually found in military aircraft applications.

There are two main types of pump design driven often by installation or structural constraints and these are:

- the vertical configuration usually installed on the lower skin of the fuel tank, referred to here as the 'skin-mounted' pump;
- the spar-mounted pump.

The skin-mounted pump is usually configured as a cartridge-in-canister arrangement (refer back to the schematic of Figure 4.7) to allow removal and replacement of the cartridge, containing the pumping element and the motor, without having to drain the fuel tank. The photograph of Figure 6.31 shows a typical cartridge-canister pump for a regional transport aircraft application.

Figure 6.31 Cartridge-canister pump for a regional transport aircraft (courtesy of Parker Aerospace).

The skin-mounted arrangement has the advantage of the fuel head in the tank which can significantly improve pump performance at altitude over the spar-mounted alternative. Furthermore, the skin-mounted pump is much better able to handle air and vapor by providing appropriate venting of the pump discharge back to tank.

On the negative side, the large penetrations in the tank skin required to accommodate the canister installation is often a challenge for the aircraft structural stress engineer. Also the presence of the canister base may inhibit pump-down capability somewhat.

The spar-mounted pump requires a snorkel inlet connecting the pump inlet to the optimum location in the bottom of the fuel tank in order to achieve the best pump-down performance.

This arrangement has some significant installation advantages. Here the motor remains outside the fuel tank which is preferred from a safety perspective whereas the skin mounted pump motor is completely submerged in the tank and must demonstrate adequate design safety and explosion-proof capability. There is, however, a fundamental problem with the spar-mounted design concerning priming and starting since the fuel inside the pump can drain away from the pumping element following shut-down. This is usually addressed using an air- and vapor-removal element referred to as a 'Liquid ring' and this is described in detail later in this section. Also, as previously mentioned, the additional lift required to move the fuel from the snorkel inlet to the pumping element will have a negative effect on altitude performance.

The photograph of Figure 6.32 shows a spar-mounted transfer pump for a large commercial transport aircraft.

In addition to the two pump configurations discussed above, an in-line mounted pump design is sometimes employed; however the prospect of having to drain fuel tanks to access a line-mounted pump for maintenance makes this approach unattractive for most situations.

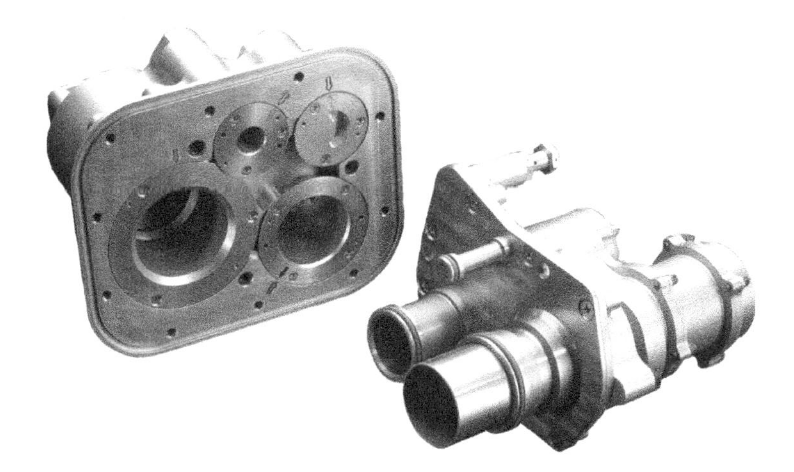

Figure 6.32 Spar mounted transfer pump for a large commercial transport (courtesy of Parker Aerospace).

The centrifugal pump is a high flow-low pressure device which is particularly suitable for the aircraft fuel system application. Maximum (dead-head) pressures for typical feed and transfer pumps are in the 30–50 psi range. Flow rates required to support the largest of today's turbofan engines at take-off power can be as high as 30,000 pph (about 80 US gallons per minute); however, the largest flow rate demand is usually for the jettison function and may require two or more pumps operating together to generate the flow rates necessary to meet the design jettison times.

The following paragraphs address the main technology issues associated with motor-driven pumping equipment including the pumping and air removal elements, motors, electrical power supply and control.

6.5.2.1 Pumping Element Technology

Centrifugal pumping element technology is essentially mature and no recent changes in the basic technology have occurred for decades, however, developments in the area of motor design, power supply and control have challenged to pump designer to identify the optimum pumping element solutions for a wide variety of motor-pump concepts.

The centrifugal pump requires a positive pressure at the pump inlet in order to prime and begin pumping. This requirement is referred to as the Net Positive Suction Head requirement (NPSHr). This positive pressure is necessary to overcome losses in the inlet line, to fill the impeller element cavities and to prevent cavitation. The NPSHr for a given pump increases as flow increases as indicated in the diagram of Figure 6.33 which characterizes NPSHr versus flow for radial impellers, inducers and mixed flow elements.

The inducer has better suction performance and a higher efficiency at higher flows and lower pressures while the radial impeller has higher efficiency at low flows and higher pressures. For this reason, a mixed flow element design is often used with an inducer immediately upstream of a radial element in order to capitalize on the benefits of both types of pumping element.

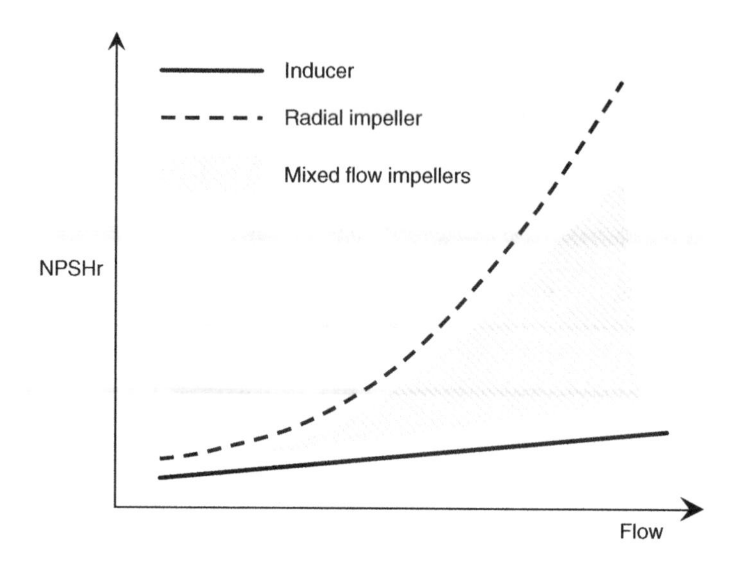

Figure 6.33 NPSHr versus flow for different pumping elements.

In the design of a fuel pump for an aircraft fuel system application, the 'Available' Net Positive Suction Head (NHSHa) must be established for the worst case operating condition which is dependent upon:

- type of fuel
- operating altitude
- fuel temperature
- lift requirements.

As a general guideline, the NPSHa must be at least 1.0 psi greater than the NPSHr.

Figure 6.34 shows an assortment of fuel pump impellers including radial, mixed flow and inducer types. The vane type of impeller shown in the lower right corner is a liquid ring impeller used for air and vapor removal. This topic is discussed in the following section.

Air and vapor removal

Large quantities of air (up to 14 % by volume) can be dissolved in kerosene fuels and this air bubbles out of solution during climb. Also at high altitudes with moderately high fuel temperatures, fuel vapor pressure can approach the ullage ambient resulting in large quantities of vapor evolution. To assist the centrifugal fuel pump handle large quantities of air and vapor, an element called a 'Liquid ring' is used to remove air and vapor and to discharge it back to the tank. The liquid ring is commonly used in spar-mounted pump applications where the pump must first consume large quantities of air from the snorkel tube as part of the priming process. Figure 6.35 shows the functional concept of the liquid ring.

The liquid ring consists of a vane type impeller located eccentrically inside a housing. For the spar mounted pump the pump axis of rotation is horizontal therefore when the pump is

Figure 6.34 Various fuel pumping impellers (courtesy of Parker Aerospace).

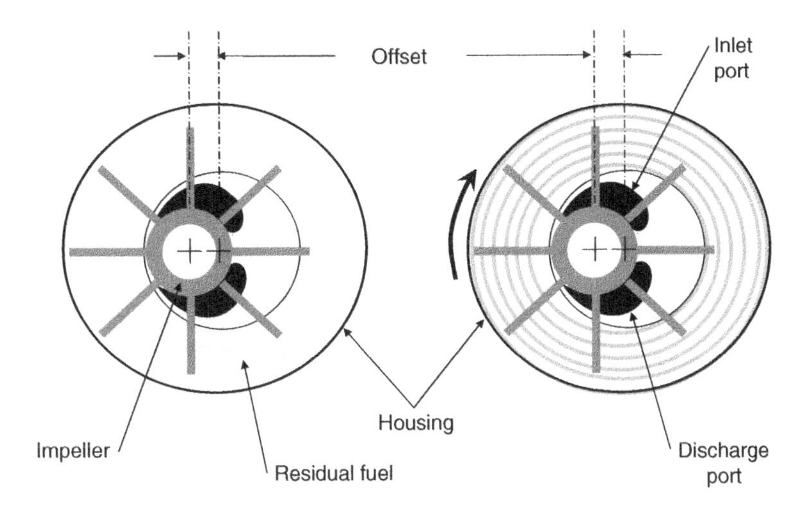

Figure 6.35 The liquid ring concept.

not rotating some residual fuel will remain within the liquid ring housing as indicated of the schematic diagram on the left of Figure 6.35. Once the pump is started, the impeller spins the fuel which is forced by centrifugal force to the outside of the housing allowing the air and vapor in the core to be pumped from the inlet port to the discharge port as shown. The inlet may be connected either to the main pump discharge or to the main element inlet; however, it should be noted that the energy added to the fuel by the main pumping element can add quantities of fuel vapor to the liquid ring in addition to the air that it is trying to remove. The discharge is vented directly into the tank.

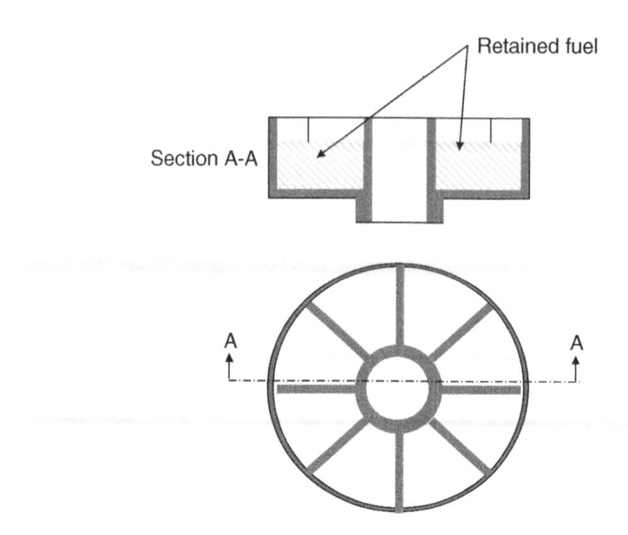

Figure 6.36 Liquid ring impeller for a skin-mounted pump.

For the skin-mounted fuel pump application, the drive shaft axis is vertical and therefore some means of fuel retention is needed for the liquid ring. This is accomplished as shown in Figure 6.36. The principle of operation is the same as described above. As mentioned before, the need for air and vapor removal for the skin mounted pump is much less of a problem and, in most cases, by providing a small additional flow capacity, and venting this flow back to tank, good performance can be achieved without the need for the added complexity of the liquid ring.

6.5.2.2 Fuel Pump Motor Technology

DC Motor Technology

There have been major developments in motor and motor driver technology over the past 25 years as indicated by the technology growth overview chart of Figure 6.37.

The old standby 28v dc brush motor pump continues to be the mainstay of the general aviation fuel pump market and this type of pump is also in commercial service in many applications primarily as the dc auxiliary pump used for APU starting or in some of the smaller regional or business aircraft for the feed and transfer functions. The limitation for commutated motors is generally considered to be 10 amps per commutator. For higher power applications double commutators are used thus limiting pump power capability for this type of pump to about 2/3 hp.

Figure 6.38 shows a typical brush motor 28v dc fuel pump. The pumping element is located at the bottom of the unit, the motor in the cylindrical section with the commutator at the top the motor. Fuel from the pump discharge is circulated through the motor and commutator assembly to provide a cooling function. The total assembly is explosion-proof.

In today's modern commercial transports, the brush dc motor-driven pumps are becoming less popular due to the demand for reduced scheduled maintenance that is necessary in brush motor designs even though a well designed pump can have removal rates in excess of 10,000

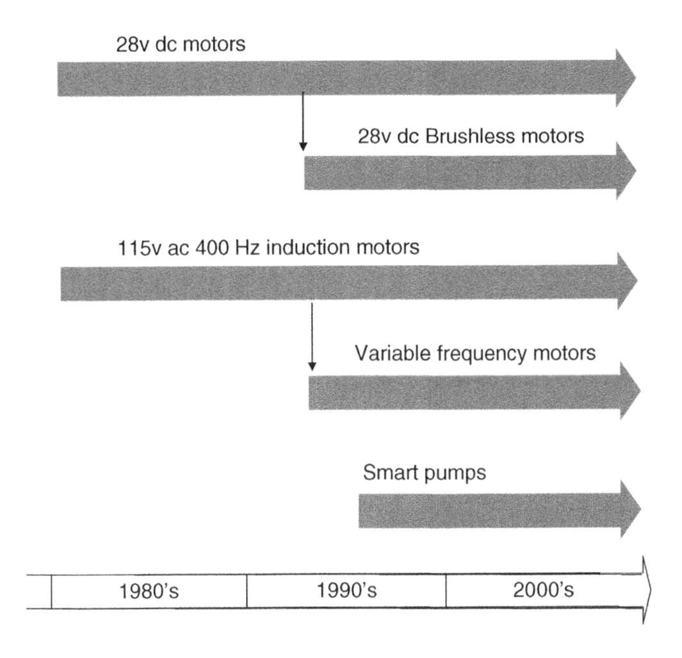

Figure 6.37 Fuel pump technology growth overview.

Figure 6.38 Typical brush motor 28 v dc fuel pump (courtesy of Parker Aerospace).

operating hours. As a result there has been a trend in the aerospace industry towards the use of brushless dc motor designs. Here the switching between motor windings is accomplished using power switching electronics and electronic control logic.

The control of the brushless dc motor is illustrated in its basic form in Figure 6.39.

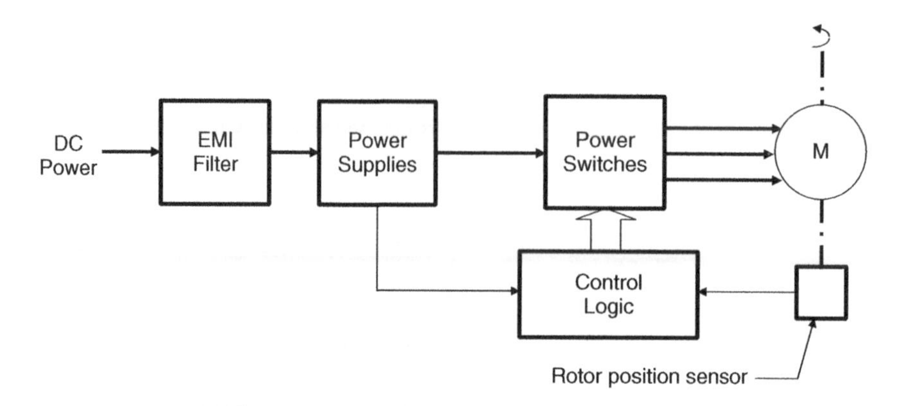

Figure 6.39 Brushless dc motor control schematic.

The power switches will be MOSFET's or IGBT's depending upon the power switching levels. Rotor angular position is required by the control electronics to facilitate the electronic commutation process. More recently, sensor-less commutation has been demonstrated using the motor current waveform to infer rotor position. Where sensors are employed, Hall devices or resolvers are both in common use.

Once electronics has been introduced to control the switching process, it becomes relatively easy and cost effective to incorporate additional functions such as:

- over temperature and over current protection
- speed control
- fault recording and reporting
- soft starting.

The last listed item involves limiting the motor voltage as the motor accelerates from rest which helps to reduce the in-rush current which is extremely high for dc motors due to the low winding resistance upon start-up.

Even though there are attractive functional benefits provided by the addition of the electronics for control and switching it may bring with it significant reliability and cost penalties which should be fully understood at the outset.

A more extensive treatment of electronically controlled 'Smart' fuel pumps is presented in Chapter 13 which addresses new technologies associated with aircraft fuel systems.

Constant Frequency Induction Motor Fuel Pumps

The constant frequency induction motor pump is probably the most common fuel pump in service today. This pump drive is both simple and robust with completely independent stator and rotor parts. The torque-speed characteristic of the typical induction motor (see Figure 6.40) results in very little speed variation with pump torque which is quite different from the simple commutator-type dc motor pump which exhibits a significant speed droop with operating load.

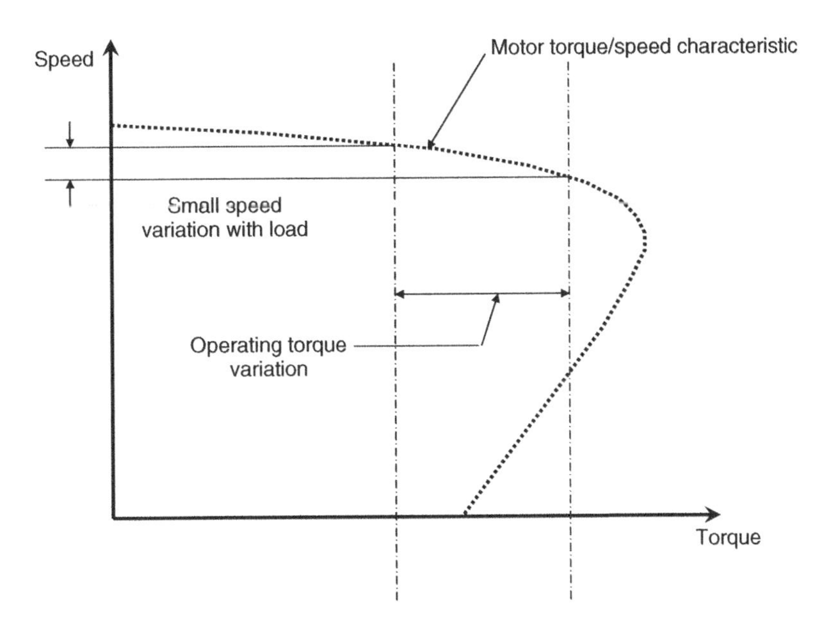

Figure 6.40 Induction motor operating characteristics.

The constant frequency motor in fuel pump applications remains an attractive solution offering high operational efficiency, low starting in-rush current and the ability to operate dry for extended periods without overheating the motor.

Variable Frequency (VF) Induction Motor Fuel Pumps

Since the early 1990s, there has been a trend towards the use of variable frequency power generation for aircraft applications Since the typical engine speed variation is approximately 2:1 between idle and maximum power, the electrical generation system has had to deal with this variation in the generation of electrical power for the aircraft. The traditional standard for ac power on aircraft has been 115 v ac, 400 Hertz, and to accommodate this standard two common techniques have been employed:

- Integrated Drive Generators (IDG's)
- Variable Speed Constant Frequency (VSCF) converters

(These alternatives were mentioned briefly in Chapter 2 in the discussion about system design drivers.)

The problem with the above approaches to power generation is the reliability of the regulation systems which involve, on the one hand, complex speed regulation mechanisms that convert the variable engine speed to a constant value for the electrical generator, and on the other hand, high power electrical components with significant failure rates due to the demanding powers and environmental conditions involved.

The alternative to this traditional electrical power generation scheme that is now becoming more 'The standard' for today's modern aircraft is to allow the ac generator frequency to vary with engine speed and have the electrical power consumers on the aircraft accommodate this situation. The result is a variable ac power system with nominally 115 v (phase to neutral) 3 phase ac power with a frequency that varies, typically from about 350 Hertz to about 750 Hertz.

The implication of variable frequency power to the traditional induction motor pump design is considerable. The challenge of optimizing the motor design to provide sufficient torque over the speed and load range without significantly over-sizing the motor is a demanding trade study.

The preferred solution to this problem is to provide a motor design that slips significantly from the synchronous speed at high frequencies so that there is a significantly smaller pump speed variation over the power supply frequency range as indicated in the pump performance curve of Figure 6.41.

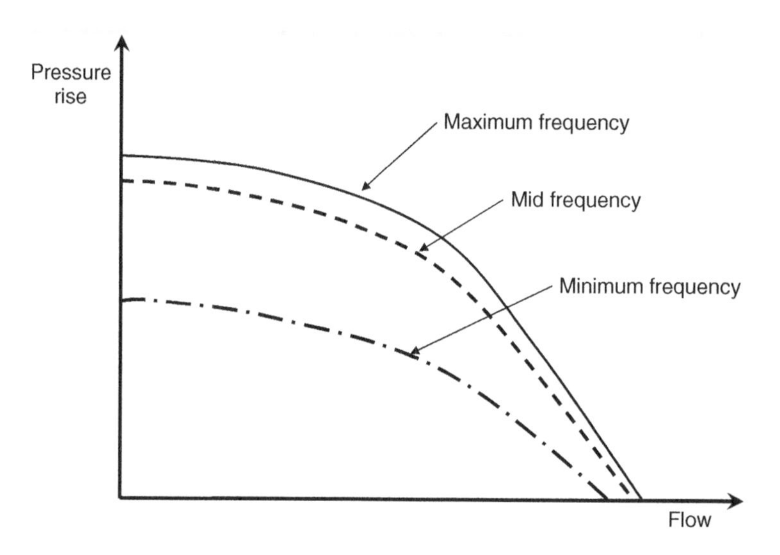

Figure 6.41 Typical high-slip VF pump characteristic.

In some cases it may be appropriate to have the highest frequency curve be actually below the mid frequency characteristic.

There are some key issues to watch for with the VF induction motor design; specifically the locked rotor current at high frequency will be small and typically not sufficient to trip a thermal fuse within the motor windings. At low frequency, however, the locked rotor current will be very high. For the same reason the dry running capability of a VF pump will be significantly limited at the lower frequencies because of the additional heat generated by the higher currents involved at these conditions.

Power Factor (PF) is another issue with the use of variable frequency power in induction motor-based fuel pump solutions. At low frequencies PF values significantly lower than 0.5 can occur and this results in a power generation penalty since the VA capacity of the generator and power lines must be increased to accommodate this issue. In general this is an aircraft-level trade-off that takes into account all power users and the simplification of the electrical generation system resulting from elimination of either IDG or VSCF equipment.

For high power fuel pumping requirements, however, the slipping induction motor solution may become untenable forcing a move to the electronically controlled pump which derives its power from the variable ac power supply. The complexities of PF correction, EMI filtration and harmonic distortion of the ac power supply make these advanced technology 'Smart pump' solutions very challenging. Once again this is addressed in more detail in Chapter 13.

6.5.2.3 Failure Accommodation

The most significant failure mode for the fuel pump is the locked rotor failure. Fuel systems are notoriously dirty and FOD inside fuel tanks is not an uncommon situation.

A locked rotor failure will result in a rapid rise in pump temperature due to this high current condition and even though the power supply will have circuit breakers the response time for the breaker maybe too slow to prevent dangerously high temperatures from developing. It is standard practice, therefore to provide temperature-sensitive fuses within the pump motor windings.

Typically all fuel pumps are mated with a pressure switch to advise the fuel management system and the crew of a pump failure. These pressure switches are usually discrete switches set typically in the range 5.0 to 10.0 psig.

In the new smart pump designs, there may be sufficient intelligence within the control electronics to infer the health of the fuel pump and even to provide prognostics to advise of an impending failure. Safety aspects of fuel pump design are covered in SAE 594 and FAR AC 25.981-1c.

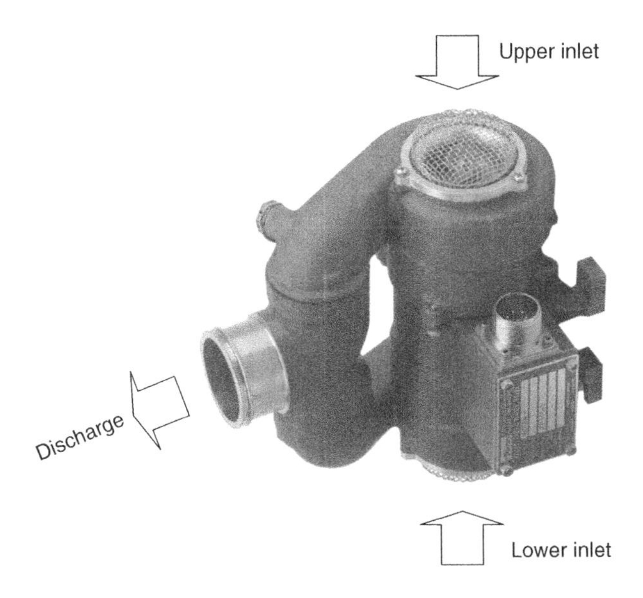

Figure 6.42 Double-ended fuel feed pump (courtesy of Parker Aerospace).

6.5.2.4 Military Fuel Pumps

As suggested in the Chapter 5 the military aircraft, particularly the fighter and attack aircraft have unique operational needs that often challenge the fuel system designer.

The need to accommodate large variations in g forces, both positive and negative for significant periods of time can be a challenge to the feed system which must continue to deliver pressurized fuel to the engine or engines.

The double-ended pump is a good example of a novel fuel pump design (see Figure 6.42).

This pump is powered by a constant frequency induction motor which drives two separate impellers feeding a common discharge. The discharge passages from each pumping element contain check valves. For a more detailed treatment of motor-driven fuel pumps the reader is directed to the Pump Handbook reference [11].

7

Fuel Measurement and Management Equipment

This chapter describes the equipment and technology associated with the measurement and management functions of aircraft fuel systems. Figure 7.1 shows how the design and development of the sensors, electronic and software fit into the overall system design and development process

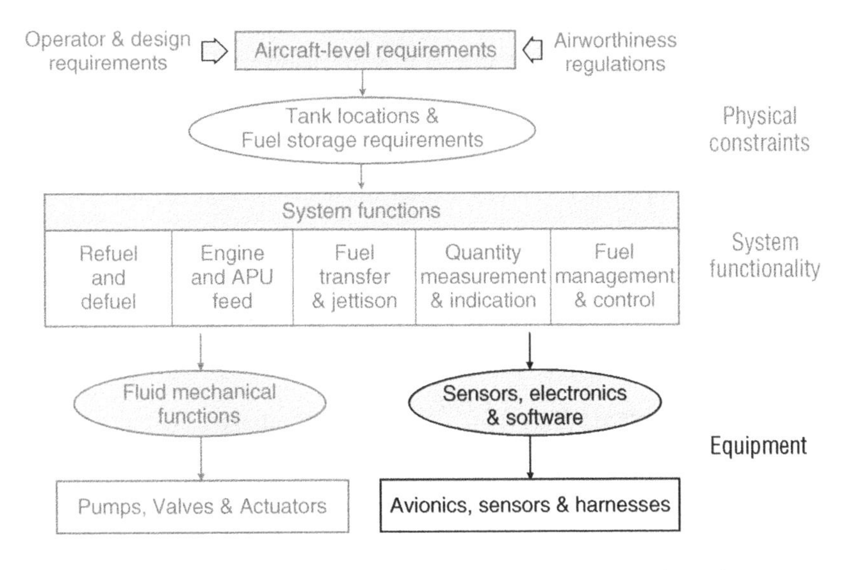

Figure 7.1 Fuel system design overview – sensors and electronic equipment.

Aircraft Fuel Systems R. Langton, C. Clark, M. Hewitt, L. Richards

7.1 Fuel Gauging Sensor Technology

There are a variety of methods that have been used to gauge fuel, ranging from successive discrete level measurement, using multiple sensors such as flotation or thermal devices, to continuous measurement using an array of full depth sensors such as capacitance devices. The most widely used to provide accurate gauging is that utilizing capacitance gauging. A recent alternative method, worthy of mention, has been developed using ultrasonics. These two approaches are now discussed in detail.

7.1.1 Capacitance Gauging

The industry has almost universally accepted this method of gauging as the way to gauge fuel quantity accurately. Although capacitance gauging dates back to a 1924 French Patent, it has been steadily improved and advanced as new technology and materials have become available over the subsequent 80 years. While the sensors are relatively unsophisticated, the long success of capacitance gauging systems is directly related to their compatibility and longevity in the relative hostile environment of the fuel tank. Capacitance gauging may be implemented in one of two approaches; ac capacitance or dc capacitance gauging. Furthermore, each of these approaches may be implemented in a variety of ways. In order to both understand and appreciate the advantages and disadvantages of the various approaches, a description of the principles involved is provided here.

7.1.1.1 Capacitance Principles

Capacitance is the physical property of an item to store charge and is developed by applying a potential difference (voltage) across a non-conducting medium (dielectric). A capacitive component (capacitor) is formed by placing a non-conducting medium between two conducting plates. The charge is configured as lines of electrical field across the dielectric. The capacitance (C) of this electrical component is expressed as the quotient of charge (q) and voltage (V) or:

$$C = \frac{q}{V}$$

In determining charge, in general, the permittivity constant for a vacuum is defined as:

$$\varepsilon_0 = 8.85415\left(10^{-12}\right)\frac{Coulomb^2}{Nm^2}\text{(from Coulomb's law)}$$

To determine the charge across a dielectric, the relative permittivity ε_r is required and is defined as:

$$\varepsilon_r = \frac{\varepsilon_s}{\varepsilon_0}$$

where ε_s is the static permittivity of the material. Hence:

$$\varepsilon_s = \varepsilon_r \varepsilon_0$$

It can be seen that the relative permittivity or dielectric constant, k as it is also known, acts as a permittivity constant multiplier. Therefore an important principle is that charge and therefore capacitance is directly proportional to dielectric constant.

The value of dielectric constant for some materials relevant to gauging, at 20 °C are as follows:

Material	Dielectric constant
Vacuum	1
Air at 1 atmosphere	1.00059
Hydrocarbon vapor	1.001 to 1.002
Water vapor	1.007
Aviation Gasoline	1.96
Aviation Kerosene (JP-4)	2.06
Aviation Kerosene (Jet A)	2.09
Ice (at -5°C)	2.85
Water	80.4

In an electrical circuit, a capacitor presents an impedance to the current (flow of charge) developed by the voltage applied to the circuit. For the case of a steady state voltage, the direct current (dc) flows in one direction in an exponentially decreasing manner until the voltage across the capacitor is fully charged and matches the applied voltage causing the current to cease to flow.

For the case of an alternating voltage, the alternating current (ac) flows as a current of alternating direction that is providing an alternating charging and discharging current. It is the alternating voltage mode of operation that is used in all fuel quantity gauging systems to excite the capacitance probes.

The terms ac capacitance and dc capacitance gauging relate to how the current developed by the excited probe is interfaced with the associated signal conditioning/processor.

7.1.1.2 The Basic Capacitance Probe

The fundamental principle of capacitance gauging is the difference in the dielectric properties of air and fuel. This phenomenon is exploited by configuring a capacitor as two concentric tubes arranged vertically or near vertically in a fuel tank. As the fuel level changes, the amount of the probe immersed in fuel changes and correspondingly the ratio of air to fuel and therefore the capacitance.

A modern, simple capacitance probe is typically configured with a nominally one inch diameter thin-walled metal outer tube and a nominally one half inch diameter thin-walled metal inner tube mounted concentrically within the outer tube to provide approximately a one quarter inch annulus for the air or fuel to accumulate (see Figure 7.2).

The theoretical capacitance of a cylindrical probe in a vacuum such as that described above is given by:

$$C = \frac{q}{V} = \frac{2\pi\varepsilon_0 L}{\ln(b/a)}$$

where: ε_0 = Permittivity Constant
L = Length of Probe
a = Outer radius of probe inner tube
b = Inner radius of probe outer tube

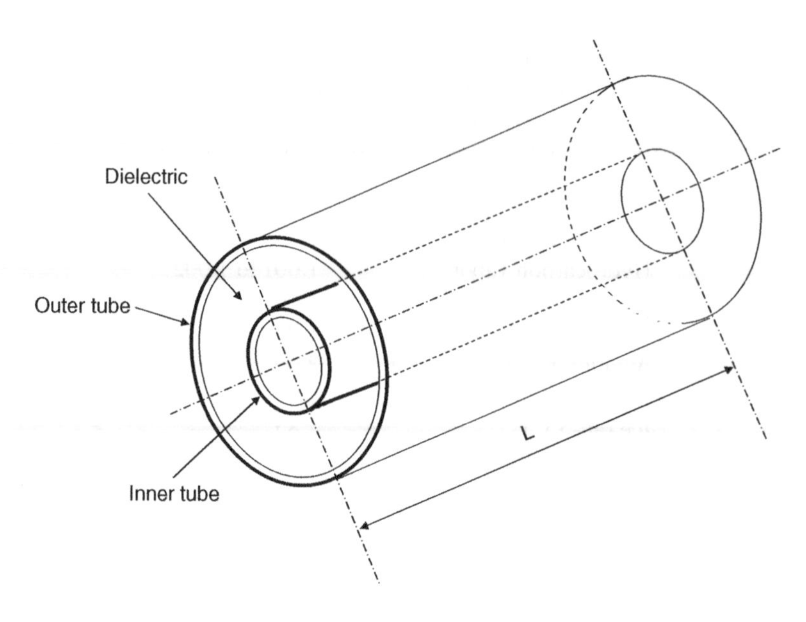

Figure 7.2 Capacitance probe concept.

In practical terms, the cylindrical probe not only develops a capacitance across the intervening dielectric between the tubes but also with other items such as adjacent structure. In other words, the electric field lines are not just contained within the probe inter-electrode gap but across dielectric gaps created in the probe design and manufacture, and its placement relative to the structure of the tank. This phenomenon is known as fringing and this effect is commonly referred to as the stray capacitance. For example, allowance for fringing effects between the probe and the top or bottom of the tank must be made. The capacitance of a probe C_P is therefore made up of two parts: the effective capacitance C_{peff} developed across the dielectric gap between the electrodes and the stray capacitance C_{Pstr}. The overall dry capacitance C_{Pdry} may therefore be expressed as:

$$C_{Pdry} = C_{Peff} + C_{Pstr} \tag{7.1}$$

In the case of a fully submerged probe in fuel of dielectric of k, the probe capacitance at full, C_{Pfull} becomes:

$$C_{Pfull} = kC_{Peff} + C_{Pstr} \tag{7.2}$$

Substituting for the constant, C_{Pstr} from (7.1) into (7.2) we get:

$$C_{Pfull} = (k - 1)\, C_{Peff} + C_{Pdry} \tag{7.3}$$

For the case of a partially submerged probe in fuel of dielectric k to a depth of n, where n is a normalized value between 0 and 1, the probe capacitance, C_{nPfull} becomes:

$$C_{nPfull} = n\,(k - 1)\, C_{Peff} + C_{Pdry} \tag{7.4}$$

The depth of immersion, n, is frequently referred to as the probe wetted length.

The change in capacitance, $C_{Pn\Delta}$ from air to partial submersion in fuel is therefore:

$$C_{nP\Delta} = C_{nPfull} - C_{Pdry} \tag{7.5}$$

Substituting for C_{Pdry} from (7.4) into (7.5) we get:

$$C_{Pn\Delta} = n\,(k - 1)\,C_{Peff} \tag{7.6}$$

In summary, Equation (7.6) reveals that the change of capacitance is:

- directly proportional to depth of immersion (wetted length), n
- directly proportional to dielectric constant, k
- independent of stray capacitance, C_{Pstr}

Figure 7.3 shows the relationship between dielectric constant and temperature for various fuels. If k were to be treated as a constant, independent of fuel type and/or temperature, a fuel gauging error of typically ±6 % of indicated contents is introduced. By introducing correction for changes in k, this error is reduced to ±2.75 %. Correction for changes in dielectric constant may be implemented by the introduction of dielectric compensation. This is achieved by using an additional probe in the tank known as a compensator. The compensator is a small probe-like capacitor mounted near the bottom of the tank to ensure total immersion and tracking of the fuel dielectric constant down to very low levels of fuel.

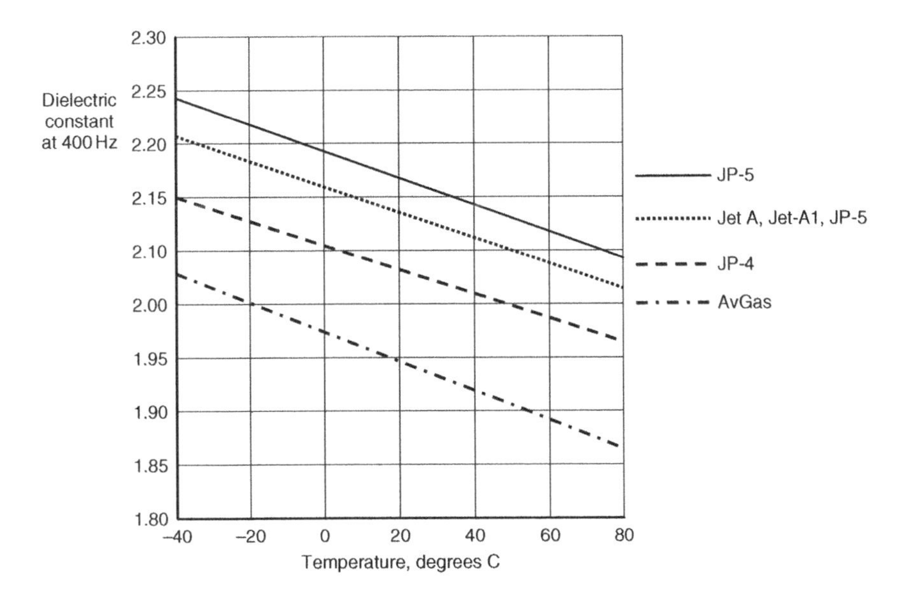

Figure 7.3 Dielectric constant vs temperature for typical aircraft fuel.

Similar to Equation (7.6) above, the capacitance of the totally immersed ($n = 1$) compensator probe, C_C with an effective capacitance of C_{Ceff} is:

$$C_C = (k - 1)\, C_{Ceff} \tag{7.7}$$

Substituting for (k-1) from (7.7) in (7.6) we get:

$$C_{Pn\Delta} = n\left(\left(\frac{C_{Ceff}}{C_C}\right) - 1\right) C_{Peff} \tag{7.8}$$

In summary, Equation (7.8) reveals that the change of capacitance is:

- directly proportional to depth of immersion, n
- independent of k.

Another method of compensation that does not require an independent compensator has been used in the industry. This technique is referred to as 'full height compensation' or 'self-compensation'. This technique was originally developed to overcome the vulnerability of earlier gauging systems to contamination effects on the compensator due to water, fungus etc. This approach requires the use of profiled or characterized probes and therefore this type of gauging technique is not in common use in modern gauging systems where software has replaced the need for the more expensive profiled probes.

Specification MIL-G-26988C reference [12] (now an obsolete US Military Specification) defined fuel quantity gauging system accuracy in terms of classes as follows:

- Class I gauging defined the accuracy requirements of uncompensated systems as ±4% of indicated quantity ±2% of full contents.
- Class II gauging referred to systems with dielectric compensation defining accuracy requirements as ±2% of indicated quantity ±0.75% of full contents.
- Class III gauging referred to dielectric-compensated systems with direct density measurement defining accuracy requirements as ±1% of indicated quantity ±0.5% of full contents.

Full height compensation gauging systems can achieve accuracies similar to the Class II requirements defined above and this gauging technique uses the relationship between fuel dielectric constant and density as explained in the text that follows.

It has been established that fuel dielectric constant k is directly proportional to fuel density D and much data has been gathered on this relationship for many different fuels from many sources and locations all over the world.

Figure 7.4 shows the best line fit between dielectric constant and density taken from the same Military Specification MIL-G-26988C reference [12]. While this direct proportionality exists, it is important to allow for the fact that D changes at a greater rate that k.

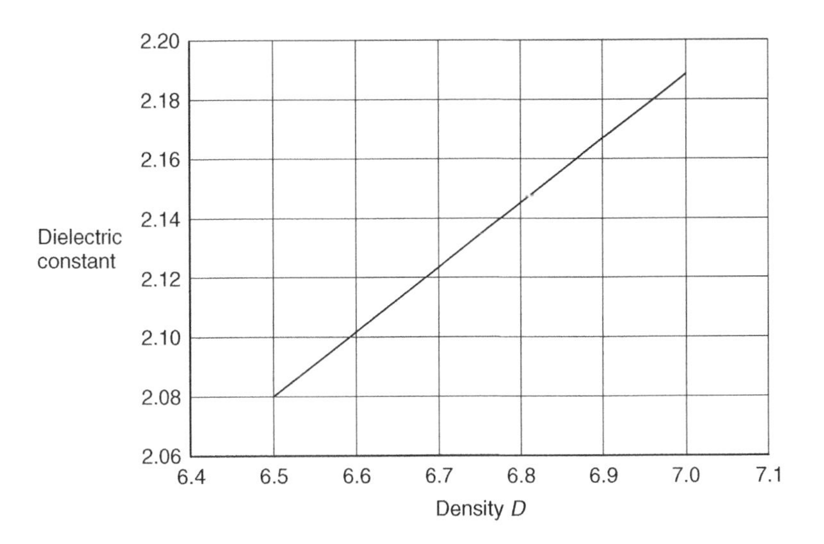

Figure 7.4 Dielectric constant vs density.

From the Military Specification the relationship between k and D is expressed by the following equation:

$$\frac{(k-1)}{D} = A\,(k-1) = B \tag{7.9}$$

Where A and B are constants.

Rearranging we get:

$$(k-1) = \frac{BD}{(1-AD)} \tag{7.10}$$

This slightly non-linear relationship can be modeled as:

$$(k-1) = aD^n \tag{7.11}$$

A probe fully immersed in fuel is naturally subject to the fuel dielectric over its entire length. Equation (7.11) is modeled by scaling the fully immersed probe output in terms of inferred density rather than dielectric constant. The probe therefore provides an output proportional to mass that does not require further compensation.

The implementation of full height compensation gauging is accomplished by locating a series capacitor of approximately twice the value of the probe dry capacitance on the probe and logarithmically modifying the necessary inner tube profiling over and above that necessary to track volumetric changes. This arrangement is designed to provide the best probe accuracy when fully submerged. At other levels, a small error is introduced which peaks at the 50 % immersion point.

The 'self-compensation' implementation arranges the compensation for best accuracy at two-thirds immersion to minimize the maximum error at other levels of immersion. A disadvantage of this technique is that the series capacitor required reduces the probe output by as much as 66 % leading to a corresponding reduction in signal to noise ratio.

7.1.1.3 Probe Excitation and Signal Conditioning

Capacitance gauging systems evolved using a sinusoidal voltage waveform for probe excitation as this was readily derived from the aircraft 400 Hz supply. With the advent of advanced semiconductor and microprocessor based signal conditioning, it became relatively easy to introduce changes in excitation frequency and amplitude as well as waveform to improve gauging performance. The reasons for these changes are more readily understood through some brief probe electrical circuit analysis.

Referring to Figure 7.5, consider a wetted probe as having a capacitance C_P. The effects of leakage resistance due to conductive fuel contamination effects may be represented by a parallel resistance R_L. In addition, a series resistor R_P represents the probe connection resistance of a few milliohms. Since this is very small it may be ignored in the analysis.

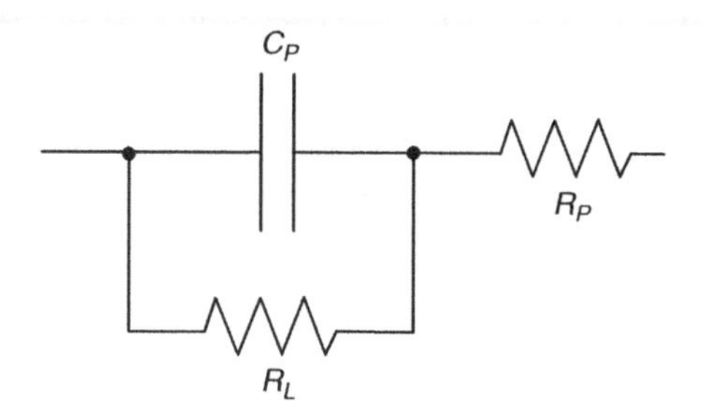

Figure 7.5 Probe equivalent circuit.

Consider a sinusoidal probe drive voltage waveform of the form

$$V_P = K\,Sin\omega t$$

Where K is a constant and ω is the frequency in radians per second, i.e. $2\pi f$ where f is the frequency in Hertz.

From Ohm's law, the current I_P produced by the sinusoidal voltage V_P through the resistance R_L is:

$$I_P = V_P/R_L \tag{7.12}$$

The pure reactance or impedance Z_P of the capacitor C_P is:

$$Z_P = 1/\omega C_P$$

$$\text{or,} \quad Z_P = 1/2\pi f C_P \tag{7.13}$$

From (7.13), it can be seen that reactance is inversely proportional to frequency.

Using Ohm's law for a sinusoidal voltage V_P applied to the capacitor C_P, then:

$$V_P = I_P.Z_P = I_P.1/2\pi f C_P \tag{7.14}$$

Rearranging for I_P we can say:

$$I_P = V_P 2\pi f C_P \tag{7.15}$$

From Equation (7.12) it is apparent that for resistance, current is independent of frequency. However, from Equation (7.15), it is apparent that for capacitance with any given voltage, the current is directly proportional to frequency. Since in the real world, the current will result from both the capacitance of the probe, and any resistance caused by fuel contamination, increasing the frequency of the excitation voltage, increases the probe signal while the leakage remains constant. Therefore by raising the frequency of the probe excitation voltage significantly the effects of resistive leakage due to fuel contamination can be minimized.

Another aspect of Equation (7.15) is that for any given frequency, the current increases with capacitance. It is therefore an important consideration in the design of a probe to make the capacitance as large as possible compatible with size and weight requirements. The resulting larger current reduces the impact of leakage and provides for more accurate signal conditioning measurement through improved signal to noise ratio.

One final aspect of Equation (7.15) is that if the product of excitation voltage and frequency is kept constant, then the current is directly proportional to capacitance. Therefore, by ensuring that any change in excitation voltage is matched by a compensating change in frequency, the current to capacitance relationship is preserved. This feature has been used in earlier gauging systems where it was found easier to keep the voltage-frequency product constant rather than try to control each parameter separately over the operating temperature range.

Leakage current can be eliminated by taking advantage of the fact that the reactive and resistive components of the return signal are in quadrature to one another, and by configuring the signal conditioning to sample the return signal in a phase synchronous manner (synchronous demodulation), as illustrated in Figures 7.6 and 7.7, errors due to leakage current can be eliminated. It can be seen that by sampling the return signal in quadrature with the excitation, the reactive component is at a maximum when the resistive or leakage component is at zero. Conversely, the signal conditioning can also be configured to sample the return signal in phase with the excitation to actually measure the leakage and provide warning of excessive leakage due to abnormal levels of conductive fuel contaminants, such as water.

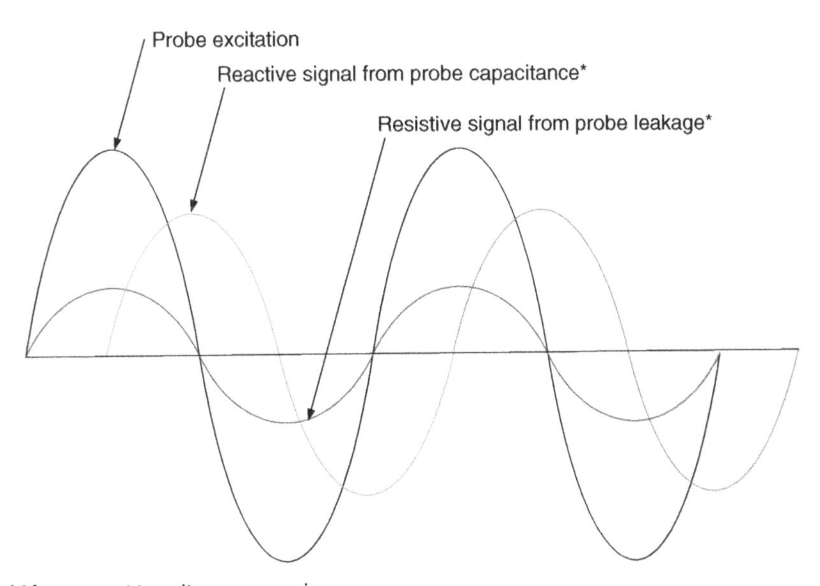

Figure 7.6 Sinusoidal capacitance probe excitation and signal waveforms.

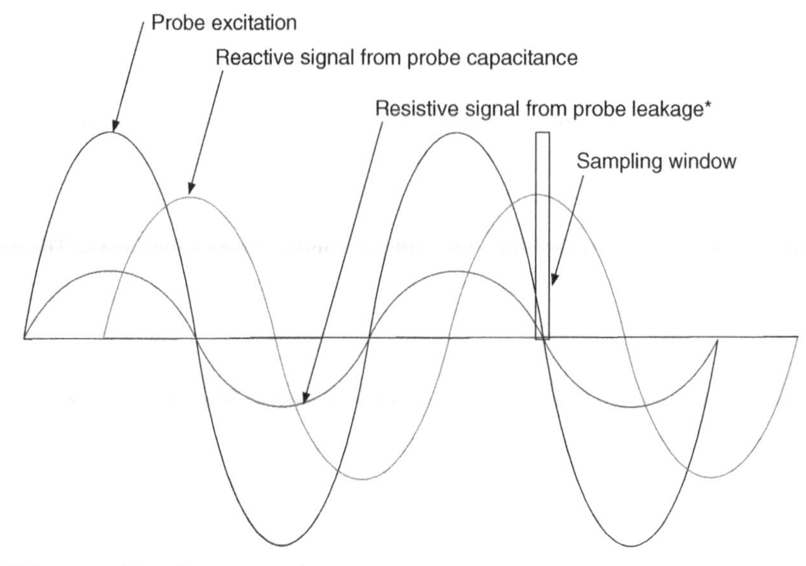

Figure 7.7 Excitation and signal waveforms with sampling.

Modern, sinusoidally excited, ac capacitance gauging systems use a combination of higher excitation frequency and phase synchronous detection. Generally the excitation frequency is in the range of 5 to 15 KHz.

Another approach to probe excitation, illustrated in Figure 7.8, was developed and patented in the 1970s using triangular excitation. This again is based on the principle of synchronous demodulation. Figure 7.8a shows the triangular excitation references [13] and [14] and Figure 7.8b shows the individual reactive (square wave) and resistive (triangular waveform) components of the return waveform following signal conditioning.

The composite waveform formed from these two components is shown in Figure 7.8c. By arranging a sampling window about the point at which the resistive component is zero, a waveform is produced (Figure 7.8d) that is directly proportional to capacitance independent of resistance as only the slope and not the area of the waveform is affected by varying leakage effects.

7.1.1.4 Capacitance Gauging Approaches

The two implementations of capacitance gauging employ either ac or dc capacitance probes. These probes may be both considered as passive probes in that they have either no electrical components or a limited number of capacitors, resistors or diodes. Another method is to use active or smart probes which feature some limited integral signal conditioning featuring more complex electrical components. A key point, in considering the merits of these different approaches is the Life Cycle Cost of the system; the fact that the probes and associated wiring are installed in a relatively hostile environment that is difficult to access and maintain should always be a prime consideration along with intrinsic safety. The principles and key points associated with these approaches are now described.

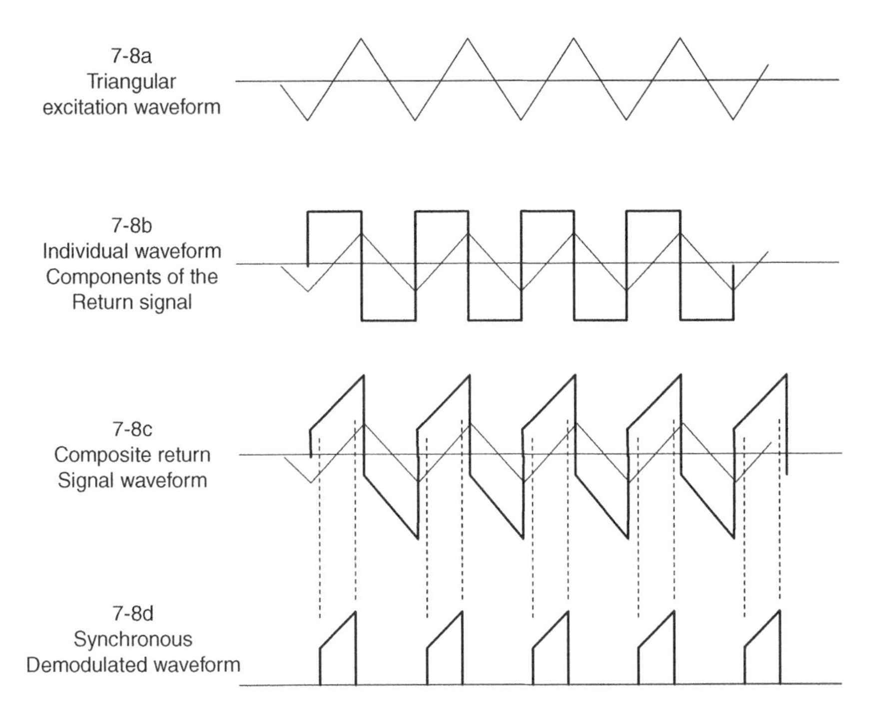

Figure 7.8 Triangular waveform excitation and response after signal conditioning.

AC Capacitance Probes

The simplest implementation of the capacitance probe is provided by the ac probe as these probes are usually configured with no electrical components. Modern ac probes are generally of the linear type comprising two lightweight aluminum concentric tubes of constant diameter, separated by insulating spacers. The outer tube, typically 1″ in diameter is configured as the low impedance (lo-Z) electrode. The inner electrode is the high impedance (hi-Z) electrode and, for a 1″ probe, is typically about 0.5″ diameter providing an internal circumferential gap of 0.25″ to safeguard against water droplets electrically bridging and short circuiting the gap. The tubes are typically polyurethane coated to readily promote water shedding. The design incorporates dual end caps to protect the tube ends, prevent contamination ingress, and provide a lightning gap. The probes are fitted with mounting brackets that are keyed to prevent incorrect installation on the aircraft. It is important, that the probes can freely drain, such that inaccuracy in system performance is minimized. The probes, therefore, incorporate drain holes to allow the purging of water by drainage. A terminal block is located on the outer tube to provide terminations for the interconnecting electrical harness comprising the high and low-z connections and an anchor point for the high-z shield. The terminals are dissimilar-sized to safeguard against incorrect interconnection with the harness. The terminal board within the block is designed to shed water and observe the required lighting gap. Probes for high accuracy gauging applications are either manufactured in the factory to high precision or include calibration adjustment to provide a repeatable dry capacitance value. This allows faulty probe replacement without the need for on-aircraft system adjustment. A typical probe is illustrated in Figure 7.9.

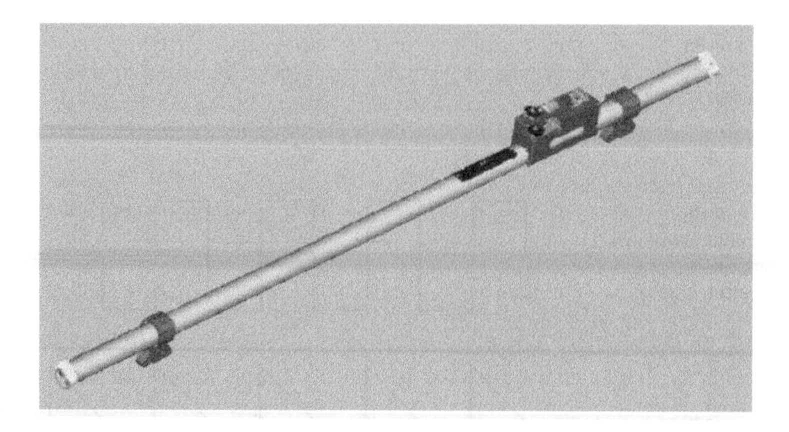

Figure 7.9 Typical ac capacitance probe (courtesy of Parker Aerospace).

The key points of an ac capacitance probe gauging system compared to other capacitance systems are:

- lowest weight
- highest reliability
- lowest cost
- most rugged
- allows reactive and resistive (leakage) current components to be readily detected
- high level of intrinsic safety
- highest HIRF compatibility
- requires shielded cable harnessing
- shield continuity is essential.

DC Capacitance Probes

The difference between DC and AC gauging is that with DC gauging the probes are fitted with an electrical network to rectify the probe output signal. The basic concept is illustrated in Figure 7.10.

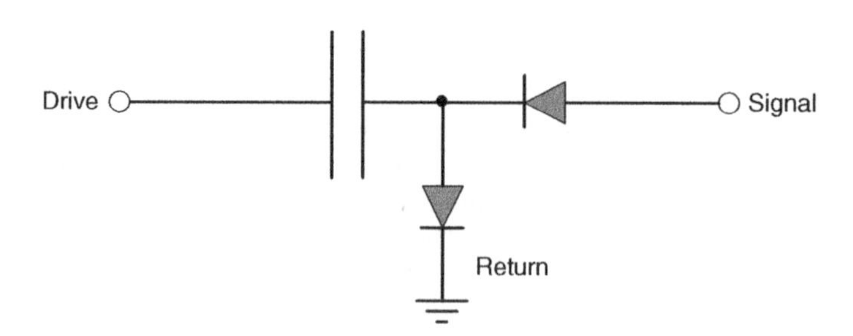

Figure 7.10 Basic dc probe concept.

The primary reasoning behind the development of DC gauging was to simplify both the in- and out-tank installation by eliminating the need for shielded wiring and thereby providing a lower weight and more reliable installation, the use of shielded wiring being precluded by the fact that the probe diode network eliminates capacitive coupling between the drive and signal in the wiring and connectors.

The DC gauging approach does however introduce the following considerations which must be allowed for in the design:

- While the wiring reliability is increased, the reliability of each probe is decreased through the introduction of electrical components mounted in the probe terminal block
- Each diode has a current and temperature dependent voltage drop (typically 0.5 to 0.8 volts) before it can conduct current
- In rectifying the ac voltage into a dc current, the reactive and resistive currents are merged into one making it difficult to identify the signal current attributable to the reactance of the capacitor from that due to the resistance of any leakage current created by water contamination etc.
- The rectifying circuit will not only rectify the excitation voltage but also any incident radio frequency waveforms
- The electrical network must be protected from lightning induced effects.

Given these considerations a number of networks and techniques have been developed in DC gauging.

The network of Figure 7.10 provides a current related to the overall capacitance of the wetted probe which includes the dry capacitance. This network may be augmented to provide a current related to the change in capacitance due to immersion in fuel only by generating a reverse current related to the dry capacitance and arranging for the currents to be additive as shown in Figure 7.11.

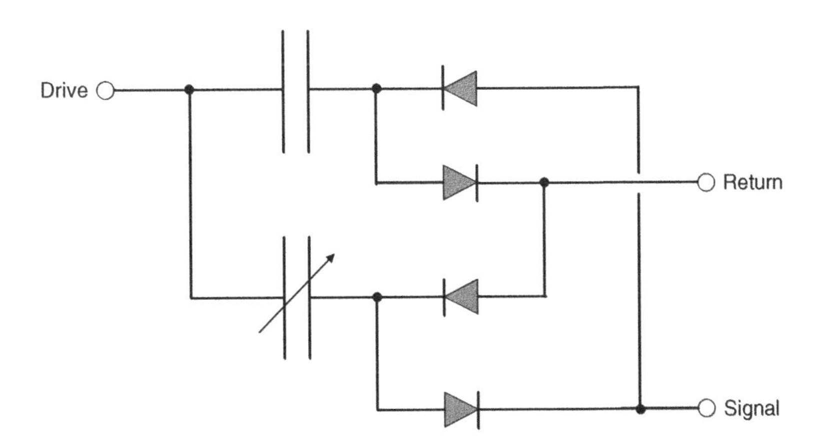

Figure 7.11 dc network providing immersed probe capacitance.

This network provides several benefits:

- It allows precision factory adjustment of the dry probe capacitance to allow probe replacement without recalibration
- By selecting a specific diode type, thermal effects on diode forward voltage drop are matched since they are all in the same environment when installed
- For military applications, it allows drop tanks to be released without any impact on total gauging contents, other than the disposed contents

As explained above, two key design dc gauging considerations are diode voltage drops and the merging of the rectified reactive and resistive current components. The effect of diode voltage drop can be significantly reduced by significantly increasing the amplitude of the excitation voltage. Typical excitation voltages for gauging systems vary depending on waveform from about 1 to 10 volts peak-to-peak; however the Airbus A320 gauging system, for example, operates with a 30 volts peak excitation to reduce the diode drop to an absolute minimum. The effect of reactive and resistive current merge can be addressed in several ways.

At the cost of increased circuit and software complexity, the industry has addressed the above issues. By periodically and sequentially doubling (or halving) the excitation voltage at constant frequency the diode voltage drop may be eliminated as follows:

From the previously established relationship:

$$I_P = V_P 2\pi f C_P \tag{7.16}$$

For voltage V_{P1} of frequency f, applied to a probe network with diode forward voltage drop V_D,

$$I_{P1average} = 2\pi f C_P \left(V_{P1} - V_D\right) \tag{7.17}$$

Therefore:

$$I_{P1average} = 2\pi f C_P V_{P1} - 2\pi f C_P V_D \tag{7.18}$$

Rearranging we get:

$$2\pi f C_P V_D = 2\pi f V_{P1} - I_{P1average}$$

Likewise for voltage V_{P2} of frequency f, applied to a probe network with diode voltage drop V_D,

$$2\pi f C_P V_D = 2\pi f V_{P2} - I_{P2averege} \tag{7.19}$$

Therefore, from (7.18) and (7.19),

$$2\pi f C_P V_{p1} - I_{P1average} = 2\pi f C_P V_{P2} - I_{P2average} \tag{7.20}$$

Rearranging we get:

$$2\pi f C_P V_{P1} - 2\pi f C_P V_{P2} = I_{P1average} - I_{P2average} \tag{7.21}$$

From this we can now say:

$$C_P = \left(I_{P1average} - I_{P2average}\right) / \{2\pi f\,(V_{P1} - V_{P2})\} \tag{7.22}$$

Hence C_P is independent of the diode voltage. Typically V_{P1} is arranged to be twice that of V_{P2}.

Referring back to Figure 7.5, the overall impedance of a capacitance probe in fuel comprises both reactance and resistance, the latter created by the leakage due to fuel conductivity. While the reactance is dependent on probe excitation frequency, the resistive component is independent. By changing the excitation frequency, the resistance will remain constant. Therefore resistance can be determined by using sufficiently different alternative excitation frequencies.

The key points of a dc capacitance probe gauging system compared to other capacitance systems are:

- lowest harness wiring weight
- high level of intrinsic safety
- no dependence on harness shield continuity

Before concluding this subsection on dc probes, the issue of thermal compensation should be addressed for completeness. In similar fashion to an ac gauging system, a dc capacitance gauging system is generally configured with a number of probes and a compensator. However the thermal properties of the diodes mounted on the probes provides an opportunity for the application of temperature compensation techniques and some systems use this approach to eliminate the need for a dielectric measuring compensator. This thermal compensation technique is based on the fundamental relationship between dielectric constant (k), density (D) and temperature (T) which can be expressed by the following equation:

$$\frac{(k-1)}{D} = G\,(1 - HT)$$

where G and H are constants.

Since this technique is both proprietary and mathematically complex it is not possible here to describe this concept further; however, it is important for the reader to understand the importance of thermal compensation in dc gauging systems and that subtle methods for its accommodation have been developed in the industry.

Active or Smart Probes

A modern trend in complex aircraft systems has been the move to remote data acquisition using a number of data concentrators to acquire sensor information locally and provide serial data bus communication with the central processing function. The primary advantages of this approach are that a great deal of sensor wiring may be eliminated to provide significant system weight savings and improved performance in the presence of HIRF. In line with this trend, some fuel gauging systems have been developed with remote data concentrators located outside the fuel tank on the tank wall. One difficulty with the capacitance probe is that its current output is uniquely different to that of the variety of other sensors employed on the aircraft. Also the intrinsic safety requirements of the probe interface have to be adopted within the data concentrator. In order to reduce the number of types of LRU utilized on the aircraft, efforts have been made to standardize the data concentrator to interface with a variety of sensors. These issues have made the direct and efficient interface of the probes to a generic remote data concentrator problematic. The smart capacitance probe has been developed as a solution to this dilemma.

A smart probe features electronics directly mounted on the probe in a hermetically sealed housing which are powered by an intrinsically safe, regulated dc power supply input from outside the tank. Oscillation is established with the capacitance of the probe and a digital waveform is output with a period directly proportional to capacitance. The output waveform is readily interfaced to a generic remote data concentrator or even a central processor. The active probe wiring interface generally comprises a shielded twisted three wire cable comprising power input and signal output.

Figure 7.12 is an example of an active capacitance probe.

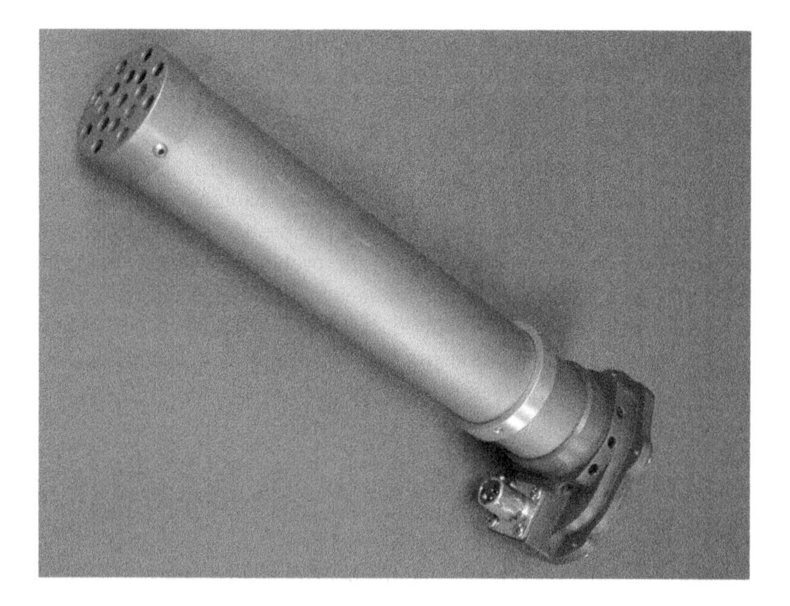

Figure 7.12 Active capacitance probe (photograph reproduced with permission from AMETEK, Inc.).

The advantages of the active or 'Smart' probe are its independence from wiring harness shield continuity and much easier signal conditioning interface. The added complexity, however, results in a slightly higher weight and cost and a lower reliability than the traditional capacitance probe. Electrical power must also be introduced into the tank.

7.1.1.5 Fuel Volume and Mass Measurement In Capacitance Systems

As discussed in the previous section, an individual probe is only able to measure fuel height or fuel level at a specific point in a fuel tank. In coping with operational conditions, tank geometry, and the accommodation of failures, a number of probes per tank are typically required to gauge the overall contents adequately. This section describes the different approaches as to how fuel height measurement may be converted into fuel volume and/or mass.

Aircraft main feed tanks, being located either within the wing or fuselage, are, by their very nature, irregularly shaped and require a number of techniques to accurately gauge the fuel contents. These include the ability to measure fuel height at several tank locations to ensure uninterrupted measurement, with no dead spots, for all levels. Additional tanks such as auxiliary tanks may be more of a regular shape and therefore easier to gauge. Military aircraft tanks, though generally smaller than their commercial counterparts, are often the most difficult to gauge as they typically have the most irregular shape (see Chapter 3).

At this point, only the measurement of fuel height has been discussed. Fuel height will only have a linear relationship with fuel volume for a theoretical tank geometrically linear about the probe. In the real world the fuel height measurement is nonlinear with the fuel volume. In addition to the shape of the tank, nonlinear allowances have to be made for the volume of the internal structure, such as ribs and stringers for an integral tank, since this reduces the internal volume and affects the relationship between fuel height and volume. Also allowance must be made for the volume and location of each fuel system component (pump, valve, etc.) within the tank.

Coupled with tank shape irregularity is the additional significant effect of the contained fuel surface changing at any one point with aircraft attitude as the fuel surface follows the imposed attitude and acceleration forces. The extent of this movement is subject to the type of maneuver and degree of coordination of any turn that may be involved. To allow for fuel surface attitude effects during ground maneuvers, take-off, cruise and approach, it is therefore necessary, as with tank geometry, to ensure uninterrupted measurement, with no dead spots, for all levels. Gauging accuracy requirements are typically stated in terms of maximum error for given ranges of aircraft attitude in terms of pitch and roll for each phase of flight. To account for gauging component failures, degraded accuracy requirements may also be stipulated for each flight phase.

An additional complexity is the need to accommodate effects of fuel slosh in the tank brought on by significant and sudden changes in aircraft attitude. This can be mitigated to some extent, by compartmentalizing the fuel by using a larger number of smaller tanks or by introducing baffles within the tank to limit the rate of fuel movement about the tank. This latter technique will introduce different levels between the baffles until the attitude stabilizes and the fuel level equalizes and may require the fuel surface to be measured separately within the different compartments.

In summary, fuel tank quantity gauging is dependent on a number of criteria to determine the relationship between fuel contents and fuel height measurement, which include:

- internal tank geometry
- internal tank structure
- internal tank components
- aircraft attitude
- fuel slosh and inter compartment leveling
- gauging accuracy requirements with or without failure.

The processes used to analyze the tank with respect to the above criteria are usually referred to as 'Tank Studies'. The process begins typically with the supply of tank drawings from the aircraft manufacturer. These are generally supplied as a set of computer files, rather than actual drawings, to provide the numerical master geometry (set of coordinates mapping the tank structure within the overall aircraft) of the tank to be gauged. In the case of a wing tank, data is required for both the in-flight and on-ground conditions to indicate the effect of wing deflection in flight and allow a gauging system to be designed that optimizes the characteristics of the wing tank in both the loaded and unloaded conditions. The data files enable the tank to be iteratively analyzed on the computer using tank studies programs for the method of gauging to be implemented. These programs are usually customized programs that have been exclusively developed by each particular gauging supplier. Tank studies are discussed in Chapter 11 under 'Modeling and design support tools'.

The objective of the Tank Studies program is to provide the optimized number of probes and their locations that are just sufficient to practically gauge the aircraft tank consistent with all the requirements and method of gauging. One of the primary tasks is to maximize the gaugable fuel, and by implication, minimize the ungaugable fuel. An important consideration in probe placement is in making the in-tank installation both electrically immune and mechanically compatible with the tank structure and fuel system components. It is necessary to minimize all fringing effects between each probe and structure by providing adequate clearance of typically 0.5″. Allowance should be made for panting (movement) of the tank surfaces with temperature, vibration and shock. The ideal probe location may be impractical in that it cannot be mounted in the required computed position because of the absence of adjacent structure such as ribs or stringers at that location. Furthermore the computed position may interfere with a proposed optimally located fuel system component such as a feed pump. Another important consideration is making the installation immune to water as much as practically possible. The fuel system designer takes every precaution to minimize the presence of water in the tanks but the fuel gauging designer should take care to place the probes away from areas of potential water collection to safeguard against the onset of major gauging problems caused by probes electrically short circuiting and thereby failing to operate. The lower ends of all probes should be kept away from area where wedges of water may become trapped between the lower tank surface and structure such as ribs and stringers. Particular care should be taken in the placement of a collector tank installation where the presence of water is most likely to be. Every effort should be made to make the tank installation as repeatable as possible with a minimum of differences from the calibration aircraft to each of the successive production aircraft so as to reduce the impact of any production tolerance on the overall gauging error analysis.

In summary key considerations in probe placement are:

- to define a minimum number of probes consistent with method of gauging and the requirements to minimize weight and cost, and maximize reliability;
- to maximize the gaugable fuel (and thus minimize ungaugable fuel);
- to maintain probe clearances to structure to minimize fringing and water bridging effects, and to safeguard against panting;
- to establish probe locations that are compatible with structure for mounting provisions;
- to ensure that probe locations do not foul fuel system components;
- to avoid potential water collection areas;
- to ensure repeatability in production.

Methods of Capacitance Gauging

There are a variety of ways of computing fuel volume and/or mass. The traditional method had been to use an array of probes in which each probe is characterized or profiled to provide an output directly proportional to immersed volume from its wetted length. By arranging for all the probes to be connected in parallel, the total resulting capacitance is the sum of the individual capacitances.

This is demonstrated mathematically in the following text.

The relationship between capacitance C charge (q) and voltage (V), $C = q/V$ can be rearranged as follows:

$$q = CV$$

Consider applying a constant voltage, V to a group of capacitors C_1, C_2 and C_n connected in parallel as shown diagrammatically in Figure 7.13. The individual charges are:

$$q_1 = C_1 V, q_2 = C_2 V \text{ and } q_n = C_n V \text{ respectively}$$

The total charge q_t is the sum of the individual charges, i.e.

$$q_t = q_1 + q_2 + q_n = C_1 V$$

Hence:

$$C_t V = C_1 V + C_2 V + C_n V \text{ or}$$
$$C_t = C_1 + C_2 + C_n$$

The total tank probe array capacitance is therefore directly proportional to fuel contents.

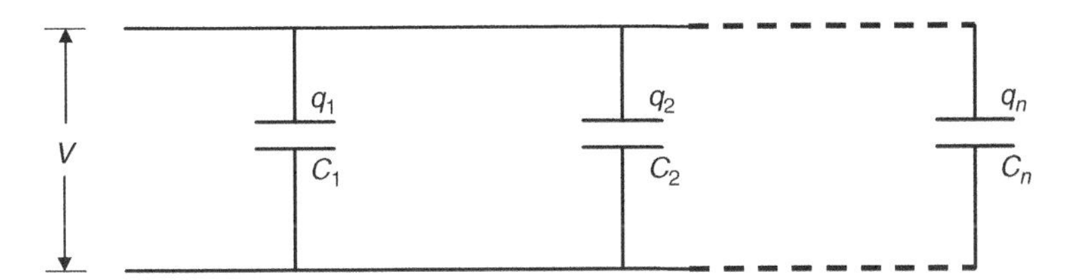

Figure 7.13 Capacitance probes connected in parallel.

Probe Characterization

In order to linearize the output of a probe in terms of immersed volume, it is necessary to characterize or profile the capacitance probe so that the rate of change of capacitance with wetted probe length immersion directly relates to the rate of change of volume. This is most commonly accomplished by one of three techniques:

- mechanically, by varying the diameter of the inner tube of the probe;
- electrically by varying the relationship between the inner and outer tubes of the probe;
- electronically by correcting the volumetrically non-linear probe output within the signal processor software.

Mechanical probe characterization features an all metal probe comprising an inner tube of varying diameter concentrically located within an outer tube of constant diameter. There are a number of methods of producing the inner tube which are in increasing order of performance:

- Mechanical assembly of differing diameters of tube, a method in which the sections of tube are riveted together. These designs have proven rugged but principally suffer from the fact that they can only approximate the necessary rate of change of capacitance particularly at the transition between sections of probe with differing tube diameters as the characterization can only provide a stepped approach. Also the capacitance change at these transitions is often impacted by fringing effects created by the rivets.
- Swaging, a method by which the inner tube is progressively reduced in diameter through the process of plastic deformation by simultaneous circumferential hammering with linear drawing. It is possible to manufacture a single piece inner tube providing the required rate of change of capacitance is maintained or changes in the same direction. Otherwise different swaged sections will need to be riveted together.
- Electro-deposition, an investment method by which the inner tube is made by metal plating a mandrel of the necessary varying diameter and then subsequently dissolving the mandrel to leave a thin walled tube.

The comparative ratings of these methods are summarized in Table 7.1.

Table 7.1 Comparison of probe fabrication methods.

Method Advantage	Mechanical Assembly	Swaging	Electro-deposition
Volumetric linearity	Steps at diameter change	Better	Best
Fringing impact	Worst	Better	Best
Mechanical strength	Lowest	Best	Better
Single piece inner tube	No	Possible if rate of change of volume does not reverse	Yes
Reliability	Lowest	Better	Best
Weight	Highest	Better	Lowest
Cost	Medium	Lowest	Highest

Electrical probe characterization features an all composite (usually fiber-glass) tube probe comprising an inner tube of constant diameter concentrically located within an outer tube of constant diameter. The outer surface and inner surface of the inner and outer tubes respectively are plated with varying amounts of copper to provide a varying capacitance in accordance with the volume requirements. The treated probe surfaces are akin to a circular printed circuit card and consequently these probes are often referred to as printed circuit probes. The electrically characterized or printed circuit probe has been used for many years in a number of applications. These installations provide for the lowest weight solution while providing very good electromagnetic compatibility. Also this type of design can allow for low manufacturing tolerance by having the ability to adjust the dry capacitance by rotating the inner tube with respect to the outer tube. One significant drawback is the reliability of the installation. Whereas all metal probe systems often can last 30 years without replacement, these probes have a tendency for the printed circuit to separate from the composite probe material.

Electronic probe characterization features an all metal linear probe (inner and outer tubes of constant diameter), the characterization being performed outside of the tank in the interfacing signal conditioner or processor. This can be performed by a series of electrical circuits but the current and most common way is to make the correction by software.

A comparison of these probe types is provided in Table 7.2.

Table 7.2 Comparison of probe types.

Method Advantage	Mechanical (Electro-deposition)	Electrical	Electronic (Software)
Accuracy	Steps at diameter change	Better	Best
Probe array return wiring requirements	Summed or Individual	Summed or Individual	Individual
Compatibility with tank installation changes	Re-design required	Re-design required	Best. Changes may be made in software
Reliability	Medium	Lowest	Best
Weight	Highest	Better	Lowest
Cost	Highest	Medium	Lowest

7.1.2 Ultrasonic Gauging

The fundamental distinction between ultrasonic and capacitance gauging is that ultrasonic gauging uses a technologically different suite of in-tank sensors that is accompanied by changes in both the signal conditioning interface and software within the processor. Once fuel height and the associated fuel parameters have been accurately determined, the calculation of fuel quantity is very similar to that of a capacitance system. Ultrasonic fuel height measurement relies on the phenomenon that sound energy can be transmitted through liquid and be reflected at an interface with that liquid. A key consideration in the measurement is that the velocity of sound in fuel is inversely proportional to temperature, with some further variation due to fuel type as illustrated by the data spread in Figure 7.14.

Many of the examples presented in the remainder of this section are courtesy of Parker Aerospace.

The basic principle of ultrasonic fuel gauging is its dependence on two measurements:

- the speed with which the ultrasound travels through fuel, as measured by a velocimeter;
- the round trip time for sound to travel upwards through fuel from the transmitting transducer to the fuel surface and downwards back to the receiving transducer, as measured by a probe.

Unfortunately it is not practical to use ultrasonic techniques to locate the fuel surface from the top of the tank by measuring the round trip time downwards from the top of the tank through the ullage. This is because very high levels of energy are required to reliably transmit through the air/vapor mixture, particularly at low fuel levels and this would be incompatible with system intrinsic safety requirements.

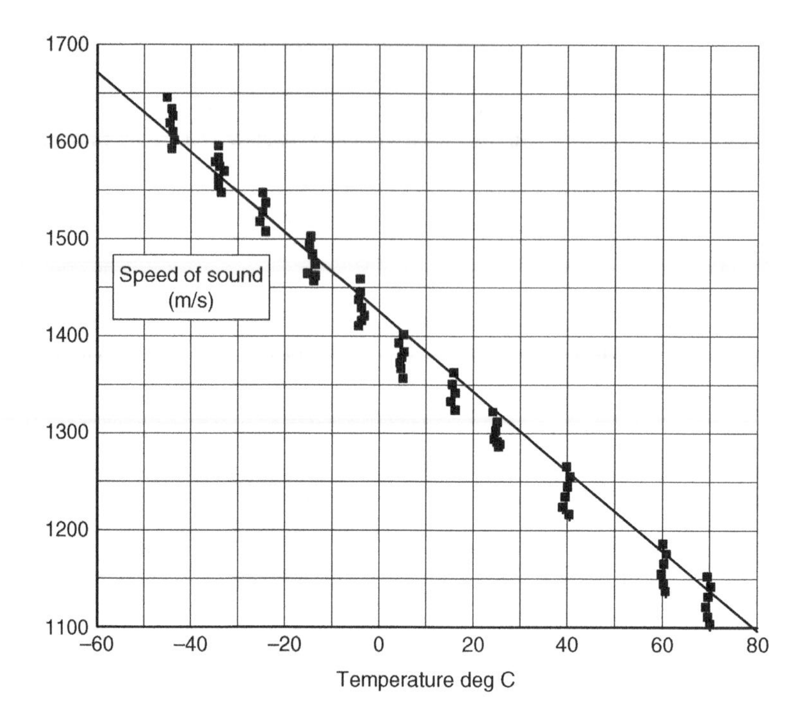

Figure 7.14 Speed of sound versus temperature (Ref ARINC 611).

Figure 7.15 illustrates the principles of the ultrasonic velocimeter and ultrasonic probe where the velocimeter acts as a speed of sound calibration using a fixed target and the probe is used to measure the fuel height in the tank.

The time chart shown in Figure 7.16 shows how fuel height can be obtained with this arrangement with the parameter definitions as follows:

T_T = Round trip time to target, T_S = Round trip time to surface
D = Known distance to target and L = Unknown distance to surface

The velocity of sound in fuel (*VOS*) can be derived from the velocimeter via the equation:

$$VOS = 2D/T_T \tag{7.23}$$

Similarly, the unknown distance to the surface (L) is defined as follows:

$$L = (VOS)(T_S)/2 \tag{7.24}$$

Substituting for *VOS* from (1) in (2) yields:

$$L = D(T_S/T_T) \tag{7.25}$$

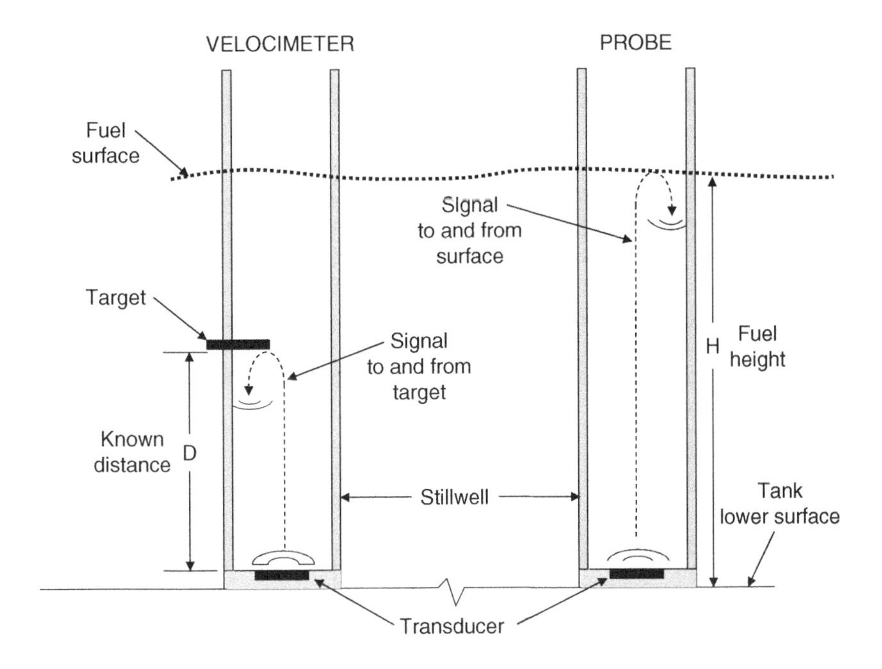

Figure 7.15 Operating principles of velocimeter and probe.

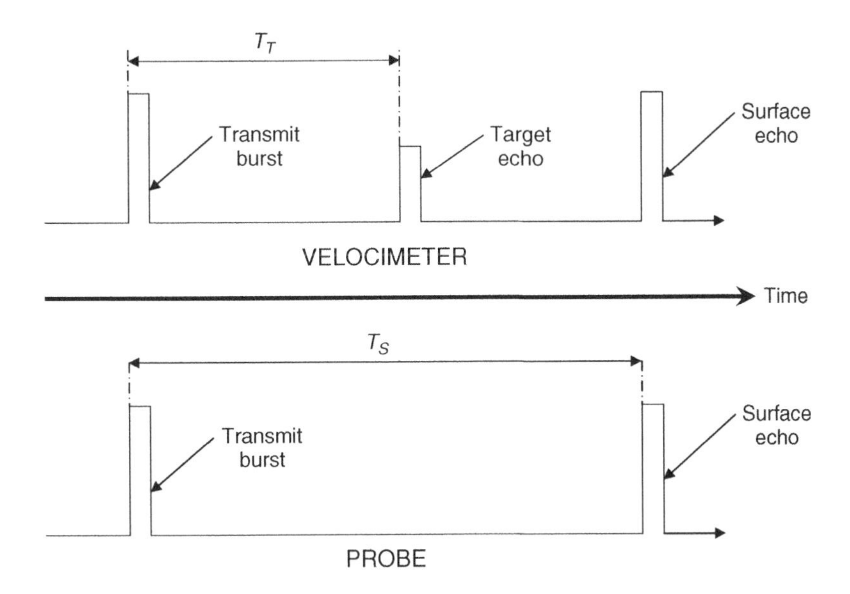

Figure 7.16 Ultrasonic fuel height measurement.

7.1.2.1 Signal Conditioning

The basic method of interfacing with an ultrasonic velocimeter or probe is illustrated in Figure 7.17. Under processor control, the piezoelectric crystal of the probe is excited at a particular amplitude and duration. The excitation comprises a pulse or series of pulses to cause the crystal to resonate and transmit ultrasound.

The excitation frequency is determined by the resonant frequency of the crystal selected to meet the accuracy requirements. The ultrasound signal or echo reflected by a target or the fuel surface is then received back at the same crystal, now acting as a receiver. The reflected signal is then amplified, as necessary, prior to passing through a threshold detector to remove unwanted signal.

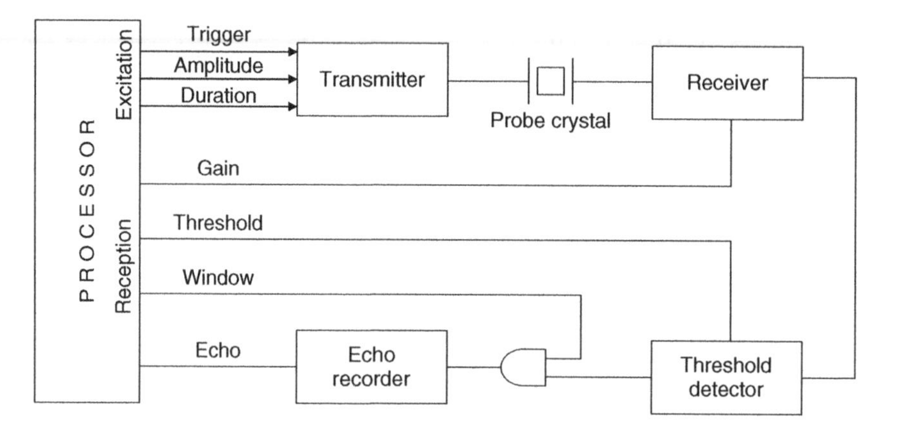

Figure 7.17 Method of ultrasonic signal conditioning.

The resulting signal is then gated with a timing window and recorded in memory as an echo for subsequent processing. Typical waveforms for a transmit burst of pulses at 1 MHz and first received echo are shown in Figure 7.18.

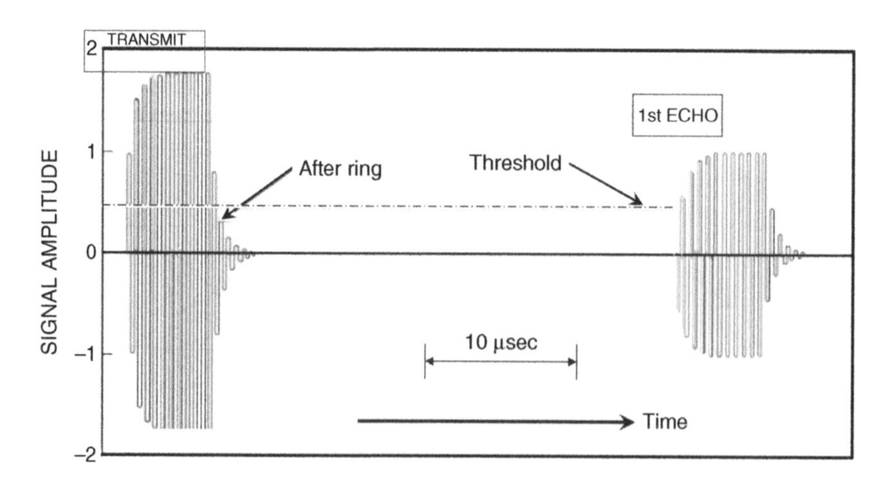

Figure 7.18 Typical burst and echo waveforms.

7.1.2.2 Physical Considerations

The fuel tank environment presents additional challenges in the application of ultrasonics to liquid measurement. Sound is attenuated when propagated through a medium. Attenuation is caused by the combined effects of sound scatter, where sound is reflected away from its original path, and adsorption, where sound is converted into other types of energy.

For a longitudinal ultrasound wave generated by a piezoceramic crystal, the wave amplitude changes according to the following relationship:

$$A = A_0 e^{-\alpha x}$$

where: A is the amplitude at distance x from the source of A_0

A_0 is the initial amplitude at $x = 0$

α is the attenuation constant for a wave traveling in the $\boldsymbol{x}$ direction

It can be seen that with increasing distance, sound amplitude decays exponentially and therefore when sound is directed upwards to the fuel surface, only a portion of the energy is reflected back downward by the fuel/air interface. The amount of sound energy reflected back to the crystal can be determined in this way.

This phenomenon does not take into account the following issues that are specific to aircraft fuel measurement.

- The fuel surface ripples and sloshes with motion to cause varying degrees of reflection.
- The fuel surface attitude changes with respect to aircraft attitude when uncoordinated.
- The fuel reflection coefficient changes that occur with altitude as the density of ullage air and fuel temperature changes.
- The fuel outgases with increase in altitude until trapped air is boiled off.
- The fuel turbulence such as that encountered in operations such as refuel may generate large bubbles.
- The presence of undissolved water in the tank.

The purpose of the stillwell is to reduce the effects of ripples and slosh, but the surface is still dynamic and certainly cannot be guaranteed to reflect all of the available incident sound energy back in the direction it came. Furthermore, aircraft attitude itself causes sound to be reflected in a zigzag way as shown in Figure 7.19.

At attitude, the reflected sound comprises two components; the direct path (primary signal) and zigzag (secondary signal) sound. As can be seen from Figure 7.19, the reflected path, caused by the zigzag effect, is longer than that of the direct path. This causes two returns to be generated sequentially; the primary signal followed by the zigzag signal. Another effect is that the primary signal progressively reduces until the incident sound path is at 22.5° from perpendicular to the surface. With increasing angles above 22.5°, the signal progressively increases again.

As the aircraft climbs in altitude, the fuel will outgas to release any air in solution and generate many fine bubbles throughout the fuel, including that in the stillwell. This process ceases when all the air has been released. In the intervening period, this process causes a degree of both sound adsorption and reflection with the net effect of reducing the signal return. Once again, suitable signal conditioning gain changes can accommodate this effect.

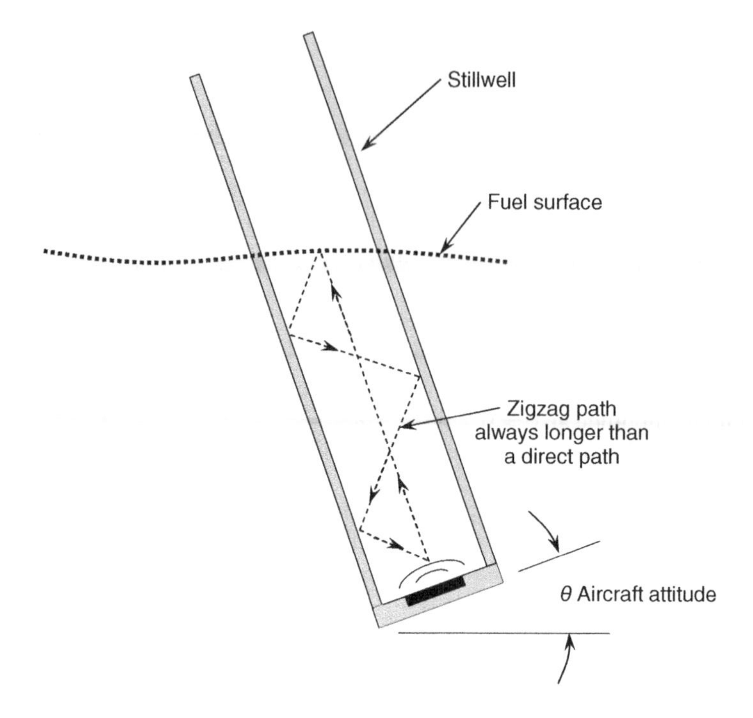

Figure 7.19 Attitude-induced zigzag reflected waveform.

The presence of large bubbles in the stillwell must be avoided at all times as, unlike outgassing, they will cause premature signal returns as they are sufficiently large to reflect significant ultrasound. These bubbles are likely to be generated by turbulent fuel such as that encountered during operations such as refuel. Software can be designed to identify erroneous readings caused by this effect but avoidance is the best policy. The first level of protection is the incorporation of a bubble shroud in the transducer assembly that allows free fuel flow but eliminates these bubbles. The second level of protection is to not locate the probes in turbulent areas of the tank wherever possible.

As with capacitance systems, it is necessary to locate the ultrasonic probes away from areas likely to encounter the presence of undissolved water in the fuel. This water will pool, as it is denser than fuel, at the lowest part of the tank, particularly in collector tank areas. Free water may also form a wedge against a rib. Whereas a capacitance probe would short if subjected to sufficient water to cover the bottom of both electrodes, an ultrasonic probe would generate a premature return signal of the water/fuel interface, assuming the probe connections have not been shorted.

7.1.2.3 Ultrasonic Probe Design

The ultrasonic probe is configured as a transducer assembly at the lower end with a stillwell attached and mounted vertically above it. The probe may be constructed from metal and/or composite materials. The overall length of the probe, for a given location, is the same as an equivalent capacitance probe, barring any necessary mounting clearances.

The transducer assembly features a piezoelectric ceramic disk that acts as a transceiver to both generate and receive ultrasound. The thickness and diameter of the crystal determine the resonant frequency of the crystal. Typically a crystal with a resonant frequency between 1 and 10 MHz is selected. The transducer assembly comprises the disk and a resistive discharge network, mounted directly on to the disk, to safely dissipate any abnormal energy created by temperature or mechanical shock, a mechanical labyrinth or bubble shroud, and provision for the electrical connections to the in-tank harness. Care must be taken in the mounting of the disk within the transducer to ensure that resonance is not impeded. Also as the resonating disk will emit ultrasound not only up the stillwell, but downwards into the assembly to cause unwanted reflections, sound absorbent material is required to be located under the disk.

The purpose of the stillwell is to both collimate the sound generated and received by the transducer, and provide a 'sheltered' area to make measurements. The stillwell protects measurements from major phenomenon such as fuel slosh or large bubbles. The design of the stillwell and transducer assembly has to be such that fuel can readily enter the stillwell so that the level follows that outside the stillwell but prevents the ingress of large bubbles caused by turbulence that may be created by operations such as refueling. This is achieved by incorporating a labyrinth-type baffle in the transducer assembly.

To help eliminate false measurement, it is important that any spurious ultrasound reflections created within the stillwell are kept to a minimum at all times. This is achieved by ensuring the inside surface of the stillwell is smooth by uniformly coating or lining the surface with acoustically suitable material. Also careful attention to the probe mountings should be made as the mechanical interface with the outside of the stillwell can lead to internal reflections. To that end, the lower mounting bracket should be fixed to the bubble shroud and the upper moveable mounting bracket(s), with damper(s) located on the stillwell.

A typical probe and an assembly view are shown in Figures 7.20 and 7.21 respectively.

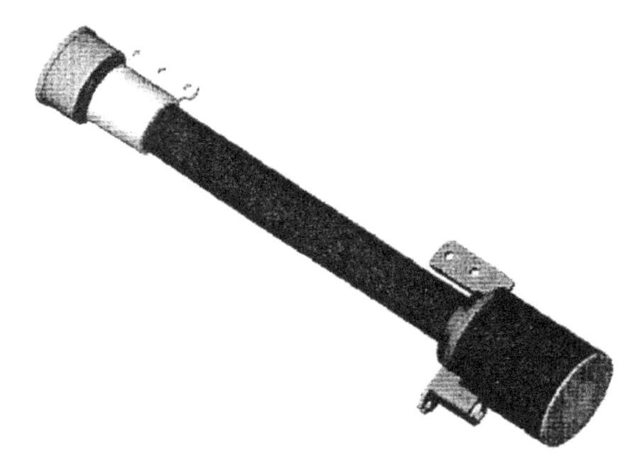

Figure 7.20 Typical ultrasonic probe (courtesy of Parker Aerospace).

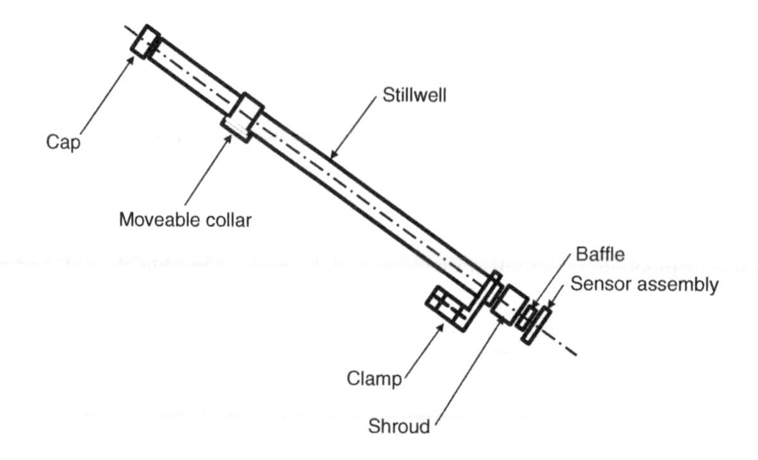

Figure 7.21 Ultrasonic probe assembly.

7.1.2.4 Ultrasonic Velocimeter

This is very similar to a probe except that a number of targets are fitted at predetermined intervals up the stillwell. These targets penetrate from the outside and through the stillwell wall into the ultrasound to provide an obstacle and so generate a reflection or echo back to the crystal. A number of targets are necessary to ensure that the speed of sound may be accurately determined by the processing for varying fuel levels and degrees of stratification. Sizing of the targets has to be such as to provide a recognizable echo by the signal conditioning process yet not prevent echoes being generated by subsequent targets through significantly blocking their exposure to the stillwell ultrasound. By using a separate velocimeter, rather than a probe fitted with targets, it avoids the issue of further optimizing target size and signal conditioning

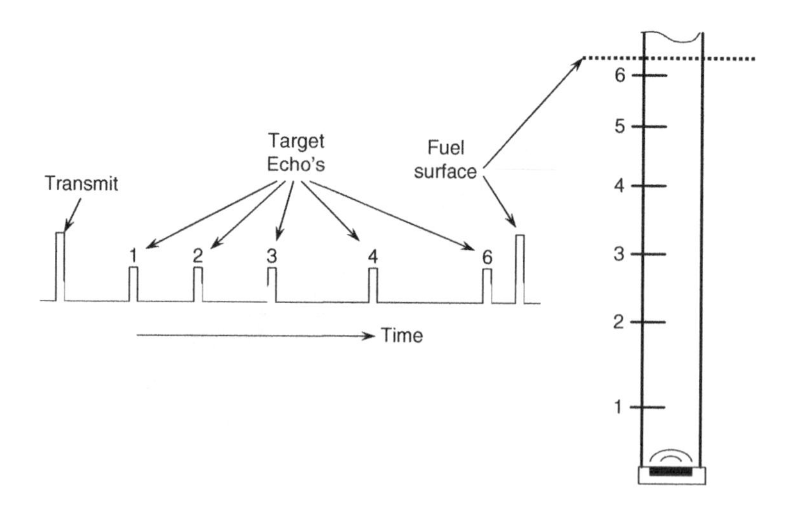

Figure 7.22 Velocimeter with multiple targets.

necessary to distinguish fuel surface echoes from multiple target echoes. Location/spacing of the targets is arranged to be nonlinear so that secondary reflections of a lower target are not confused with the primary reflection of an upper target as illustrated in Figure 7.22.

7.1.2.5 Installation Considerations

Probe Installation

The number and location of the probes is determined, as for a capacitance system, by conventional tank studies to provide optimized wetted probe length data. This process generally results in the same number of probes that a capacitance probe system would utilize. There may be some differences due to the need for greater clearance of structure caused by the larger diameter of the bubble shroud and the necessity to avoid direct sources of major fuel turbulence.

Velocimeter Location

In concert with tank studies, velocimeter location is determined by the following three principle requirements:

1. measurement of the velocity of sound for almost all fuel levels and attitudes to compensate for temperature stratification and fuel type variation;
2. mapping of fuel density measurement against speed of sound to provide accurate fuel mass measurement;
3. avoidance of areas of high fuel turbulence and/or water collection.

Depending on the complexity of the tank, 1 or 2 velocimeters are normally sufficient to meet requirement (1). Requirement (2) determines that a velocimeter be located adjacent to a densitometer. Requirement (3) determines that a velocimeter should not be located near refuel or pump outlets. Taking all this into consideration for a dihedral wing tank, it is recommended that a velocimeter be located in the inboard collector area next to the densitometer, if available, but away from sources of turbulence and areas of water collection. If necessary, a second velocimeter should be located in the outboard area of the tank.

7.1.2.6 Signal Processing

As explained in Section 7.1.2.2, there are a number of physical phenomena that the signal conditioning must adapt to. This is accomplished by processor software control of the transmitting and receiving circuitry using a set of algorithms unique to ultrasonics.

Referring back to Figure 7.17 above shows the signal conditioning interface for a single probe. The drive and return interfaces may be both readily multiplexed to facilitate a multiple probe and velocimeter system. Each probe is addressed sequentially to excite and receive the resulting signals. An advantage of ultrasonic over capacitance probe systems is that a probe array may be excited by one of two signal conditioning circuits and the return signals interfaced to both to provide a fault tolerant, dual redundant tank gauging system.

The software performs an essential role in optimizing an ultrasonic system to measure accurately under all conditions. For example, in Figure 7.17, the processor has the ability to control the excitation of the probe(s) in terms of timing, amplitude and duration. Amplitude control is particularly important to increase the excitation that is necessary to compensate for the attenuation effects of the longer ultrasound round trip times associated with increasing depth of fuel and/or longer probes.

Gauging of low levels presents a challenge for ultrasonics in that the round trip time reduces with reducing level to the point where the drive and return signal become intermingled and therefore indistinguishable. To overcome this, the excitation amplitude and duration are both reduced. This has the effect of reducing the ringing effect of the transducer to allow it to be switched more quickly to the receive mode. Also software techniques may be employed on the receiving side to identify the second or third instead of the first reflection of the primary signal/echo.

On the receiving side the software has the ability to control the time during which the return signal or echo is sought (window), adjust the gain and set a threshold only above which a signal may be accepted. These controls, along with those on the excitation side, allow adjustments to be continually made for the dynamically changing aspects of the ultrasound signals with fuel surface and level changes within the tank.

An advantage of ultrasonics is its ability to gauge accurately in the presence of fuel stratification. As mentioned elsewhere in this book, fuel in a tank is subjected to changes of temperature with the ascent and descent of the aircraft. In particular during ascent, the fuel at the bottom of an integral wing tank, will become colder than that in the body of the fuel in that tank due to the cooling effect on the tank outside surface as the wing travels through the air of steadily reducing temperature. Another example is provided by an aircraft that has recently landed in a hot climate, after a long flight at high altitude, only to be refueled with warm fuel (most likely of different characteristics to that already onboard) from a hydrant or fuel truck. In both cases the fuel stratifies into layers of fuel of different temperature. A multiple target velocimeter provides a method to measure the speed of sound over a number of submerged targets and develop a profile of the fuel from the bottom of the tank to the fuel surface at a series of defined intervals. As the velocity of sound (VOS) changes approximately 0.3 % per degree Celsius, it is essential for high accuracy measurement, that stratification effects are taken into consideration in determining fuel height. The application of the appropriate VOS to each probe measurement can be applied by the software. Furthermore given the fuel density to VOS relationship, density as measured by a densitometer located at the bottom of the tank (for maximum coverage) can be mapped to each stratum of fuel as monitored by the velocimeter.

7.1.3 Density Sensor Technology

Direct density measurement is an essential factor in the accomplishment of high accuracy fuel quantity (mass) gauging. Traditional tank units measure wetted length in order to establish a surface plane within the tank thus determining the fuel volume. To arrive at mass the following basic equation must be applied:

$$Mass = Volume \times Density$$

Direct density measurement by densitometer has evolved from the use of sensors based on the principle of buoyancy to those based on vibration to achieve the highest accuracy attainable.

Another method of note is based on the adsorption of gamma radiation but has been the subject of environmental issues.

The two principal types of vibration sensor are the vibrating cylinder and vibrating disk versions. The vibrating cylinder densitometer features a thin walled metal cylinder that is held under tension within a rigid housing. This arrangement acts as a toroid with the fuel being sensed located in the central hole. The spool is electromagnetically deformed by one of three sets of coils arranged internally around the spool. Cylinder resonance is sustained by drive circuitry within a fuel signal conditioner or processor. The spool's resonant frequency is determined by the elasticity of the spool wall and the mass of the volume of fuel located within the spool, thus as the density of the fuel in contact with the spool varies, so does the resonant frequency.

With the spool resonant frequency measured and the mass and elasticity of the spool known, the mass of the volume of fuel within the spool may be determined. Since the spool volume is known, the density of the fuel is computed as follows:

$$D = K_1 + K_2T^2$$

Where: D is the fuel density
K_1 and K_2 are densitometer constants
$T$is the time period of the output waveform

An alternative densitometer design uses a vibrating disk instead of a cylinder. In this concept the one half of the disk is mounted within a metal block while the other half is exposed to fuel as indicated in Figure 7.23.

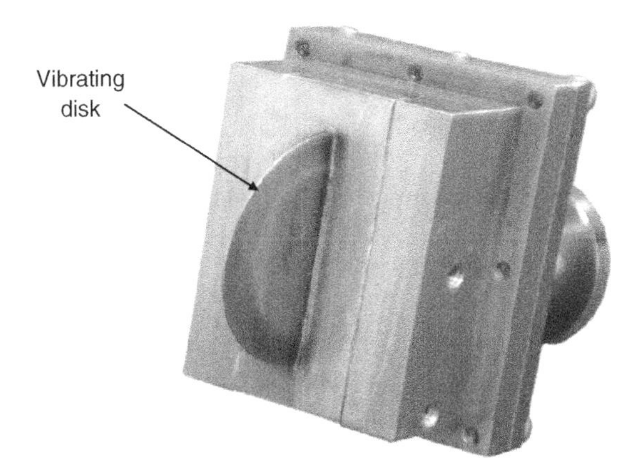

Figure 7.23 Vibrating disk densitometer (courtesy of Parker Aerospace).

The operating concept of the vibrating disk is the same as for the vibrating cylinder. The vibrating disk uses a vibration mode which is sufficiently high to make the device relatively insensitive to environmental vibration. A high level schematic of the vibrating disk densitometer is shown in Figure 7.24.

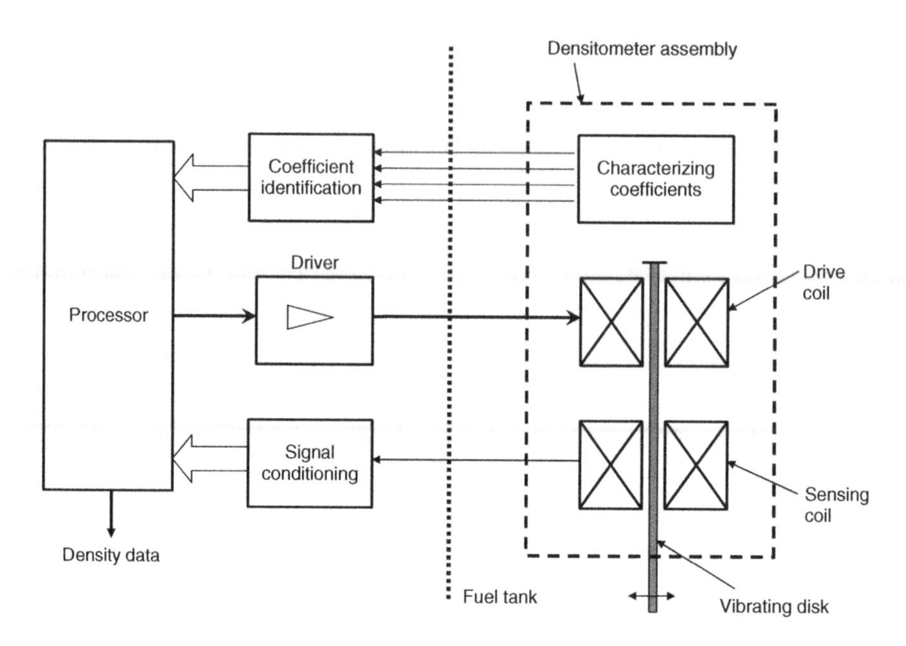

Figure 7.24 Vibrating disk densitometer schematic (courtesy of Parker Aerospace).

In similar fashion to the vibrating cylinder, the disk is electromagnetically maintained in resonance by integral drive and sensing coils operating in concert with the signal conditioning electronics.

The vibrating disk densitometer has some distinct advantages over the vibrating cylinder type device including:

- an 'air-point' (an operating frequency in air); this characteristic is guaranteed by design, not by selection;
- the design which does not require an internal vacuum, thus minimizing the possibility of leaks and associated failure mode;
- the design has a higher tolerance to external vibration.

Every manufactured vibrating cylinder or disk has unique vibrating characteristics and requires calibration to establish a pair of calibration coefficients. These coefficients are stored within the densitometer unit. This is mandatory in order to achieve accuracy and to allow sensor replacement without the need to replace or recalibrate the associated signal conditioning LRU. These coefficients are represented by resistor networks for highest reliability.

Both the vibrating cylinder and vibrating sensor densitometers require electronic signal conditioning to read and maintain resonance of the sensor, as well as read the calibration coefficients. In some cases the associated electronics are contained within the densitometer itself, and are thus referred to as 'Active' densitometers. Here the electronics are housed in a sealed enclosure fitted to the sensor which allows a processed data interface with the fuel quantity processor. While the active densitometer approach is designed with an intrinsically safe interface within the fuel tank, it is no longer preferred since the introduction of SFAR 88 for the primary reason that power is required to be routed into the fuel tank.

The currently preferred approach is having a 'Passive' densitometer with the signal conditioning electronics located within a system data concentrator or fuel quantity processor.

An important operating feature of both of the above vibration type sensors is that small air bubbles, created within the fuel from phenomenon such as turbulence created during refuel or fuel outgassing as the aircraft climbs, become attached to the sensor element during vibration. This has been shown to be less of a problem for the vibrating disk sensor than the vibrating cylinder. In this latter type, the bubbles become trapped in the cylinder causing inaccurate measurement.

A solution to this problem has been to periodically but briefly switch off the signal conditioning so as to release any bubble accumulation within the sensor (or more radically, to acquire fuel density information only while on the ground during the refuel process (see below).

In the case of both the vibrating cylinder and vibrating disk densitometers, it is necessary for the unit to be mounted in an enclosure to shelter the sensor from the undesirable effects of bubble formation. Figure 7.25 shows the enclosure for the vibrating disk version.

Figure 7.25 Vibrating disk densitometer enclosure (courtesy of Parker Aerospace).

Since the densitometer plays a key role in high accuracy gauging systems, the use of the densitometer requires very careful implementation. Given the bubble formation issues particularly associated with outgassing mentioned above and that a densitometer mounted low down in the fuel tank only measures the fuel at the bottom of the tank, the industry is now increasingly using the densitometer only on the ground as part of a Fuel Properties Measurement Unit (FPMU) where the dielectric constant and temperature as well as the density of the uplifted fuel during refuel are measured in close proximity to one another by a group of sensors housed in a sheltered environment. The parameters of the fuel are then mapped to the dielectric (compensators) and temperature sensors distributed around all the tanks to provide a more complete and accurate picture of fuel density across the aircraft.

7.1.3.1 Fuel Density Acquisition Alternatives

In lower accuracy systems, 'inferred density' measurement methods may be adequate. There are two principle methods of inferring density from other parameters:

1. from the fuel dielectric constant in capacitance gauging systems;
2. from the velocity of sound in ultrasonic gauging systems.

In the case of capacitance gauging systems a dedicated sensor referred to as a compensator or reference unit is used. The principal behind the measurement is based on the Clausius Mossati relationship:

$$D = \frac{(k - 1)}{\{A + B\,(k - 1)\}}$$

Where: D is the fuel density
k is the fuel dielectric constant
A is a constant related to a type and batch of fuel
(for Jet A this is $\approx 1.0 \pm 10\%$)
B is also a constant based on the type of fuel (for Jet A this is 0.3658).

Thus with knowledge of fuel type and dielectric constant k density can be obtained albeit with a level of uncertainty that contributes about $\pm 1.2\%$ to the overall accuracy error.

Measurement of dielectric constant is typically via a compensator probe which is designed for installation near the bottom of the tank to ensure that it remains fully immersed. Great care must be taken to avoid water contamination since this will render the compensator unusable. A more effective method that is sometimes used is to use gauging probe that is fully immersed as an alternative to the compensator (e.g. on the ground or early in the flight when tanks are close to full). This approach has the advantage of providing additional accuracy by measuring dielectric constant over the full depth of fuel in the tank thus taking into account any stratification effects that a conventional compensator has no knowledge of.

The second alternative method of inferring density is by ultrasonically measuring the speed of sound within the fuel. This is achieved by measuring the speed of sound over a defined distance using a velocimeter. A velocimeter utilizes a piezoelectric crystal to both transmit and receive sound. The crystal is electrically energized or 'pinged' by signal conditioning electronics to send a sound pulse through the fuel over a known short distance of several inches to a target that reflects the sound back to the crystal now acting as a receiver; the time taken for the sound pulse to return is used to determine the velocity of sound. The velocity of sound measurement requires correction for temperature which is usually accomplished via a co-located temperature sensor.

The primary advantage of this ultrasonic approach is that it provides greater accuracy of inferring density over that by dielectric measurement with a total error contribution of error $\pm 0.5\ \%$ compared with $\pm 1.2\%$ for the dielectric constant approach.

Disadvantages of the ultrasonic method are its vulnerability to bubble effects and, perhaps more importantly, the intrinsic safety of the piezoelectric crystal which requires a resister network to discharge any potential static charge build-up arising from environmental shock or temperature effects.

7.1.4 Level Sensing

Most fuel quantity gauging systems employ both gauging probes and level sensors. While fuel quantity is measured in mass terms, being the measure of stored energy and hence range of the aircraft, knowledge of fuel volume is also required to prevent over filling of the fuel tanks on hot days when the tank volume limit can be exceeded before the maximum specified fuel mass is reached. FAR Part 25 regulation reference [15], (specifically 25.969) specifies a minimum expansion space of 2 % of the tank volume be provided to prevent over filling and to accommodate fuel thermal expansion.

There is also a requirement mandating high level protection against over-pressurization during pressure refuel due to a refuel valve failure. Provision of a tank high level sensor and associated valve control independent of the gauging system satisfies this requirement. This arrangement also provides the refuel 'Pre-check function' where the refuel shut-off function is verified.

Level sensing is also used to warn the crew of a low fuel state in a feed tank or collector cell. For integrity reasons this function must be independent of the gauging system for that tank/cell.

Data fusion advocates have promoted the use of common sensors for both management and gauging particularly on larger systems; however recently there has been a move to eliminate all potential sources of common mode failures (e.g. water contamination) and to require the use of different technologies for gauging and low-level sensing. For convenience of installation the level sensors are normally mounted on gauging probes but this may not always be possible due to sensor size, location, tank shape, fuel system component interference etc, in which case a standalone configuration is utilized.

There are many ways of detecting fuel level and various types of level sensors are described in the following paragraphs.

7.1.4.1 Magnetic Float Activated Reed Switch

This sensor comprises a vertically mounted, thin circular stem made from a non-magnetic material or plastic that houses one or more encapsulated electrical reed switches located at the one or more levels to be detected. A circular float containing a permanent magnet is arranged to slide up and down the stem with increasing and decreasing fuel level, the magnet closing the contacts of a reed switch when adjacent to it. The sensor requires one wire per switch plus a common wire. The opening and closing contacts may be used to directly activate or deactivate a relay or a signal conditioner to operate a fuel system component. It is possible to measure to about $\pm 1/8$ in. The primary disadvantages of this type of level sensor are:

- relatively large size due to float and difficulty of location
- performance in turbulent fuel
- float stiction at attitude
- float durability
- hysteresis.

7.1.4.2 Thermistor

The thermistor sensor is based on the use of a material exhibiting a resistance that changes significantly and predictably with temperature. In aircraft level sensing applications, a semiconductor material whose resistance decreases with temperature (negative temperature coefficient) is normally utilized. A typical sensor comprises two thermistors. One thermistor is exposed to the air or fuel, in accordance with the fuel level, and is operated at relatively high current in order to provide a self-heating effect. The other thermistor provides a reference at lower current within an evacuated capsule. By immersing the sensor in fuel, the heating effect is conducted away and the increased resistance relative to that of the reference is detected by remote electronics. The sensor is sensitive, has a fast response time, high accuracy and small in size. However the intrinsic safety issues associated with the necessary heating current of between 20 and 30 milliamps are the main disadvantage of this sensor.

7.1.4.3 Zener Diode

The zener diode level sensor works on the principle of the zener diode reference voltage sensitivity to temperature. The sensor comprises two zener diodes assembled in a small cylindrical housing, one operating at a relatively high current to produce a self-heating effect, and the other at a lower reference current producing negligible heating effect. In a manner similar to the thermistor, upon immersion the fuel cools the heating effect created within the diode operating at the higher current. Remote signal conditioning electronics monitor the two-diode assembly as it is immersed or uncovered from fuel, to derive a switching signal based on the current change in the heated diode with respect to the reference diode. As with the thermistor, this sensor is fast, accurate and small. It also has the same disadvantage as the thermistor relative to intrinsic safety requiring a heating current of between 20 and 30 milliamps. A fuel quantity probe with multiple zener diode level sensors is shown in Figure 7.26.

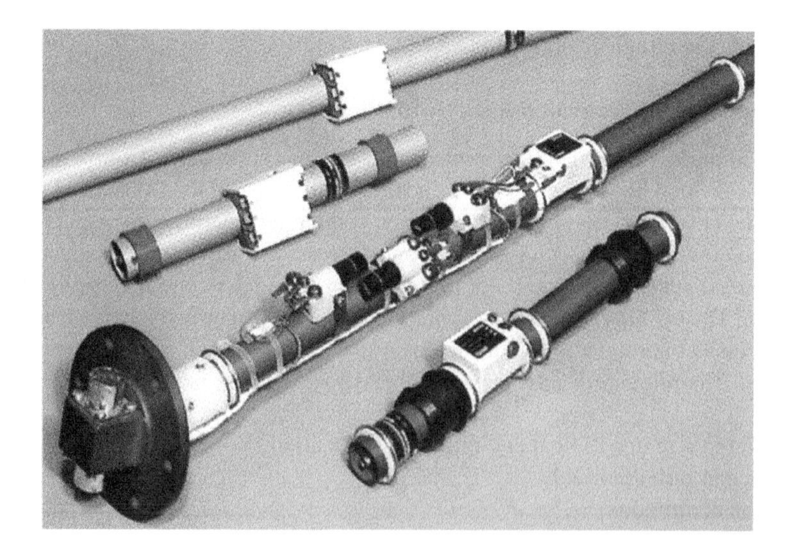

Figure 7.26 dc probe with multiple zener diode level sensors (courtesy of GE Aviation formally Smiths Aerospace).

7.1.4.4 Capacitance Level Sensor

This sensor is virtually identical in shape and size to the compensator used in a capacitance based fuel quantity system, as described previously. The sensor is designed to detect level at a point prior to total immersion. Consequently the interface with the signal conditioning within a capacitance based gauging system is simplified as long as independence of measurement is maintained where mandated. The primary disadvantages of this type of level sensor are its relatively large size and therefore difficulty of location. Inaccuracy can occur due to variations in fuel dielectric.

7.1.4.5 Optical Level Sensor

This category of sensor utilizes the principle of light refraction. A typical sensor utilizes an infra-red LED source of light that is projected downwards into a short, 0.5 inch diameter, translucent, vertical shaft. The lower tip of the sensor is configured as a point in which the light is totally internally reflected back up the shaft to an infra-red activated receiver when the probe is uncovered, or refracted into the fuel and away from the receiver when immersed. Remote electronics are used to detect the change in output from the optical receiver. This sensor has the advantages of good sensitivity and accuracy. However the sensor is prone, as with all tank located optical sensors, to contamination, such as the progressive build-up of fungal growth, to inhibit its optical properties.

7.1.4.6 Ultrasonic level sensor

Ultrasonics may be used to measure fuel level by transmitting sound through fuel by measuring the time of travel from the transmitter to the fuel surface and back to the receiver. The approach is similar to the fuel surface acquisition technology discussed in detail in Section 7.1.2 above.

7.1.5 Secondary Gauging

In the event of a total failure to determine the quantity of fuel within a tank or tanks by the primary gauging system, a back-up method of gauging is normally mandated in order to dispatch or recover the aircraft in accordance with the Master Minimum Equipment List (MMEL). This role is performed by the secondary gauging system comprising (in most applications) of a number of Magnetic Level Indicators (MLI) commonly referred to as drip-sticks. These units allow the fuel level at a specific location within a tank to be externally determined. An MLI (see Figure 7.27) comprises an assembly that is mounted vertically and internally to the lower skin of an integral or bladder tank with external access through the lower skin. In the case of a bladder tank, the MLI also has to penetrate both bladder and structure for external access. The MLI features a graduated rod, with a magnet at the upper end that is free to slide downwards within an outer tube when unlocked from the stowed position. A float with matching integral magnet is located on this tube and is free to move up and down the tube to follow the fuel surface. In the event of a tank(s) primary gauging system failure, the level of fuel may be determined during ground operations by unlocking the associated drip-sticks and allowing the graduated rods to slowly fall until the rod magnet is attracted by the float magnet. The level of fuel is then determined from the exposed graduations.

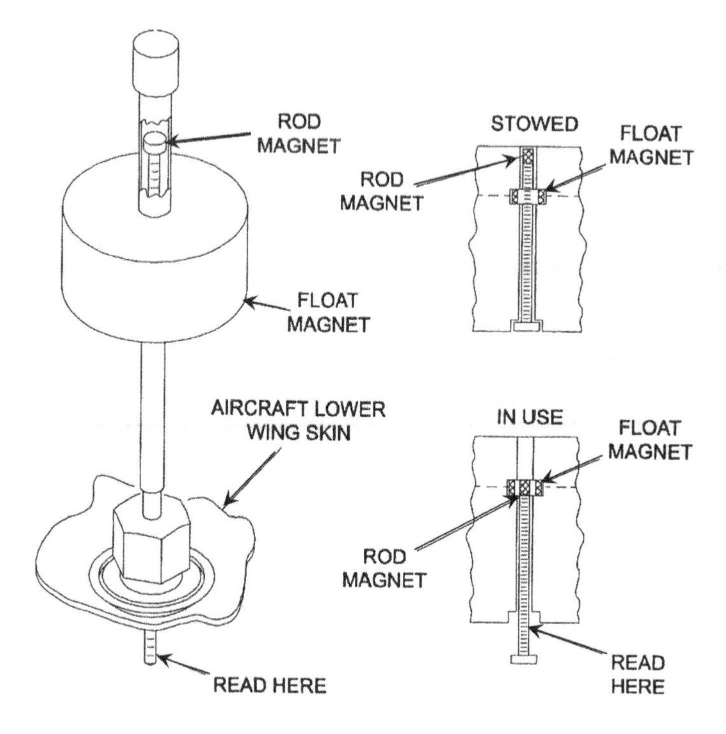

Figure 7.27 Integral tank Magnetic Level Indicator (MLI) details.

The number of MLIs is determined by tank studies with the requirement to measure the fuel in the tank at ground attitudes (typically less than ± 3° pitch and roll). Depending on tank shape, generally one or two MLIs are required per tank.

In order to determine the fuel quantity within a tank the following process is undertaken after the fuel level has been allowed to settle following completion of the refuel process:

- The pitch and roll of the aircraft is established using the attitude indicator(s), typically located in the landing gear (undercarriage) bay.
- The density of the uplifted fuel is established from the supplier/fuel farm.
- The fuel level(s) from the MLI(s) are determined.
- Quantity is established from the aircraft quantity conversion table handbook using the attitude, density and level data.

Fuel quantity determined by the MLI method is typically accurate to an error of approximately ± 5 % of tank total capacity.

Alternative, more accurate secondary gauging systems are available having accuracy of about ± 2 %. These systems use electronic gauging technology and therefore require both in-tank sensors and an electronic unit. Long-range versions of the McDonnell Douglas (now Boeing) MD-11 and Boeing 747-400 and 777-200 aircraft use this more-accurate technology to provide secondary gauging of the aircraft auxiliary tanks. These systems feature gauging probes with compensation to provide fuel height/level information independent of the primary fuel quantity indicating system. The secondary probes for each auxiliary tank are routed to a separate, maintenance bay where the electronic processing unit is installed. This unit also contains a display on which individual level measurements may be selected for quantity calculation using the above process. This approach allows relatively quick auxiliary tank installation or removal using secondary gauging with integral sensing rather than externally accessed MLIs.

7.2 Harnesses

The reliability of the harnessing and associated interconnections to both the fuel gauging and the signal conditioning/processing equipment cannot be over-emphasized. Historically harnesses have tended to be considered as the Achilles heel of gauging system reliability. These harnesses are basically built into the aircraft and are much more difficult to replace than other items such as a processor, indicator or even a sensor. Remembering that a system is only as good as its weakest link, both the in-tank and out-tank harnessing needs to be extremely robust, particularly in the hostile tank environment, as these harnesses are required to last for very many years for the reason that their replacement is such an extremely time-consuming and costly process. Careful selection of environmentally robust and compatible materials, wire gauge, wire runs, rework provisions and connectors will pay dividends in terms of harness longevity. There are different considerations for in- and out-tank harnesses and these are now discussed.

7.2.1 In-Tank Harnesses

The role of the in-tank harnessing is to connect all the sensors to the tank wall connectors for routing outside the tank through the tank wall. The initial design consideration is the impact of the chosen sensor technology e.g. ac or dc capacitance, ultrasonic etc. This determines the initial shielded wiring requirements. For instance, the shield is of the utmost importance in an ac capacitance system as continuous shield continuity between the probe and processor is essential to eliminate capacitive coupling between the drive and signal wires thus ensuring correct operation. For the same reason, separate Hi-Z and Lo-Z tank wall connectors should be used for the interface with the out-tank wiring.

Certification requirements for High Incident Radiated Frequencies (HIRF) are being pushed to higher and higher levels by airworthiness authorities further challenging the design of the shielded wiring.

Aircraft tanks are also increasingly using composite rather than metallic materials. Composite tanks do not provide the same level of protection to HIRF or Lightning and so a more robust electromagnetic shielding approach is required to compensate for the lack of a ‘Faraday shield’ provided by an all metallic tank. The effects of lightning strike on a composite tank require special design considerations. This includes keeping all wiring to a minimum that runs parallel to the spars to minimize strike-induced current.

In dc Capacitance systems, the diode networks on each probe will rectify any HIRF thus injecting a bias on the output signal causing a gauging error. This phenomenon can be eliminated in a metallic tank by the use of filtered tank wall connectors. In a composite tank, the wiring (drive, signal, return) will also require shielding and the probes may require an additional outer tube to act as a shield. Ultrasonic in-tank harnessing, while requiring some shielding, is more immune to HIRF because the probe signals are pulsed and may be more readily extracted from noise.

An important system consideration in the design of the in-tank harnessing is the observance of the fault survivability requirements. Whereas a small aircraft may be able to tolerate loss of all the probe signals from a particular tank, this type of single point failure may well be unacceptable in a larger aircraft. Consequently, a large aircraft tank will usually have two harness routings where, for example, alternate probes are connected by alternate harnesses. Thus a harness failure will result in loss of only half the probes so that adequate gauging is still available at cruise attitudes. In order to protect this requirement in a wing tank located in a rotor-burst zone, it will be necessary to have one harness with its tank wall connectors on the leading edge and the other harness with its connectors on the trailing edge.

Installation, reliability and serviceability are key considerations with in-tank harness design. The in-tank environment is extremely harsh because of the impact of the environmental conditions coupled with the effects of fuel and contaminant build-up. In particular, consideration has to be given to the combined effects of prolonged exposure to sulfur, water and fungus. Common to all in-tank harnessing installations are a number of features that are recommended:

- All sensor and connector terminations should be electro-chemically compatible with the wiring material.
- Wiring should have PTFE conductor and jacket insulation.
- Nickel-plated copper wire should be used, 22 AWG recommended.
- All wiring terminations should be crimped and soldered.
- Ferrules are recommended for shield termination.
- Wherever possible all wiring should be routed along the top of the tank with branches down to the probes.
- Prevention against wire chafing is achieved by running as much wiring as possible through conduit and/or braiding them in 'Nomex®' type material.
- Each probe connection should be dissimilar in size to prevent false connection safeguards.
- All probes should feature differing terminal block locations to prevent false branch connections.
- All probe and harness connections should feature captive hardware (screws etc) for ease of installation and FOD prevention.
- Probe wiring branches should feature water shedding drip loops formed by an initial downward direction followed by an upward connection to the probe terminal block.
- Drip loops should allow for future rework to avoid harness or branch replacement.
- All harnessing should be clamped at the probe terminal block or tank wall connector.
- In-tank wing harnesses should be run inside metallic tanks as much as possible to protect them from HIRF and lightning, the tank wall connectors being located by the wing root.
- Tank wall connectors should be of the fuel resistant jam-nut type.
- Tank wall connectors should feature provisions for receiving 'poke-home' harness termination pins.
- Adaptor hardware to enable external access to the tank wall connector and its rear in-tank harness connections is a desirable feature.
- The harness wiring to the tank wall connector should feature provision for rework and/or external connector removal.

7.2.2 *Out-Tank Harnesses*

The role of the out-tank harnessing is to connect the tank wall connectors to the normally remote signal conditioning/ processor. Once the wiring is outside the tank, it may not always be possible to keep all the wiring inside the protective shield of the aircraft structure. Any exposed harnessing may require double shielding to protect it from electric and magnetic fields created by the effects of HIRF and lightning. In order to maximize reliability, the number connectors between the tank wall and the signal conditioner/processor should be the minimum compatible with the aircraft construction. This is particularly the case with ac capacitance systems where shield continuity must be preserved through each connector, a technique that requires careful design and implementation.

A requirement introduced in the recently released SFAR 88 reference [10] is to both separate and clearly identify fuel gauging wiring runs from other aircraft wiring runs. Separation eliminates any possible tank ignition threat caused by a major electrical issue such as chafed wire damage resulting in gauging harnessing wires shorting with the wires of another system. Identification, such as pink colored wiring, aids in the subsequent installation of other aircraft systems by readily distinguishing the gauging system wiring so as to maintain its separation from the wiring of other systems.

7.3 Avionics Equipment

The fuel measurement and management functions in today's aircraft rely substantially on state-of-the-art avionics in support of signal processing, data concentration, computation, communication, performance monitoring, fault accommodation and displays. While the older, traditional fuel gauging systems involved limited amounts of analog electronics for signal conditioning, processing and displays, the enormous growth of digital electronics technology has significantly influenced the avionics approach to fuel measurement and management.

The challenges for new aircraft fuel quantity gauging system designs include:

- more demanding airworthiness requirements (e.g. SFAR 88) which impact avionics as well as other system equipment;
- adoption of more composite materials in aircraft structures leading to an increased threat from lightning and HIRF;
- the application of data concentration;
- integration of the measurement and management functions within a common hardware and software solution;
- integrated modular avionics open architectures that use common core hardware designs operating partitioned software (Examples of this approach include Airbus A380, Boeing 787, Typhoon, F-22 Raptor and the new F-35 Joint Strike Fighter);
- glass cockpit (integrated multi-function displays).

7.3.1 *Requirements*

Performance requirements have increased in a number of areas leading to the need for more capable avionics delivering higher accuracy gauging, improved availability and integrity, superior fault survivability and enhanced built-in-test (BIT). The fuel measurement and management avionics of today has to interface with significantly more sensors that are all separately addressable and may have to withstand multiple failures. Furthermore the signal conditioning of these sensors, as well as the sensors themselves, have to operate to higher certification standards. This is due to enhanced safety implications of the higher levels of HIRF and lightning resulting from the increasing use of composite materials in aircraft construction that require more substantial protection of wiring and avionics equipment (see Chapter 9).

7.3.2 Data Concentration

As with other systems, there is a trend to locate electronics nearer to the sensors and save out-tank wiring harness weight by conditioning and concentrating sensor data prior to transmitting it by data bus to the central processor. Consequently a number of data concentrator designs are emerging for either the tank wall or an intermediate location between the sensors and the central processor.

This trend does have the advantages of achieving significant installed system weight savings and of eliminating the intrinsically safe tank sensor interfaces from the central processing function and simplifying the design of the latter. The main disadvantage is having to design a sufficiently rugged mechanical and electrical package that is able to meet the necessary reliability requirement in an environment that is typically much harsher than the avionics equipment bay. A good compromise is to locate the data concentrator nearer to the sensors but still within the protected and pressurized environment of the fuselage. For data concentrators that are installed outside the pressurized cabin area, the mechanical design becomes much more challenging since the vibration and temperature requirements can be much more severe. For example, ambient temperatures below –40 degrees, which is the lower operational limit for most electronics, are quite realistic and special considerations within the design (e.g. provision for internal heating) may be necessary to ensure continued operation in this extreme environment. An alternative approach is to design a processor with integral signal conditioning and data concentration for an intermediate location nearer to the sensors and away from a central electronics location.

7.3.3 Avionics Integration

Over the last 25 years, there has been a progressive move to the adoption of integrated avionics. This trend was initiated by military aircraft designers in the need for systems to increasingly interact with one another particularly in the area of flight and propulsion control to make possible the advanced flight handling characteristics of the latest fighter aircraft. This was then followed by the integration of other systems, including the fuel system, into Utilities Management Systems. The final step of integrating all the systems was taken with the development of Vehicle Management Systems to control all the aircraft systems. All of these developments have only been made possible by the major advances achieved in micro-electronics, data bus communications and software development. Naturally, much of this technology has now been adopted on the latest commercial aircraft designs to reduce the life cycle cost of avionics by adopting similar, if not common, processor hardware designs operating partitioned software.

An example of the integration of the fuel system with other systems avionics is provided by the Airbus A380 which has an integrated modular avionic (IMA) system. The IMA features a number of similar Core Processor and Input/Output Modules (CPIOM) that inter-communicate by Asynchronous Fully DupleX (AFDX) data bus, as outlined in Figure 7.28. More detail on this system is provided in Chapter 12.

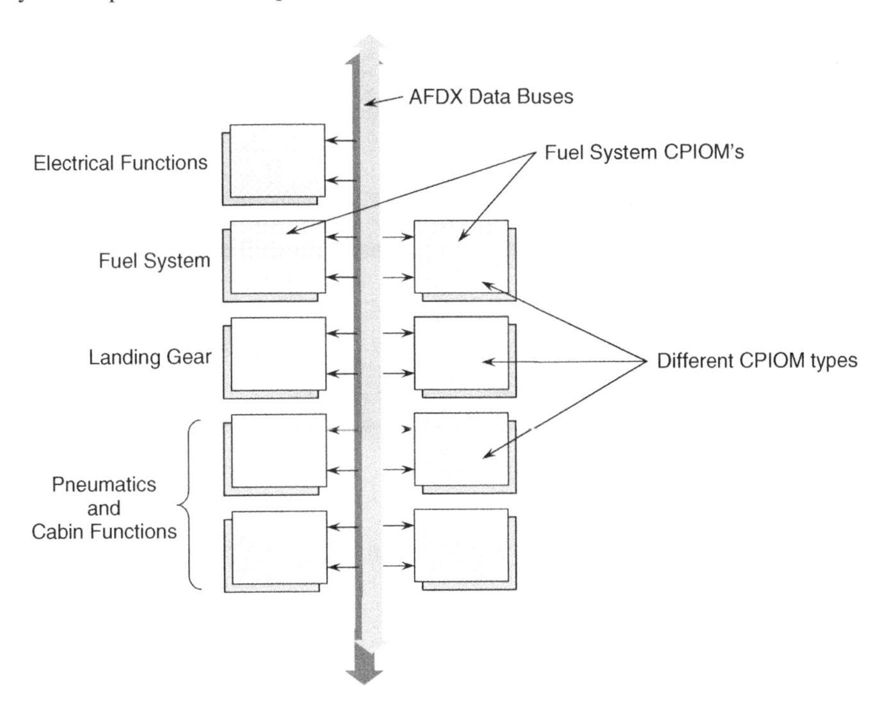

Figure 7.28 A380 Integrated Modular Avionics (IMA) concept.

Integrated avionics have been introduced on both regional jet and business aircraft. Limiting factors in the integration of the fuel system on these aircraft and, in particular, the fuel quantity gauging system with the other systems, have been both the intrinsic safety and battery refuel requirements. The special intrinsic safety demands of fuel gauging sensor interfaces have often led to reluctance to compromise common module designs by imposing these requirements throughout the integrated avionics. In the case of battery refuel, the limited battery capacity is typically insufficient to power the entire integrated avionic suite for any extended period such as that necessary for aircraft refuel. For these reasons the fuel quantity system has not always been integrated with the rest of the avionics.

7.3.4 Integration of Fuel Management

The integration of fuel management into the fuel gauging system significantly enlarges the signal conditioning interface of the processor given the many discrete inputs and outputs from the aircraft necessary to monitor and control the various fuel system components. Figure 7.29 is a photograph of A340-600 Fuel Control and Monitoring Computer of which there are two per aircraft. The computer is housed in an ARINC 600 4MCU case to accommodate all the necessary electronics and the large rear connector.

Figure 7.29 A340-500/600 Fuel Control and Monitoring Computer (FCMC) (courtesy of Parker Aerospace).

7.3.5 Fuel Quantity Display

The universal adoption of the glass cockpit has eliminated the need for standalone dedicated fuel quantity displays in the cockpit. Fuel quantity and other related data is computed by the processor and sent by several data busses for viewing on the central integrated LCD displays such as the Engine Indication and Crew Alerting System (EICAS) displays. In the case where the system also provides the fuel management function, the processor is required to generate the data for, and in some cases, compile a synoptic display for the EICAS in which the operating status of the entire fuel system is diagrammatically represented.

The refuel panel remains an area where dedicated fuel quantity displays are required. The panel usually provides means to display total fuel quantity, pre-selected total fuel quantity, individual tank quantities and pre-selected tank quantities. Refuel panels may be configured with a separate indicator unit for each tank, a combined total fuel indicator with selectable display or an integrated display with simultaneous displays of all the fuel tanks. Figure 7.30 shows two examples of these refuel panel types; one is for the Gulfstream V two-tank business jet aircraft and the second example represents, perhaps the most complex panel used on the Airbus A380-800 aircraft.

The refuel panels are located in a harsh location and the indicators therefore have to be ruggedly designed to survive the environment and are packaged and sealed accordingly. Normally the indicators are microprocessor based units that act in a slave fashion to receive information by data bus from the fuel quantity processor, for formatting and display. A key consideration

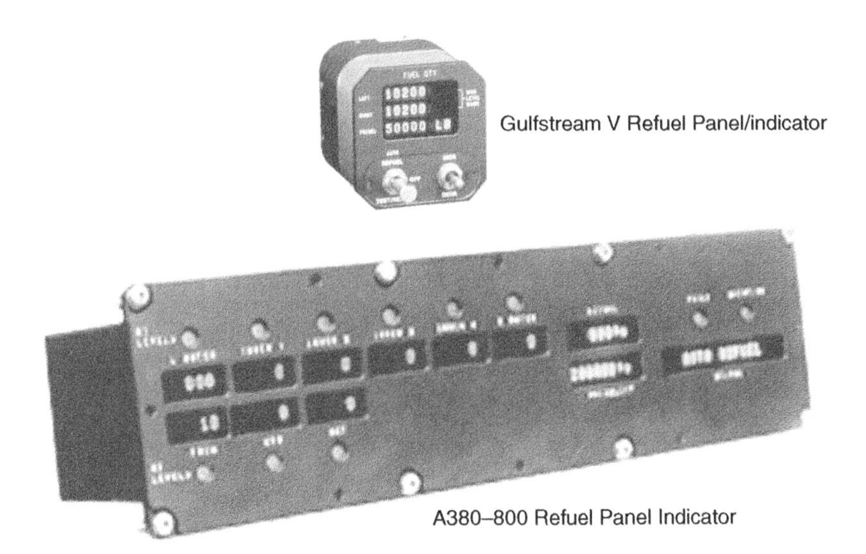

Figure 7.30 Examples of refuel panel indicators (courtesy of Parker Aerospace).

is the choice of a suitable display element for the environment. The best choice, for reasons of readability and reliability, are LED based displays using either 7-segment or dot matrix format. The latter format enables a more comprehensive alpha-numeric indication to be adopted so as to allow units of measurement, tank designation and BITE messages etc. to be integrated within the display.

8

Fuel Properties

This chapter addresses the properties of aircraft fuels that are significant to the design and functionality of the overall system and to the components that make up the system. Specifications for jet fuels were first introduced in the early 1940s in order to establish and control limits to key properties including composition, volatility, specific energy, thermal stability, lubricity and corrosion characteristics.

The United States introduced the JP series of fuels in support of military aircraft programs under the control and oversight of the American Society for Testing and Materials (ASTM). In 1951 JP-4, a gasoline/kerosene blend was introduced and became the mainstay of US Air Force for the next several decades. Shortly thereafter in 1952, JP-5 was introduced with a lower flash point and reduced volatility specifically for use on US Navy ships.

Commercial jet fuels were introduced in 1951 with the introduction of Jet A and Jet A1. These fuels had a maximum flash point temperature of 100 degrees F for improved safety. A commercial jet fuel defined as Jet B was introduced in 1953 with a significantly lower freeze point for operation in extremely cold climates.

Table 8.1 shows the evolution of the complete JP series of fuels together with the specific properties and target applications.

Other countries outside the United States and Europe have also developed their own commercial aviation fuels however, the specifications for these fuels are generally similar to those defined by MoD (Def Std) and ASTM.

8.1 The Refinement Process

Most of today's jet fuels are refined from crude oil which, in its basic form is made up of many components from light gasses to heavy materials. Separation of the various hydrocarbon components is achieved using the distillation process.

In the refinement process crude oil is pumped into a distillation column which is maintained at decreasing temperatures throughout its length as indicated by the schematic of Figure 8.1.

Aircraft Fuel Systems R. Langton, C. Clark, M. Hewitt, L. Richards

Table 8.1 Evolution of the JP series of jet fuels

Fuel	Year	Key Attributes
JP-1	1944	Kerosene, Freeze point −77 Deg F, Flash point 109 Deg F minimum, Limited availability
JP-2	1945	Experimental, unsuitable viscosity and flammability
JP-3	1947	High vapor pressure fuel, boil-off losses and vapor lock problems at high altitude
JP-4	1951	Gasoline/kerosene blend with Reid vapor pressure restricted to 2 to 3 psi to reduce boil-off/vapor lock problems
JP-5	1952	Kerosene with 140 Deg F flash point developed for the US Navy
JP-6	1956	Developed for the XB-70. Similar to JP-5 but with a reduced freeze point and improved thermal stability
JPTS	1956	Highly refined kerosene developed for the U-2 with low freeze point (−64 Deg F) and thermal stability additive package (JFA-5)
JP-7	1960	Kerosene developed for the SR-71. Low vapor pressure and excellent thermal stability for high altitude Mach 3+ operation
JP-8	1979	Jet A-1 kerosene with icing inhibitor, corrosion/lubricity enhancer and anti-static additive

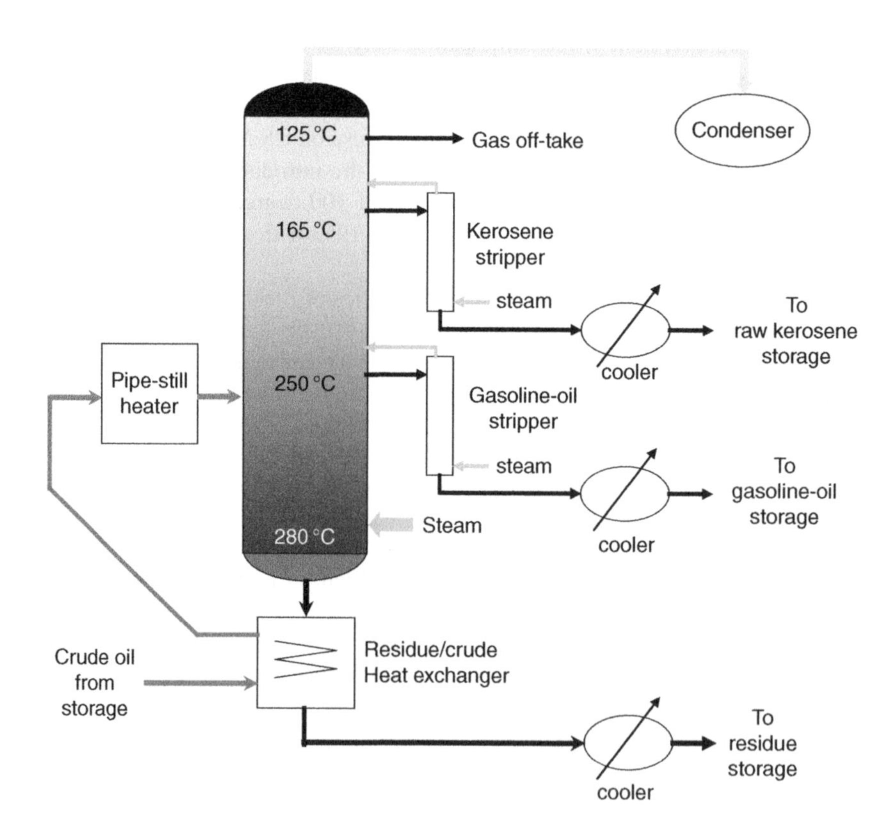

Figure 8.1 Simplified schematic of a distillation column.

The lightest fractions such as propane and butane rise to the top of the column and are drawn off as gases.

Gasoline being a little heavier does not rise as high and is drawn off at the side of the column. Kerosene and diesel are heavier products and are drawn off lower down the column.

The distillation process alone, however is not able to support the market needs for gasoline and kerosene fuels and so to maximize the yield, a number of additional refining processes known as conversion and up-grading are use by the refining industry.

Fluid catalytic cracking is the most widely used conversion process where the higher boiling point hydrocarbons are broken down into a number of separate hydrocarbon constituents. The hydro-cracking conversion process is similar to catalytic cracking since a catalyst is used. In this case, however, the reactions take place under high pressure hydrogen. This process yields a large percentage of fuels in the kerosene/diesel boiling-point range.

Upgrading includes:

- A process called 'Sweetening', which is used to remove certain sulfur containing compounds referred to as 'Mercaptans'. These compounds are undesirable since they are corrosive and contain an offensive odor. The 'Merox proccss' (mercaptan oxidation) is thc most commonly used sweetening method used today.
- 'Hydroprocessing'. This is a generic term for processes that remove undesirable components including olfins, sulfur and nitrogen compounds by breaking down the molecules.
- Clay treatment is used to remove surfactants which can be introduced by the Merox process referred to above. Merox fuels are therefore usually treated this way.
- Each fuel that is produced is made up of a number of hydrocarbons. For jet fuels, most of these hydrocarbons can be classified as follows:

Paraffins

These fuels are very stable in storage and do not attack materials usually found in fuel systems. They have the highest heat of combustion per unit weight but the lowest per unit volume.

Napthenes or Aromatics

These fuels are known to cause swelling of elastomeric seals.

8.2 Fuel Specification Properties of Interest

8.2.1 Distillation Process Limits

The distillation process specification limits seek to achieve a balance between providing sufficient volatility to achieve vaporisation and combustion in the engine but not so high as to create risk of vapor lock in the aircraft fuel system. Widening the range between these two key parameters would increase availability but would have a negative impact on volatility (if the initial boiling point is lowered) or freeze point (if the end point is raised).

8.2.2 Flashpoint

The flashpoint is the temperature at which the vapor above a pool of fuel will ignite when a flame is applied.

8.2.3 Vapor Pressure

Fuel vapor pressure is an important characteristic when considering pump performance and gravity feeding. It is particularly important when considering wide cut fuels such as JP-4 but not so significant with Jet A.

Vapor pressure is a parameter that affects fuel tank flammability since it is the vapor that burns rather than the liquid fuel itself. Furthermore for vapor to burn, an appropriate ratio of vapor to air (i.e. oxygen) must be present.

If there is insufficient vapor then the mixture is too lean. Conversely, with an excess of vapor the mixture is too rich for combustion. For kerosene fuels the lean limit is 0.6 % vapor to air by volume while the rich limit is 4.7 %.

Figure 8.2 shows the flammability limits for Jet A fuel in terms of the energy required in millijoules to cause ignition for fuel temperature plotted against altitude. Outside of zone 6 it requires a substantial level of energy (more than 25 joules) to ignite fuel vapors. The boundary of all zones can be extended slightly by the presence of static charge within the fuel or by fuel sloshing. Static moves the rich boundary limit to the right while sloshing moves the lean limit to the left.

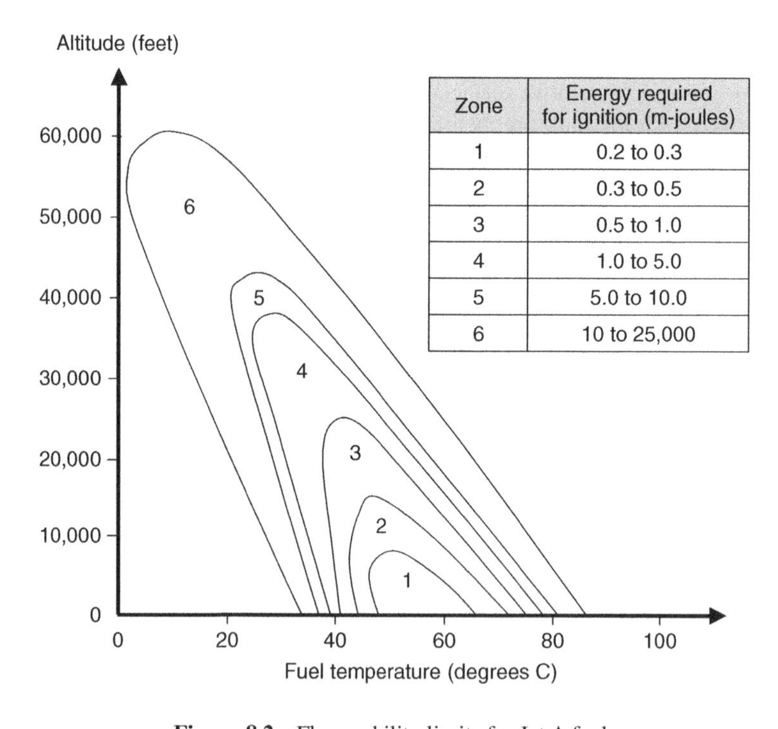

Zone	Energy required for ignition (m-joules)
1	0.2 to 0.3
2	0.3 to 0.5
3	0.5 to 1.0
4	1.0 to 5.0
5	5.0 to 10.0
6	10 to 25,000

Figure 8.2 Flammability limits for Jet A fuel.

Fuel vapor pressure is perhaps most important in its effect on fuel pump performance at altitude. As ullage pressure is reduced with increasing altitude, vapor evolution increases as the fuel vapor pressure is approached until boiling occurs when ullage pressure equals the vapor pressure of the fuel. This is particularly critical in military applications where relatively volatile fuels (e.g. JP-4) are used to take advantage of the lower freeze point desirable for very high altitude operation. Therefore to prevent excessive vaporization at altitude many military aircraft have closed vent systems to maintain a positive pressure margin between fuel vapor pressure and ullage pressure.

The vapor pressure issue is much less significant in commercial aircraft operation where Jet A is perhaps the most common fuel in use.

This point is illustrated by Figure 8.3 which shows a graph of true vapor pressure against fuel temperature for JP-4 and Jet A fuels.

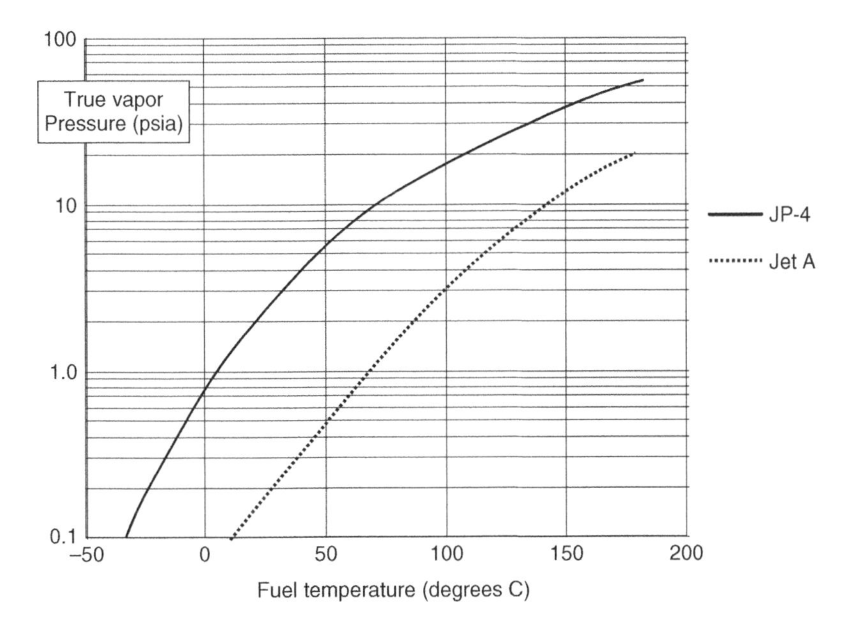

Figure 8.3 Vapor pressure versus temperature.

8.2.4 Viscosity

The viscosity of jet fuel is a measure of its ability to flow and it increases as temperature decreases. While there is a general relationship between freeze point and viscosity, there can be significant scatter in measured viscosity as temperature approaches the freeze point. The standard specification test for viscosity is to measure the time taken for a specific fuel volume to flow through a capillary tube.

As viscosity increases to the point where flow transitions from turbulent to laminar, flow losses increase substantially and burner performance is degraded. For this reason, engine manufacturers require fuel viscosity delivered to the engine to not exceed a predefined upper limit. Twelve centistokes is a common specified limit.

Figure 8.4 shows a graph of kinematic viscosity versus temperature for JP-4 and Jet A fuels.

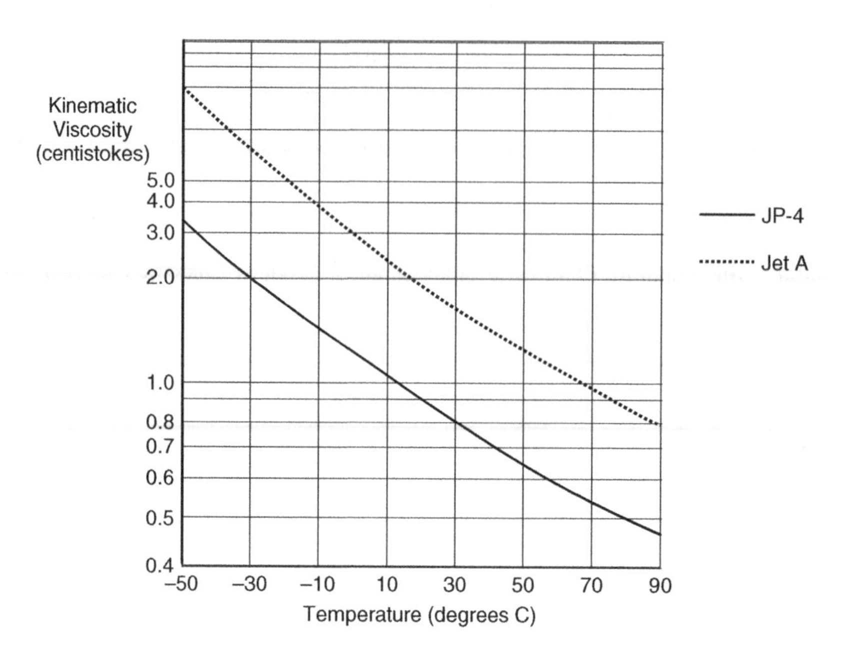

Figure 8.4 Viscosity versus temperature for JP-4 and Jet A.

8.2.5 Freeze Point

Since fuel is a mixture of different hydrocarbons each with its own freeze point, fuel does not turn into a solid at a specific temperature. Instead wax crystals of solid fuel from the hydrocarbons with the higher freeze point precipitate out. With further cooling, the liquid fuel turns into slush and eventually into solid fuel.

From a fuel system perspective it is crucial to ensure that is not trapped within tanks due to its inability to flow or be pumped. Wax formation within fuel tanks can also lead to blockage of pump inlet screens with equally disastrous results.

Testing for the freeze points of a specific fuel involves warming up a fuel sample and determining the temperature at which the last crystal of solid fuel disappears.

Engine manufacturers also typically specify a minimum fuel inlet temperature relative to the freeze point.

8.2.6 Density

Density is an important fuel property from a fuel system perspective because it can vary with temperature more than 25 % over the typical operating range of most aircraft. As has been mentioned before, the fuel load required to fulfil a particular mission is proportional to the fuel mass since this is a measure of the energy stored. Because of the large variation that can occur over temperature, volume limitations may determine the maximum fuel load in hot climates where the uplifted fuel is relatively warm. Conversely, fuel mass will determine the maximum fuel load available in low temperature refuel situations.

As already discussed, density can be inferred from permittivity (refer back to Figure 7.4) or by direct measurement where a higher gauging accuracy is required (refer again to Chapter 7 for a detailed discussion of this methodology.

8.2.7 Thermal Stability

When exposed to high temperatures, fuel has a tendency to oxidize and form gums and varnishes. This can cause clogging of fuel filters, fuel metering equipment, combustion nozzles and heat exchangers. The issue of thermal stability is a major challenge within the engine fuel system where high fuel temperatures occur. Within the aircraft fuel system, aircraft that can operate at supersonic Mach numbers, where recovery temperatures can be very high (Concorde and military attack aircraft are good examples of this), fuel thermal stability becomes important. Also, the use of fuel as a heat sink to cool hydraulics or avionics can lead to the development of high fuel temperatures with resultant thermal stability issues.

8.3 Operational Considerations

8.3.1 Fuel Temperature Considerations – Feed and Transfer

Environmental temperatures during typical aircraft operation vary considerably from ground high temperature soak to high altitude cruise conditions. The issues associated with the colder end of the operating spectrum are illustrated by Figure 8.5.

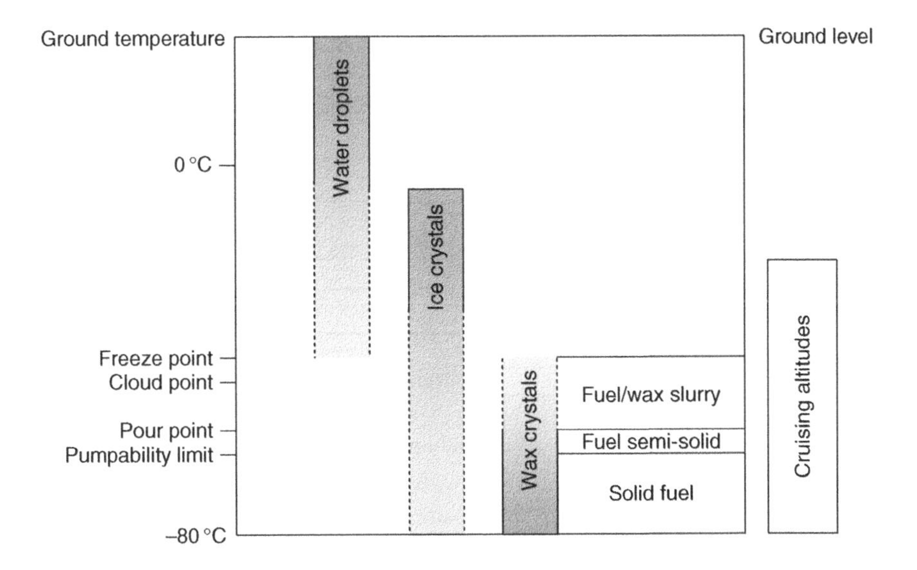

Figure 8.5 Operational fuel states.

As indicated in the figure, dissolved water and water droplets are typically present as a result of relatively warm and humid vent air entering the ullage during descent with very cold tanks. This water turns into ice crystals when the fuel temperatures fall below freezing. The fuel system design must ensure that these ice crystals do not adversely affect system performance. Typically this requires special purpose icing tests to be completed as part of the certification process (see Chapter 11 for a detailed treatment of this subject).

As the freeze point is reached, wax formation begins which ultimately affects fuel pumpability. Since these conditions can occur during typical cruise conditions, the measurement of fuel bulk temperature becomes a critical fuel management system requirement and the crew is required to initiate certain procedures such as increase Mach number (to increase recovery temperature) and/or the reduce altitude (to move into a region of higher ambient air temperature) when low fuel temperature limits are reached.

When flying through cold air with recovery temperatures significantly below the fuel freeze point, wax will begin to form on the tank walls where fuel is on one side and ambient air is on the other. This wax formation will continue to thicken over time as indicated in Figure 8.6.

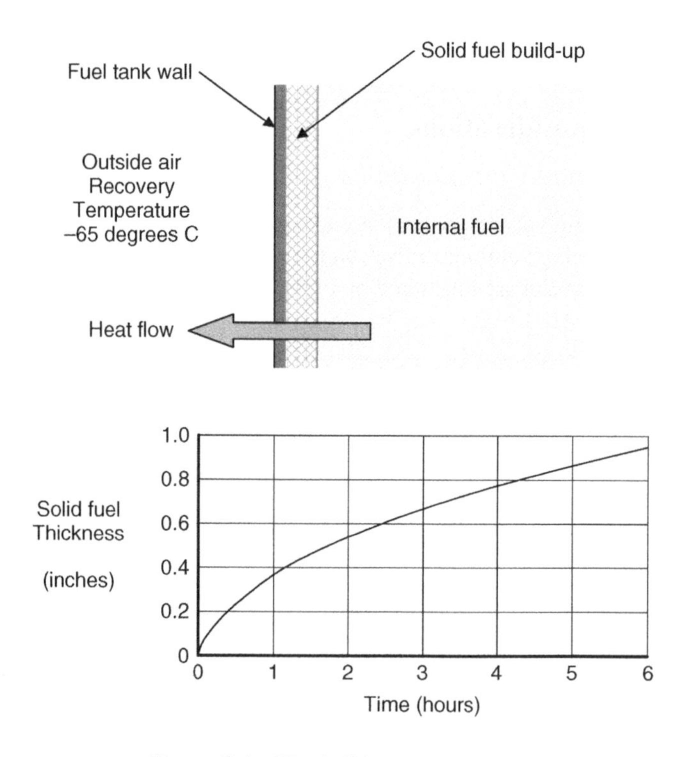

Figure 8.6 Wax build-up on tank walls.

The example shown is for metal tanks. Composite materials have a higher thermal resistance than aluminium and therefore wax formation rates will be much slower.

8.3.2 Fuel Property Issues Associated with Quantity Gauging

Fuel gauging in-tank sensors can be significantly affected by variations in fuel properties as discussed previously. The inherent properties of kerosene fuels such as permittivity and density can be accommodated by the use of sensors that measure these parameters; however, there are other aspects of fuel properties that can also significantly impact the gauging function including:

- water contamination
- dissolved air and outgassing
- anti-static additives
- sulfur
- microbial growth.

Each of the above issues are discussed in the following text.

8.3.2.1 Water Contamination

The presence of significant quantities of water in fuel tanks can have a serious effect on a gauging system. The water contamination is produced primarily by condensate that is formed as the aircraft descends from cruise altitudes after a long, cold soak. As the aircraft descends, warmer, more moist air at lower altitudes flows into the tanks through the vent lines to balance the pressure between the ullage and the outside are altitude. The moisture in this incoming air condenses onto the cold aircraft structure and eventually mixes with the fuel. This water exists either in suspension within the fuel or collects as a puddle at the bottom of the tank and can be problematic to a gauging system in various ways.

The in-tank gauging equipment should have design features such as coatings and finishes to shed water. Electrical terminals should have features to prevent water bridging and shorting the terminals. Interconnecting harnesses should feature drip loops. For a capacitance system, measurement is dependent on dielectric constant. The dielectric constants of fuel and water are vastly different; approximately 2.1 for fuel and 80 for water. Water suspended in fuel will therefore increase the dielectric of the fuel between the electrodes of any affected probe or compensator so as to increase the effective capacitance and impact measurement accuracy accordingly. Water in suspension is particularly prevalent in a collector tank where the actions of the pumps will cause a continuous mixing of fuel and any free water that tends to collect there. A good scavenging system is important to reduce the risk of fuel pooling.

Free water will collect in pools away from pump activity such as forming a wedge outboard of a lower dihedral wing surface stringer and/or rib. It is vital to keep the bottom of probes and compensators away from potential pooling areas otherwise the water will cause an electrical short. Another problem is the formation of ice during a very long flight which may not even melt between flights. Compensators are particularly prone to being covered in ice and, together with careful location, should incorporate umbrella shields to protect them from dripping condensate and ice formation between the electrodes. Some fuels have icing inhibitors added in the region of 0.1 to 0.15 % by volume. The addition of biocide to fuel (see microbial growth) can also inhibit ice formation.

Ultrasonic gauging systems are impacted by the presence of water in a different way. For an ultrasonic system, measurement is dependent on the velocity of sound. Water in suspension has an effect on increasing the velocity of sound for the affected sensors. Significant free water collection may cause a premature signal return from the water/fuel interface.

In summary, every effort should be taken to both minimize the presence of water and harden the gauging tank installation to the effects of water. As a general guide, the gauging system should tolerate conductivity levels of 1000 units (Siemens/cm) without any impact on accuracy.

8.3.2.2 Air in fuel

The presence of air in fuel can provide problems for the gauging system. Dissolved air can be released from fuel (outgassing) as the aircraft climbs in altitude. Turbulent fuel such as that encountered in the tank near an inlet diffuser during the refuel process can contain larger bubbles of air. It is necessary for the gauging system to be tolerant to the presence of air in these forms.

The fine bubbles generated by outgassing are attracted to vibrating elements such as those encountered in a vibrating cylinder or disk densitometer, causing inaccurate density measurement. These can only be shed if the densitometer is switched off periodically or not used in flight. Ultrasonic probes are affected by outgassing in that the signal is attenuated by the bubbles.

Larger bubbles resulting from turbulence will upset the accuracy of measurement of most gauging sensors and therefore the sensors should be shielded or located well away from all sources of turbulence.

8.3.2.3 Anti-static additives

A very important safety issue with fuel systems is the need to dissipate static charge within the fuel. As fuel is fundamentally non-conducting, without taking precautions there would be a significantly hazardous accumulation of charge, particularly in modern aircraft refueling operations where larger and larger quantities of fuel are being uplifted at high flow rates. The presence of a static dissipation additive to provide some conductivity in the fuel allows the charge to gradually and safely leak away. Typically a concentration of 3 mg per liter of fuel is used. The fuel quantity system, however, has to operate with this additive but, as with dissolved water, the system is designed to work unimpaired at conductivity levels up to 1000 units.

8.3.2.4 Sulfur

Most kerosene fuel contain relatively large amounts of sulfur and sulfur products. The presence of free sulfur and hydrogen sulfide are key factors in causing fuel to be corrosive. A major issue is the possibility of the generation of sulfides by chemical reaction with the materials used in the tank components. Historically a key gauging system reliability issue has been the impact of the blackening and subsequent erosion of the connections between the harness wiring and the probes. These connections are very susceptible given the very small currents that flow through them. This problem has been largely resolved by the determination of materials and practices necessary for longevity of the in-tank gauging installation. An installation comprising a harness assembled from nickel plated copper wire with crimped and soldered terminations that is connected to electrochemically matched fuel sensor terminals has proved to be the most reliable. All screw terminals should be tightened with sufficient torque to ensure a reliable connection featuring an anaerobic contact.

8.3.2.5 Microbial Growth

A very important phenomenon with fuel tanks is the potential for fungus to grow. The aircraft environment is ideal given the right combination of air, moisture and fuel in the right climate. The fungus grows in undissolved water at interfaces with the fuel and, if unchecked, produces a green or black substance in the form of a sludge or slime, which tends to adhere to most surfaces. Growth is particularly encouraged at temperatures between 25 and 35 degrees Centigrade, especially in tropical climates. The growth process is not entirely understood and differs for various fuels with tank construction, shape and location. The predominant fungus has been established as 'Cladosporum resinae'. It has been found that the growth mechanism is self-propagating in that it actually feeds on the fuel and that, as the material builds up, further water is trapped to encourage other types of fungal growth.

The consequences of unchecked microbial growth are extremely significant not just for the gauging system, but for the entire fuel system. First of all the physical properties of the fungus are such as to cause restriction or blockage of filters, screens, sump drains, valves, pipelines etc. with potentially catastrophic results. Secondly, if this was not enough, the fungus causes Microbially Induced Corrosion (MIC). This serious effect is caused by the fungus feeding of the carbon within the fuel to produce further fungus and a range of organic compounds including acids that enter into solution with the fuel and can result in significant tank and component corrosion if precautions are not taken. Thirdly, the fungus is conductive because of its water content. This is particularly important in the case of capacitance gauging probes as their accuracy performance will degrade with the resulting change in dielectric, due to the compromising effect of the fungus, and eventually short and fail.

From the above discussion, it can be seen that it is absolutely essential to limit the formation of fungus. Two key prevention measures are to add a fungus growth-inhibiting biocide to the fuel during manufacture and to limit the amount of water present. Recognising the ability for fungus to grow in fuel is not just limited to aircraft fuel tanks and that each step in its journey from the refinery to the aircraft refuelling point is susceptible, precautions have to be taken in the storage, transportation and pipeline transfer to airport fuelling trucks and hydrants. Therefore at every stage on its route to the aircraft, fuel is passed through filter water separators. Once on board the aircraft, tank design and maintenance must be such as to limit the growth and effects of fungus. In addition to eliminating leaks, tanks are normally sealed with a polysulfide compound protection to stifle fungus growth opportunities in all voids in the structure. All fuel components should be protected against corrosion, fuel quantity probes having water shedding coatings. Water produced from condensate etc should be effectively scavenged by the fuel system. In view of the potential consequences of unchecked fungus growth and particularly in tropical locations, it makes good safe and economic sense to instigate an aircraft clean tank program in which periodically all residual water should be extracted using the tank water drains and fuel samples should be taken and tested for the presence of fungus. In the case of a positive fungus test, an application of biocide may be used to kill the fungus in the tanks. Fungus growth may be inhibited with an application of biocide with a concentration typically between 50 and 300 ppm.

9

Intrinsic Safety, Electro Magnetics and Electrostatics

This chapter describes the precautions that must be taken in the design and implementation of aircraft fuel systems to ensure that they remain intrinsically safe in every aspect and continue to operate reliably and safely both during and after exposure to high energy electromagnetic fields or lightning strikes.

Susceptibility to Electro-Magnetic Interference (EMI) has always been a concern of the airworthiness authorities and the qualification of avionics and electronic equipment installed in aircraft involve tests that demonstrate the ability of this equipment to operate satisfactorily in EMI fields. The specific magnitudes and frequency ranges, which depend significantly upon installation and operational environment, have traditionally been defined in the old military specification MIL-STD-461.

In addition to the concerns about EMI radiated susceptibility, is the potential for electronics equipment to interfere with other on-board systems or equipment. This Electromagnetic Compatibility (EMC) issue has been controlled historically via the old military specification MIL-STD-462.

The most stringent EMI susceptibility requirements today demand that the system tolerate exposure to High Incident Radiated Fields (HIRF) which includes direct exposure to high energy radar. This involves significantly higher radiated energy levels than the earlier MIL specifications.

Lightning strikes pose a significantly different threat to the safety of aircraft fuel systems. Here there are three issues that must be addressed:

1. the ability of sensitive electronics to tolerate the effects of induced currents and voltages in exposed wiring;
2. the prevention of arcing resulting from the extremely high induced currents passing through structure, piping, couplings etc.;
3. the direct ignition of fuel vapors emanating from the fuel system vent lines.

Aircraft Fuel Systems R. Langton, C. Clark, M. Hewitt, L. Richards

Each of the issues summarized briefly above are addressed in detail in the following paragraphs.

9.1 Intrinsic Safety

As mentioned previously in this book, fuel systems have been the subject of intense scrutiny since the tragic events of July 18, 1996 when a Boeing 747, operating as TWA Flight 800, crashed into the Atlantic Ocean near East Moriches, NY, following a mid-air explosion. The subsequent investigation focused on the center wing fuel tank and its components and wiring. This led to the FAA creating Special Federal Aviation Regulation 88 (otherwise known as SFAR 88) to eliminate tank ignition sources .

A significant issue affecting the intrinsic safety of the fuel quantity gauging system is associated with the design of the signal conditioning avionics (sensor interfacing) together with segregation of the associated aircraft wiring installation.

Airworthiness Circular AC 25.981-1 released in connection with SFAR 88 reference [10], addresses this and includes the following key points:

- Transient energy introduced into tank shall be less than 50 micro-Joules during either normal operation or operation with failures. (This is a significant tightening of the energy limit requirement previously set at 200 micro-Joules.)
- No single failure shall cause an ignition source.
- No single failure combined with a latent fault not extremely remote ($< 1 \times 10^{-7}$ per flight hour) shall cause an ignition source.
- Any combination of failures not shown to be extremely improbable ($< 1 \times 10^{-9}$ per flight hour) shall cause an ignition source.
- Normal operating 'no-fault' steady-state in-tank short circuit current shall be limited to 10 milliamps.
- Operating 'no-fault' steady-state energy introduced into fuel tank shall be less than 20 micro-Joules during either normal operation or operation with a failure.
- Steady-state short circuit current shall be limited to 30 milliamps for failures of probability $< 1 \times 10^{-5}$ per flight hour.
- Signal Conditioner internal design and construction for segregation of intrinsically safe from non-intrinsically safe signals and power shall be in accordance with International Standard, Electrical Apparatus for Explosive Gas Atmospheres, ICE 60079-11 Section 6 – reference [16].
- Separate avionics connectors for power, non-intrinsically safe signals and intrinsically safe signals (tank sensor interface signals).

The use of Transient Suppression Units (TSUs) to protect the in-tank wiring from sudden unsafe levels of energy has been discussed by the regulatory authorities and within the industry. These units are designed for location at the tank-wall to provide an electrical safety barrier between the out-tank and in-tank wiring. While readily suitable for protecting older designs, they are not recommended for new designs. One of the drawbacks of TSUs is that their protective circuits may fail open in a latent manner, leaving the so-called 'protected aircraft wiring' in a non-detectable unprotected state. For this reason, TSUs must be periodically removed from the aircraft and tested for latent failures.

9.1.1 Threats from Energy Storage within the Signal Conditioning Avionics

The signal conditioning tank interface may be protected from potentially dangerous energy transfers using the design approach presented in Figure 9.1 which shows a typical well-protected multiplexer interface. As indicated in the figure, multiple series resistors are inserted into every possible path that could potentially connect a source of power to the tank I/O lines. EMI capacitors are set back behind a series resistor and a low-power-consuming integrated circuit is used to allow larger resistors to isolate the tank interface.

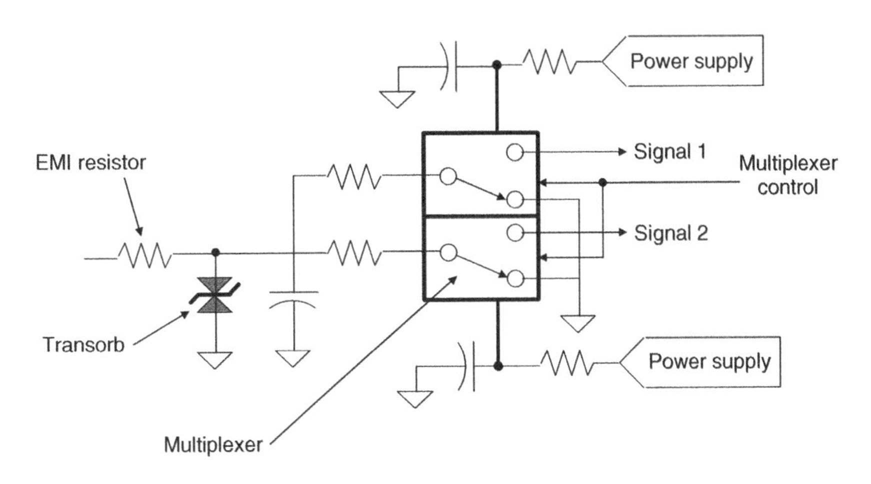

Figure 9.1 Protected signal conditioning interface.

A comprehensive approach to intrinsic safety design is well documented in the European CENELEC document EN 50020, Electrical Apparatus for Potentially Explosive Atmospheres – Intrinsic Safety – reference [17].

9.2 Lightning

Lightning is a major threat that significantly impacts the implementation of aircraft fuel systems. Following a lightning strike, large electrical currents can flow through the structure before exiting the aircraft. Any significant resistance to this current flow will create a potential difference that can result in arcing, which if this takes place in a volatile environment (e.g. a fuel tank or vent line) could result in an explosion or a fire. Careful attention to electrical bonding is therefore critical in the installation of fuel components in areas where fuel and/or vapors can accumulate. Bonding requirements that have been adhered to in current aircraft fuel system designs are again defined in the, now defunct, military specifications which specify resistance maximums between component boundaries and structure that must provide low resistance paths for lightning-induced electrical currents.

A more direct concern related to lightning strike protection relates to vent lines which connect fuel tank ullage vapors to the outside airstream. Since surge tanks (vent boxes) are often located close to the wing tips, which are considered to be a favorable location for a lightning strike, this risk could be potentially catastrophic if fuel vapors are ignited and the flame front is allowed to propagate into the fuel tanks via the vent line piping. As a result, it is standard practice for vent lines to incorporate flame arrestors close to the air inlet (NACA) scoop. These devices contain a large number of small diameter tubes that support the required vent line area while eliminating the possibility of flame propagation into the vent lines.

Similar devices are installed in drain outlets that could possibly contain fuel vapors in the presence of a failure condition where fuel is being drained overboard.

The increasing use of composite materials in aircraft construction also adds to the challenge of providing comprehensive protection against lightning strikes since the protection normally provided by a metal wing is no longer available. This issue is particularly critical for in-tank electrical equipment such as fuel probes, sensors and harnesses.

The design approach for lightning protection of electronic equipment is to perform a safety assessment process in order to establish a hardened fuel system design solution that demonstrates its ability to safely survive lightning strikes by using a combination of suitably protected components and optimally routed harnesses. The design aim is for the system to recover from a lightning strike event with an absence of latent faults and have the same level of protection available for a subsequent strike. The testing level depends upon the zone of installation of the equipment on the aircraft relative to the likelihood of a direct strike.

9.2.1 Threats from Induced Transients in Electronic Equipment

Lightning events generate large electro-magnetic fields that can induce large transients in wiring harnesses. Especially troublesome are the transients, that are coupled to the in-tank interface wiring since the energy is already in a hazardous region. Wiring specifications may require harness installations within a metallic shielding conduit or overall braid in areas of exposure to prevent coupling into the tank.

A key aspect of any safety assessment process (see Figure 9.2) is the determination of ignition threats associated with in-tank harness faults that bring an exposed conductor (one of the sensor interface wires) to within sparking distance of any grounded element of the tank structure. A catastrophic threat is said to exist if the lightning stress can transfer more than 50 micro-Joules to the arc created at the fault. It is important to note that the wire contact need not be shorted to draw an arc. All that is needed is for a fairly low voltage to start the arc, and a low impedance path to channel the energy to the arc. Dielectric failures can also create shorts to structure, and pose a threat similar to direct harness failures. The only difference is that a larger voltage is needed. The insulation elements must therefore be able to withstand the voltage stress imposed by the lightning threats. It should be noted that this part of the safety assessment covers only the energy transfer to the harness fault and electrical currents in excess of one ampere may flow during the lightning event.

Threats are divided into separate categories, normal and egregious. Egregious events are more severe but harder to generate. The main threat comes from lightning stresses, which resonate with the wiring. Harness faults act like nodes attached to tuned antennas. Estimates indicate that resonance circuits created by harness faults fall within the frequency spectrums associated with ring lightning (i.e. 1 MHz to 30 MHz). Furthermore, both the resonance loops and the lightning spectrums have high Qs, such that voltages and currents can build up well

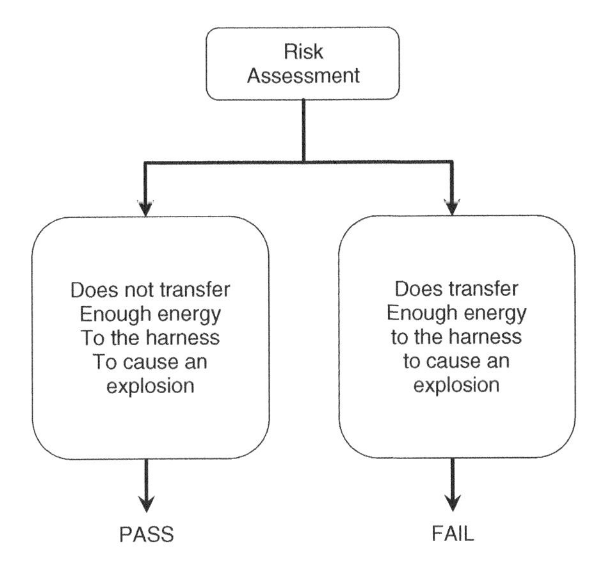

Figure 9.2 Risk assessment process.

above the normal threat levels. A 600 volt transient can induce several thousand volts and place great stress on the dielectric insulation barriers.

The following highlights some of the main issues involved.

Dielectric Breakdown at the Connector Sites

The high voltage insulation at the connectors must be 'infallible' since a breakdown from any connector pin to ground would violate the isolation. Connectors must be utilized that comply with the intrinsic safety spacing requirements for all the in-tank interfaces, including those connectors at the sensors, as well as at the signal conditioning. Proper clearances and creepage distances must be maintained, even at the pin protrusions. Solid insulators, coatings and/or casting compounds must be applied to provide complete protection from 'Punch through' and 'Flashover'.

Low Impedance Paths through the Signal Conditioning Avionics

Internal circuits are floated with respect to aircraft ground but stray capacitance can still provide fairly low impedance paths. The higher frequency components associated with the 'Ring' transients tend to pass through these parasitic paths without much attenuation.

The capacitance coupling between the electrical circuits and the enclosures should be minimized, and coupling disrupted wherever possible by inserting resistance in series with the I/O pins. Ground plane to chassis capacitance of less than 150 pF should be a design goal. This value provides good isolation with regard to long-wave transients, but it is too high for the 'Ring' transients. Series resistors should be inserted into each I/O line to ensure that parasitic paths through the ground plane cannot find low impedance paths to the tank harness. The series resistors also help to reduce (dampen) the resonances formed by wire inductance interacting with the isolation capacitance. Safety assessment indicates that at a value of at least 500 ohms is needed to cover the most egregious cases.

Stray capacitance associated with the capacitance probes also provides a parasitic path to structure. Estimates place this stray capacitance coupling as high as 30 pF. Based on historical safety assessments a 910-ohm series resistance embedded in each of the probes lines is required to bring the threat below the 50 micro-Joule safety limit.

The stray capacitance estimates for typical compensator devices are higher than for tank probes and therefore need more protection.

The stray capacitance associated with the Resistance Temperature Device (RTD), which is the most commonly used fuel temperature sensing technology used in today's fuel systems is generally inconsequential with respect to resonance threats. Some series resistance is needed but since this reduces the accuracy of the temperature measurement a four-wire RTD interface is recommended.

Dielectric Breakdown within the Signal Conditioning Avionics

Intrinsic safety spacing must be maintained inside the avionics to separate the areas of great stress from the internal electronics. Safety resistors need to be carefully selected. Two-watt high reliability surface mount components, capable of handling the high voltages and severe transient stresses imposed by the lightning threats are recommended. They should be precision components fabricated using a highly stable Metal Glaze^{TM} process, which deposits a resistive material onto an alumina substrate to be subsequently laser trimmed and encapsulated. These devices have excellent steady state and transient power capabilities, are well characterized, and have a history of providing reliable performance. The peak voltage rating for these two-watt devices is typically 3000 volts. These resistors are laid out in parallel banks, a technique which takes advantage of the resistor lead spacing, to create an isolation barrier, which is consistent with the intrinsic safety requirements. It should be noted that certain circuit board enhancements, such as slots, can add robustness to the isolation.

Internal Faults

If left unchecked, certain faults can compromise the isolation barriers. Wires can break and components can become partially or fully dislodged from their circuit board assemblies. Solid insulation, coating and/or casting compound should be used to ensure the integrity of the barrier.

Threats Coupled through Power Input

The power lines, which typically run along the spars in composite tanks, are subject to lightning levels consistent with exposed wiring. Several layers of protection can prevent the propagation of these threats into the in-tank harnesses. The first level is to provide isolation through the power supply design by transformer coupling through a dielectric barrier with a protective shield between the primary and secondary windings. The supply is attached to the power connector to keep the power pins, and primary circuits in a tight area well away from other circuits. The transients that do get through tend to be decoupled by the parasitic capacitance between the ground plane and chassis. This parasitic capacitance now helps by bypassing some of the transient to ground. Resistors in series with the I/O pins attenuate whatever is left.

Direct Paths through Lightning and EMI Protection Circuits

It should be mentioned here that the Lightning and EMI protection circuits are not connected directly to I/O lines. If they did they would complete ground loops through the parasitic capacitance between the ground plane and chassis.

The same applies to the shield connections. The shields are connected to active circuits, which provide a virtual ground to low level signals. The available energy that can be transferred to the fault is limited by current limiters.

9.2.2 Protecting the Signal Conditioning Avionics from Lightning

EMI capacitors and transient suppression diodes are used to protect the internal electronics from the potentially damaging effects of lightning. Because of the floating nature of the design, these components are not located at the I/O pins, and do not bypass the transients to ground. The protecting components are tied to the local ground plane and prevent differential voltages from appearing across the sensitive electronic devices.

Most protection networks consist of a T-connection of two series resistors with a capacitor and a transorb connected to the center node. This circuit is robust in that the individual elements help protect each other. The resistors protect the diodes and the diodes protect the capacitors.

9.3 EMI/HIRF

Electro-magnetic compatibility is required of all aircraft systems in that each system does not generate conducted or radiated interference to upset the operation of any other system, or that the system itself is not susceptible to conducted or radiated interference from all sources. EMI that is generated by most operational aircraft systems is considered a steady state condition, and it is therefore a requirement that all fuel system electronics and sensing equipment must perform without degradation when exposed during normal operation to aircraft ambient EMI levels. An EMI test susceptibility level of 20 volts/meter per RTCA DO160 reference [18], for category (U) and (R) equipment generally provides an adequate design margin. HIRF (High Incident Radiated Fields) is generated by sources external to the aircraft such as radar and telecommunication transmissions and has become the focal point of the overall subject of Electro-Magnetic Interference (EMI) because the qualification levels necessary for system certification have been raised significantly from earlier EMI specification requirements. HIRF is considered as a transient EMI condition of significantly greater amplitude in the region of 200 volts/meter and above. During these tests electronic equipment may be upset during periods of HIRF exposure but must automatically recover without damage after exposure.

9.3.1 Threats from HIRF Energy Transfer

The electro-magnetic fields associated with HIRF can couple energy to wiring harnesses. Especially troublesome are the HIRF fields inside the fuel tanks since the energy is already in the hazardous region.

A key topic of any safety assessment is the determination of the ignition threat(s) associated with in-tank harness faults that can bring an exposed conductor (one of the sensor interface wires) to within sparking distance of any grounded element of the tank structure. A catastrophic threat is said to exist if the HIRF develops a current in excess of 30 milliamps.

Like EMI, HIRF threats are also divided into the normal and the egregious categories. Egregious events are more severe; however, they occur if, and only if, the HIRF frequency matches a natural resonance. Again harness faults act like nodes attached to a tuned antennas.

Estimates indicate that resonance frequencies associated with harness faults can fall inside HIRF frequency ranges.

The following comments highlight some of the main issues.

Dielectric Breakdown

The voltages produced by RF coupling are lower than voltage transients induced by lightning, so the dielectric rating determined to be safe for lightning is a valid solution for HIRF.

Parasitic Paths

The high frequencies associated with HIRF can significantly modify the circuit properties. Many parasitic paths are created, and the architecture of the interface can change dramatically. The isolation between the floating ground plane and chassis is reduced typically to a few ohms, and becomes inconsequential. The series resistors connected to the I/O pins now act as line terminators, which help suppress oscillations.

RF coupling that induces just 15 volts into a section of the harness can force 30 milliamps through a fault site, placing the maximum acceptable RF field at about 600 volts per meter.

Threats Coupled through the Power Input

A shield in the input transformer will bypass most of the RF threat to ground. Provided that the shield and bonding straps are designed to take the RF load there will be no coupling risk to the power interface.

EMI Protection Capacitors

To avoid ground loops, EMI protection capacitors should not be connected directly to the I/O lines. The same applies to the shield connections. Shields are connected to active circuits, which provide a virtual ground to low frequency signals. The higher RF frequencies will overwhelm the shield drivers and therefore will act like resistive terminators.

9.3.2 Protecting the Signal Conditioning Avionics from HIRF

EMI capacitors and transient suppression diodes are used to protect the internal electronics from the potentially damaging effects of HIRF. The lightning protection circuits are designed to provide a satisfactory HIRF interface. As stated earlier these protection networks consist of a T-connection of two series resistors with a capacitor attached to the center node. This circuit is useful in that it provides filtering in both directions. HIRF can apply up to a half watt load on the protection resistors, which ties in with the 2 watt resistors selected for lightning protection.

9.3.3 Electrostatics

In the fluid handling area of fuel system design, care must be taken to prevent static charge build-up in the fuel itself as a result of high velocity impingement with the internal tank structure and installed equipment. History has demonstrated that sufficient energy storage can be achieved in this way to generate sparking with resultant ignition of fuel vapors within the tank.

One commonly used approach to managing the build-up of a static electric charge within the fuel involves the use of anti-static additives to facilitate leakage of any charge back to the aircraft structure over time as described in the fuel properties section of Chapter 8.

Within the basic fuel handling system design, however, there are fundamental techniques that should be followed in order to minimize the probability of the occurrence of excessive electro-static charge within the fuel during refuel or fuel transfer operations. As a general guideline, fuel discharged into the fuel tanks should occur at the bottom of the tank where it will be normally submerged in fuel. Secondly the velocity of fuel as it enters the tank should be limited to ten feet per second or less. Typically, diffusers are installed at the fuel discharge point to provide safe conditions for fuel discharge.

10

Fuel Tank Inerting

Fuel tank safety has been a perennial issue associated with the design and operation of aircraft fuel systems since the inception of powered flight, however it was not until the 1960s that meaningful studies were made into the practicality of providing a safe environment for aircraft fuel tanks by using controlled inerting of the ullage within the fuel tanks.

The need for fuel tank inerting has always been critical in military aircraft applications where fuel tank penetration by enemy fire can result in a spontaneous explosion of the fuel vapors within the ullage. The resulting over-pressure can lead to immediate destruction of the aircraft. It is this over-pressure, rather than the potential for an ensuing fuel fire that is the major threat since it can cause sufficient structural damage to destroy the aircraft. Until the 1960s the only explosion suppression technique in service with the military was the use of polyurethane reticulated foam installed within the fuel tanks. This approach prevents flame propagation and subsequent explosion within the tanks but it has many disadvantages:

- It adds weight to the aircraft and reduces fuel capacity.
- Maintenance of in-tank equipment is problematic since foam has to be removed and replaced to access the hardware.
- It has an operational life of about 5 years.

On the plus side, foam does offer full time protection from take-off to landing and there are no operational components to fail.

The following section provides a brief history of the emergence of fuel tank inerting technologies in military aircraft applications from the 1950s through to the present day.

10.1 Early Military Inerting Systems

Fuel tank inerting systems for military aircraft began to emerge in the late 1950s and 1960s. The F-86 and F-100 aircraft demonstrated gaseous nitrogen systems that provided part-time inerting of their fuel tanks. The F-86 system weighed 116 lbs and provided only nine minutes of inert fuel tank operation. The later F-100 system showed significant improvements with a

Aircraft Fuel Systems R. Langton, C. Clark, M. Hewitt, L. Richards

system weight of 42 lbs and an inerted time of 35 minutes; however, neither of these systems entered operational service. This information and much more can be found in SAE AIR 1903, reference [19].

Liquid nitrogen inerting systems emerged in the late 1960s on the SR-71 Blackbird, the XB-70 Valkyrie (which was canceled in the prototype phase of the program) and the C-5A Galaxy ultra large transport aircraft. The primary need for tank inerting in the SR-71 application was to prevent spontaneous ignition of the fuel which can achieve temperatures of over 200 degrees Fahrenheit during high Mach number operation.

The main challenge for the liquid nitrogen approach is the provision of support logistics in remote operating theaters. For the C-5A this was acceptable since the aircraft operates from only a few large bases around the world. Liquid nitrogen is stored on-board in insulated containers which must be topped up after every flight. Figure 10.1 shows a C-5A LN_2 dewar undergoing vibration testing.

Figure 10.1 The C-5A LN_2 dewar during vibration testing (courtesy of Parker Aerospace).

Since cost weight and logistics made the liquid nitrogen inerting system impractical for most fighter aircraft an alternative inerting technology was developed. In the early 1970s an inerting system approach using halon 1301 was developed, demonstrated and eventually introduced into the new F-16 Fighting Falcon aircraft, reference [20].

This alternative approach involved the use of stored liquid halon that is gasified and fed into the tank ullage. Protection provided by the halon system varies with halon concentration, for example, a 9 % concentration by volume will provide protection against a 50 caliber armor-piercing Incendiary (API), while a 20 % concentration is needed to provide protection against the greater threat of a 23 mm High Energy Incendiary HEI round.

Due to weight and space limitations, this type of inerting system was only available for relatively short periods of time, therefore the pilot had to select the halon inerting system prior to entering hostile airspace. Figure 10.2 shows a schematic of the F-16 halon inerting system. Upon selection by the pilot, ullage pressure is reduced from about 5.5 psig to 2 psig and air is vented overboard. At this time, halon is dumped into all tanks for 20 seconds to achieve an immediate inert state. From this point on a halon proportioning control valve mixes halon in a fixed proportion to the incoming air from the vent system to maintain a nominally constant halon concentration level within the tanks as the aircraft altitude increases. During descent, the vent system allows the ullage air/halon combination to spill overboard in order to maintain the required ullage pressure. To compensate for halon absorption by the fuel and fuel usage, a continuous bleed of halon into the ullage is provided via a fixed orifice. Thus halon is consumed as fuel is consumed and as the aircraft maneuvers. As a result the fully inerted operational time is limited by the amount of liquid halon carried.

The F-16 Fighting Falcon was the first aircraft to employ this type of tank inerting system and a derivative of this system was later installed in the F-117 Nighthawk.

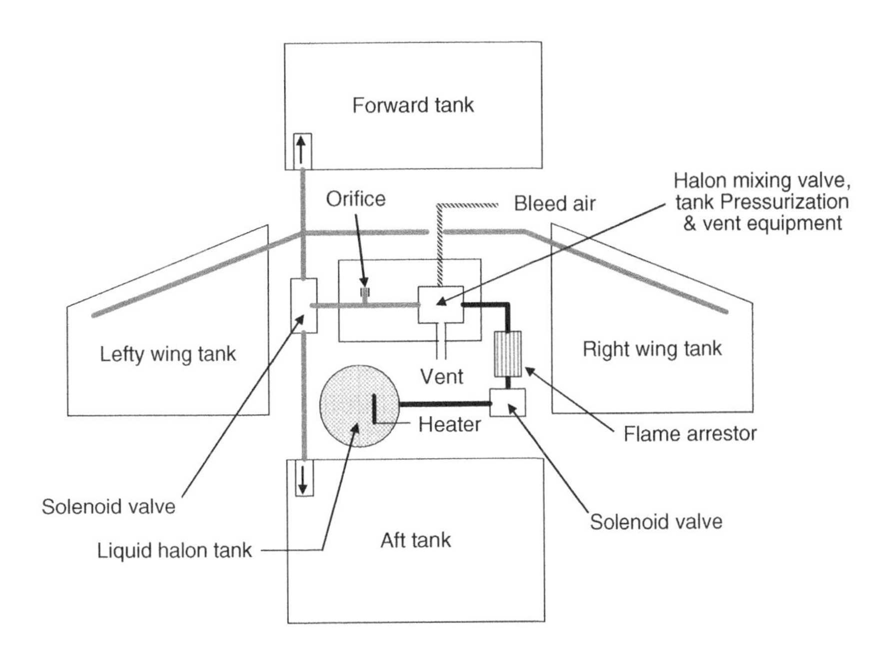

Figure 10.2 F-16 halon inerting system conceptual schematic.

An alternative halon system approach was designed for the A-6 Intruder aircraft. This system uses an electronic controller to infer, the actual concentration of halon within the ullage from measurements of ullage pressure and density. Closed loop control of halon delivery into the ullage can then be achieved. This system provides a significantly more accurate approach to halon control than the previous F-16 system and as a result the inerting time for a given halon load is substantially longer.

Since halon is a fluorocarbon with the potential for damaging the ozone layer and as a result of pressure from the EPA halon inerting systems are no longer in production.

The next generation inerting system where inert gas is generated on-board the aircraft was first introduced on the C-17 as the On-Board Inert Gas Generation System or OBIGGS. This system uses air separation technology to strip oxygen molecules from engine bleed air leaving Nitrogen Enriched Air (NEA) to displace the air in the ullage within the fuel tanks.

Figure 10.3 shows a schematic diagram for a generic OBIGGS using engine bleed air as the air inlet source.

The bleed air must be cooled via a heat exchanger to temperatures below 200 degrees Fahrenheit prior to being fed to the air separator sub-system. Filters are also necessary to protect the air separators from contaminants that can cause functional deterioration of the air separation process.

Since the OBIGGS provides less than pure nitrogen into the ullage, the issue of acceptable levels of oxygen concentration for protection against enemy action is important. Studies have shown that provided that the oxygen concentration is maintained below about 9 % by volume, the ullage will provide protection against a 23 mm High Energy Incendiary HEI round.

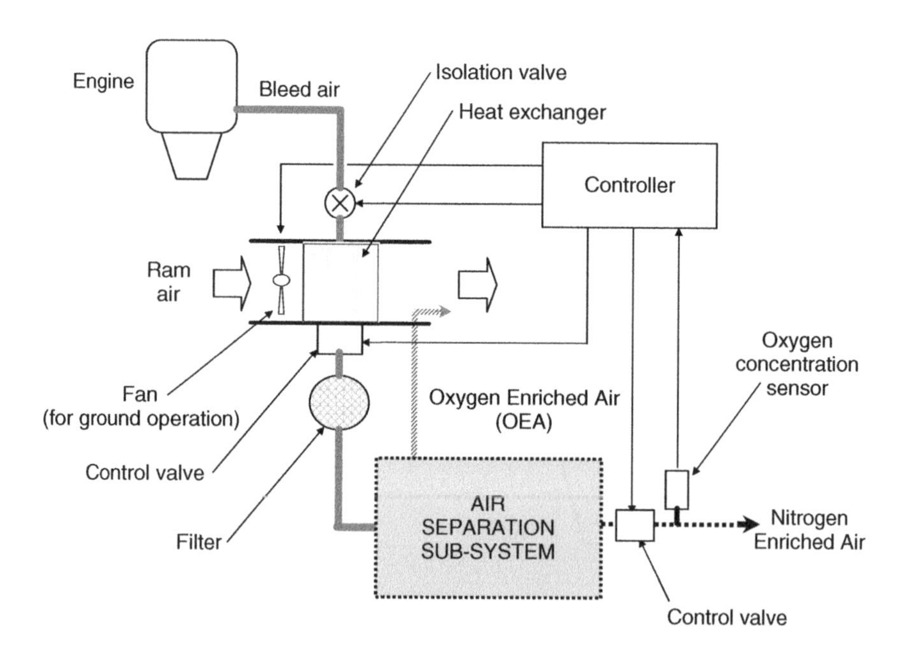

Figure 10.3 Generic OBIGGS system schematic.

The challenge for the early OBIGGS designs was its capacity to deliver sufficient NEA to the system particularly when the aircraft is in descent with low power throttle settings on the engines. In this situation outside air is rushing into the tanks via the vent system to equalize the pressure between the ullage and the outside air and engine bleed air available for NEA generation is at significantly lower pressure at low power throttle settings. When fuel tanks are close to empty the ullage volumes are large thus exacerbating the problem. On the C-17, this problem was solved by storing NEA under pressure early in the flight when fuel tanks are full and engine power settings are high and using this stored NEA when the OBIGGS NEA flow capacity cannot meet demand. The penalty of this solution in terms of weight and system complexity is considerable.

An additional issue than must be addressed in OBIGGS type inerting systems is the air (and hence oxygen) dissolved in the fuel as it is loaded into the aircraft both on the ground and (for military aircraft) during the aerial refueling process. This air will outgas into the ullage during a climb to altitude. To mitigate this problem on the C-17, a 'Scrubbing' device (see Figure 10.4) is installed in the system through which in-coming fuel passes before being distributed to the various fuel tanks in the aircraft. This device bubbles NEA through the fuel forcing the dissolved air out into the ullage and replacing it with NEA. Throughout the refuel process the ullage contents are purged with stored NEA to ensure effective inert status.

Figure 10.4 C-17 Aspi-scrubber (courtesy of Parker Aerospace).

10.2 Current Technology Inerting Systems

10.2.1 Military Aircraft Inerting Systems

The air separation technology used initially by the C-17 was the 'Molecular sieve'. This approach uses beds of synthetic zeolite material which preferentially absorbs oxygen when exposed to air under pressure. Two beds are used in this system, each bed being sequentially exposed to high and low (atmospheric) pressure. This is necessary since the oxygen capacity of a given zeolite surface area is limited. When exposed to low pressure, the oxygen is de-absorbed and vented overboard. This technique is referred to as the 'Pressure swing absorption' method of air separation.

The competing technology to the molecular sieve is the permeable membrane fiber technology initially developed by DOW Chemical Corporation. In the early 1980s the molecular sieve was capable of delivering about 8 lbs per minute of NEA at optimum air inlet conditions versus about 4 lbs per minute for the permeable membrane air separator, however in the mid to late 1980s a major breakthrough occurred in the permeable membrane fiber technology with air separation test results showing an order of magnitude improvement in NEA flow capacity over the early fibers. These new high permeability fibers are typically larger in diameter and with thinner walls.

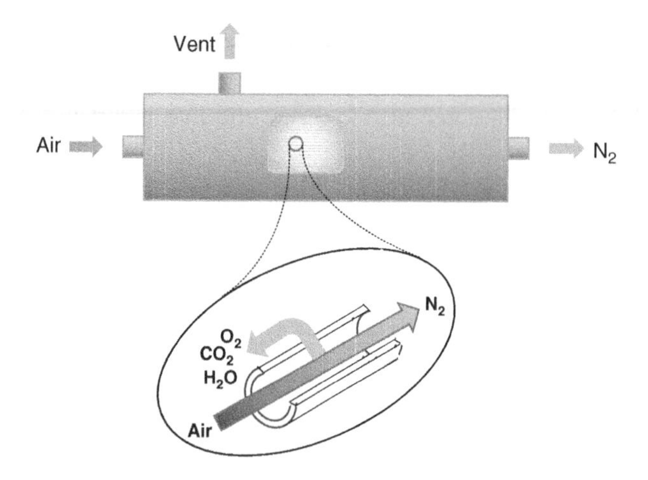

Figure 10.5 The Air Separation Module (ASM) concept (courtesy of Parker Aerospace).

These hollow fibers are specially treated to maximize performance and are assembled in cylindrical bundles as shown in the conceptual drawing of an air separation module presented in Figure 10.5.

As indicated in the diagram, air passing through the module is separated into its molecular constituents, primarily oxygen (O_2) and nitrogen (N_2), as it passes through the fibers. In this process the oxygen molecules are encouraged to migrate towards the vent (together with any entrained carbon dioxide and water vapor molecules, leaving the nitrogen to pass through to the axial outlet as Nitrogen Enriched Air (NEA). Several different fiber sources are available today; each with differing characteristics and these fiber designs and fabrication processes are highly proprietary. In all cases however, the latest fibers available have much improved performance in terms of NEA yield and flow capacity compared to earlier fiber products.

Improvements in fiber technology have continued to the present day and now the permeable membrane fiber has become the standard air separation method for all state of the art OBIGGS systems.

An upgrade of the C-17 OBIGGS to the latest permeable fiber technology was competed in 2004 and this new system is now on operational service.

The new permeable fiber air separation technology is in service on the F-22 Raptor and will be operational on the new F-35 Joint Strike Fighter. Both these aircraft use a demand system without any on-board storage of NEA. This has been made possible by the improved performance of the latest available fibers. Figure 10.6 shows a photograph of the F-22 air separation unit.

Figure 10.6 F-22 Air Separation Module (ASM) (courtesy of Parker Aerospace).

The European A400 military transport aircraft currently in development also uses an OBIGGS for fuel tank inerting using the permeable membrane fiber technology.

10.2.2 Commercial Aircraft Inerting Systems

For the commercial aircraft market, the penalty of providing some form of on-board explosion suppression system in terms of both the equipment acquisition cost and operating expense has been has long been the main obstacle to the inclusion of tank inerting systems until major improvements in air separation technology were made in the late 1980s and 1990s. Prior to these later developments, the commercial aircraft community has relied on the enforcement of strict airworthiness design standards as part of the aircraft certification process. These safeguards include requirements that embrace a number of issues relating to electrical components, mechanical components and installation within fuel tanks. Several examples follow.

In-tank Wiring

A goal of intrinsic safety is supported by specifying an upper limit to the electrical energy entering the fuel tank during normal operation, short circuits and induced currents/voltages in fuel tank wiring that may potentially lead to ignition of flammable vapors. A limit of 200 micro-joules originally established was recently superseded by a lower limit of 50 micro-joules (see comments below).

Pump Wiring

Pump designs are required to ensure that spark erosion and hot spots due to short circuits in pump wiring are avoided. Also independent protection shall be provided against the development of local high temperatures resulting from a locked rotor failure.

Pump Design
Fuel pumps must be capable of extended dry-running operation without generating local high temperatures with the potential for ignition of fuel vapors. The design must also minimize the probability of sparking due to component wear or as a result of foreign object damage.

Bonding
Installation of in-tank equipment must be adequately bonded so that electrical discharges due to lightning, High Intensity Radiation Fields (HIRF), static or fault currents can not ignite flammable vapors.

Arc Gaps
Adequate separation between components and structure must be provided to ensure that arcing due to lightning will not occur.

The above provides an indication as to how the airworthiness regulations attempt to eliminate to probability of fuel tank explosions. The above list is just an overview of some of the more critical issues involved in this important subject of fuel tank safety.

In April 2001 airworthiness regulations associated with the design of aircraft fuel systems were tightened further via the release of Special Federal Aviation Regulation (SFAR) 88 by the FAA applicable to all aircraft registered in the USA. The JAA (now the EASA) released a similar document INT/POL 25/12 applicable to all Airbus aircraft. This was as a result of the TWA Flight 800 incident in July 1996 and the conclusion by the NTSB that loss of the aircraft was most likely caused by an explosion of fuel vapors in the center fuel tank from some unknown ignition source. This ruling has had a major impact on the commercial aircraft marketplace to the point where most new aircraft seeking Type Approval by the FAA or the EASA may be expected to have some form of on-board inerting capability in order to reduce the risk of a fuel tank explosion to a newer and more stringent level of safety.

Generally, commercial aircraft store fuel in integral wing tanks. Additional auxiliary tanks located within the fuselage are typically used to provide additional operating range. From a safety risk perspective, fuselage tanks have an inherently higher risk of a fuel vapor explosion than wing tanks (see Figure 10.7) for the following reasons:

- These tanks are frequently flown with little or no fuel in them. This was in fact the case for TWA flight 800 which exploded over Long Island in 1996.
- The location within the fuselage keeps any fuel and/or fuel vapors warmer than the in the wing tanks which are exposed to the airstream.
- In some aircraft designs, air conditioning packs are located close to these tanks and reject heat into these tanks thus increasing the volatility of any fuel or vapor contents.

These considerations focused initial efforts in the evaluation of inerting systems for commercial aircraft to the center fuel tanks.

Wing tanks are considered less hazardous because the ullage is initially small and the tanks are quickly cooled by the airstream to very low temperatures. This is particularly true of metal wing aircraft. Wings manufactured partly or wholly of composite materials have a substantially lower heat transfer capability between the fuel and the outside airstream. For this reason, composite wing aircraft are therefore not considered to be significantly safer than fuselage tanks.

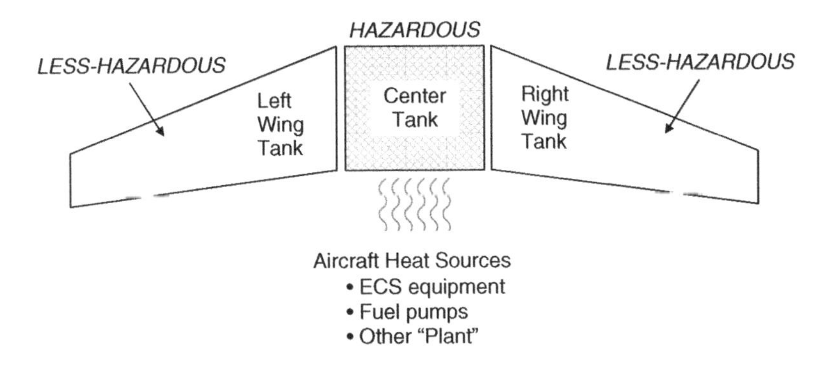

Figure 10.7 Fuel tank hazard assessment for commercial aircraft.

Both Boeing and Airbus embarked on flight test evaluation of center fuel task inerting systems following the release of SFAR 88. Boeing flew trail installation pallets comprising permeable membrane type air separators, pneumatics and control equipment on their 737 and 747 aircraft. In each case, the air separation system was sized to maintain inert conditions in the center tanks only of these aircraft on the basis that the aluminum wings would be acceptably safe without having to inert the wing tank ullage.

The Boeing 747 demonstrator (see Figure 10.8) comprised five separator modules.

This design solution was required to meet two critical operational requirements:

- To maintain inert status during the descent case with an empty center tank and a flight idle power setting.
- To meet a specified initial 'Pull-down' time to reach inert status with engines at idle power on the ground with an empty center tank.

Inert status for commercial aircraft fuel tanks has been established as 12 % oxygen concentration or less.

The new inerting system for Boeing commercial aircraft, designated the Nitrogen Generation System (NGS), is already being installed on production 737 aircraft and is planned to enter service on production 747s and 777s in 2008 and 2009 respectively.

The production NGS for the 747 will have only three separator modules. This has been accomplished by using cabin air as the air source, instead of engine bleed air, and using a compressor and heat exchanger to provide close to optimum air inlet conditions to the air separator sub-system for all engine power settings which significantly helps both the pull-down time and the descent case requirements.

This palletized assembly is supplied by Honeywell using air separator modules provided by Parker Aerospace.

In parallel with the Boeing NGS demonstration and development, Airbus tested a similar inerting system on their A320 aircraft. The photograph of Figure 10.9 contrasts sharply with the 747 demonstrator unit of Figure 10.8 since the Airbus A320 is a much smaller aircraft.

The Airbus system also uses compressed and cooled cabin air as the air source for the separation system.

Figure 10.8 Boeing 747 flight test demonstrator (courtesy of Parker Aerospace).

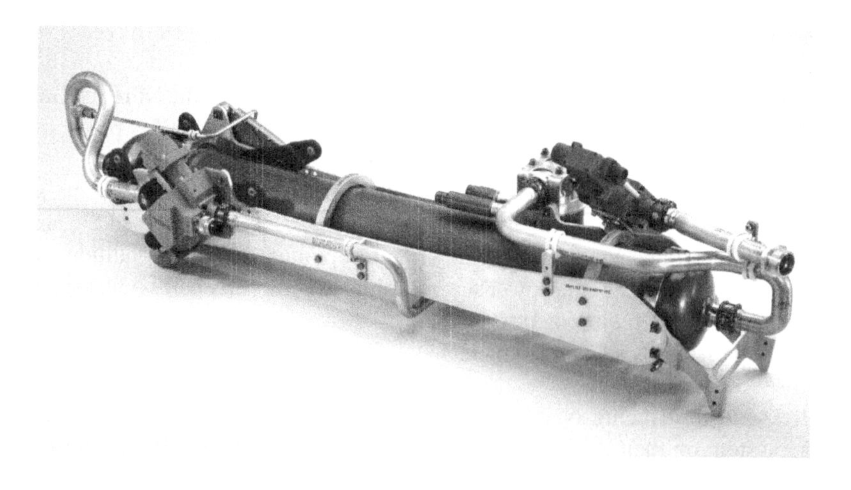

Figure 10.9 Airbus A320 Inerting system demonstrator (courtesy of Parker Aerospace).

New commercial aircraft either recently certified or in the certification process include the Airbus A380-800 super jumbo which entered service with Singapore Airlines late in 2007 and the Boeing 787 currently being prepared for first flight.

The former aircraft has no center fuel tank; all fuel being stored in the integral wing tanks which are aluminum. This version of the A380 does not have an inerting system; however, the freighter version does have provision for a center wing fuel tank. It is expected that this center tank will have to be inerted in order to satisfy the current fuel tank safety standards.

The Boeing 787 has a center fuel tank and a composite wing box and therefore has an inerting system sized to meet the requirements for the center tank plus both wing tanks. The pull down time requirement for this application depends upon the wing tank fuel quantities specified. With relatively full wing tanks, the pull down time will not be significantly more challenging than for a center-tank-only inerting system since the added ullage volume is small. On the other hand if worst case minimums are specified much more separator capacity will be needed to meet the pull down time.

10.3 Design Considerations for Open Vent Systems

All commercial aircraft have open vent systems and therefore the oxygen concentration of the ullage is dependent upon the air separation sub system performance, the variations in fuel tank quantity and the air flow into and out of the vent system throughout the flight.

Simulation techniques have been developed that can predict the inert status of fuel tanks during simulated missions.

On most flights the aircraft will take off with relatively small ullage volumes allowing the inerting system to achieve oxygen concentration levels of less than two or three percent, during the cruise phase. Therefore at the start of the descent, as outside air comes in through the vent system, the air separation system has good operational margins and oxygen concentrations will remain well below the 12 % limit.

'Short hop' flights with small fuel tank quantities become much more challenging for the inerting system. Starting at the 'pull down' limit of 12 %, the inerting system has insufficient time at cruise to reduce oxygen concentrations significantly and the descent case will be much more marginal for a given NEA flow capacity. This point is illustrated in Figure 10.10 which shows simulated inerting system performance for long-range and short-range flights for the same aircraft with the same separator capacity.

The long-range flight begins by burning center tank fuel through the climb and initial cruise. During thus time the separators are able to pull down the oxygen concentration level so that the tank ullage is almost pure nitrogen by the end of the cruise phase. This serves as a buffer during the descent when the separator system cannot keep up with the vent inflow and the peak oxygen concentration remains below the inert status limit.

The short flight begins with an empty center tank so the separator system has to work harder to lower the oxygen concentration. The short cruise duration adds to this problem and descent begins with a much higher oxygen concentration than for the long-range flight. In this example it shows the oxygen concentration level exceeding the limit during the descent suggesting that the separator capacity is insufficient.

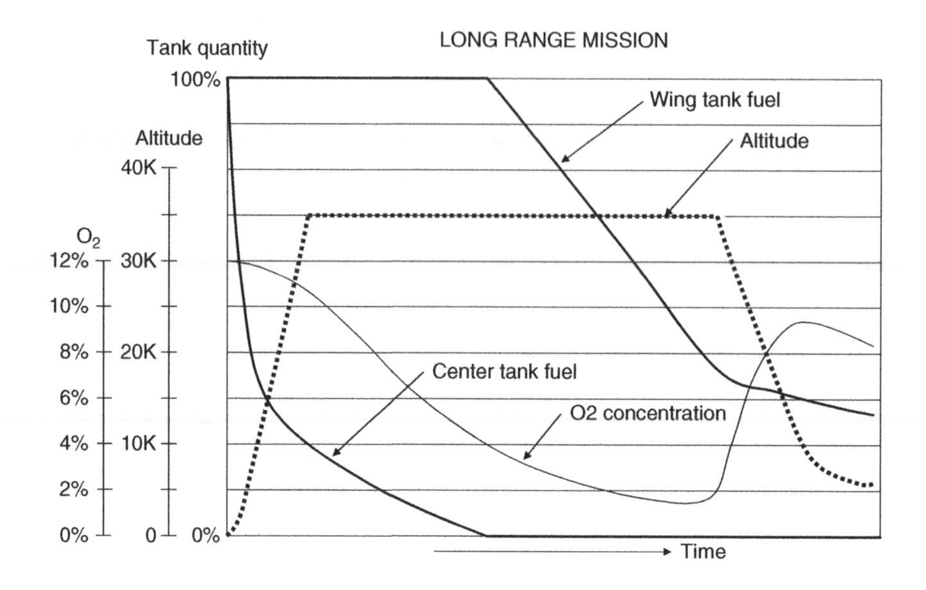

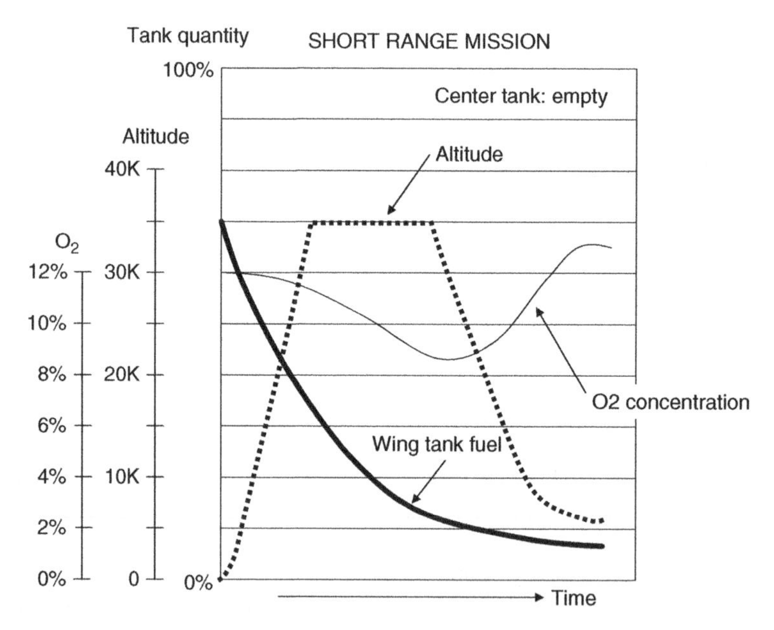

Figure 10.10 Simulation plots for long-range and short-range missions.

10.4 Operational Issues with Permeable Membrane Inerting Systems

10.4.1 Fiber In-service Performance

Permeable membrane air separation technology has been established as the system of choice for fuel tank inerting for both the military and commercial aircraft industries. For the commercial aircraft industry this technology is still relatively new and little long-term information about the operational capabilities of this type of system is known at this time.

One of the major concerns is the operational life and reliability of these new air separator modules which are the most critical functional aspect of the system. Factors that can potentially affect the performance and reliability of these plastic fiber assemblies include contaminants carried in via the air supply. Of particular concern is the effect of ozone which could cause serious corrosion of the fibers. The effects of other contaminants such as water vapor, oils and particulate are less well understood.

As a result, the newly fielded inerting systems protect their separator modules via filters and ozone converters in order to prevent early performance deterioration which could prove to be expensive in terms of operational cost and maintenance.

In an attempt to stay ahead of the fielded systems in terms of operational exposure time substantial effort is being made on endurance testing of the various fibers available with varying degrees of contamination in order to obtain meaningful life information that can be applied to the in-service environment.

10.4.2 Separator Performance Measurement

The inerting systems being introduced into service today utilize oxygen concentration sensors to monitor the separator output NEA. This is for prognostic purposes to generate a history of separator performance over time and to be able to predict the useful life of the device.

The available technology used for oxygen concentration sensing is not considered intrinsically safe for installing in the tank ullage to assess the inert status of the fuel tank directly.

10.4.3 NEA Distribution

Another area that is receiving a lot of attention in aircraft systems test laboratories is the effectiveness of NEA distribution within the ullage volume itself. This is of particularly concern in thin wing tanks with supporting rib structure that tends to inhibit natural mixing of the ullage gasses. For the center-tank-only inerting systems this may not be as much of a problem since these tanks are usually of rectangular form.

11

Design Development and Certification

Aircraft fuel systems are both unique and complex with regard to their interaction with other aircraft systems and the task of executing a successful design, development and certification program can be extremely challenging to both the aircraft designer and the equipment supplier. The design, development and certification issues associated with this process are therefore considered to be a subject that warrants address and discussion in this book.

It is recognized here that most fuel system complexities and challenges apply primarily to the modern transport aircraft and military aircraft communities, however, the lessons that can be gleaned from this chapter are considered invaluable to the practicing engineer at all levels of program involvement and management.

The contents of this chapter address the methodologies that have evolved in fuel systems design, development, certification and in-service support that have occurred over the past two decades and their implications on program management methods and, more fundamentally, the way new aircraft are designed, developed, certified and supported in the field by the aircraft manufacturer and the equipment supplier community.

The design and development disciplines that have been successfully applied to modern aircraft fuel systems form the basis for the methods and processes described herein and how they can contribute to the reduction of program risk from the perspective of schedule and certification plan management and the achievement of a high level of system maturity at entry into service.

11.1 Evolution of the Design and Development Process

Aircraft system design and development methodologies have changed radically over the past 15 years. Prior to the 1990s the standard mode of operation for the aircraft designer was to develop the aircraft preliminary system design in isolation from the supplier community and to establish specifications for specific products and equipment which are then competed and suppliers of these equipments selected. At this stage, the supplier has no visibility as to the aircraft mission or operational expectations and, as a result, provides a product solution based

Aircraft Fuel Systems R. Langton, C. Clark, M. Hewitt, L. Richards

solely on the information provided in the product specification. Schedule realities typically mean that specification requirements must change as the aircraft design evolves and as a result a renegotiation of the contract between the aircraft manufacturer and the equipment supplier with the attendant schedule delays becomes necessary to introduce these design specification changes into the various products.

The complexity of the aircraft fuel system and its interaction with so many of the typical aircraft's system and structural providers accentuates the difficulties of operating a procurement strategy in this way and the result of this component-based equipment procurement approach leads to an increase in the probability of latent operational problems emerging after the aircraft has entered service. In today's highly competitive marketplace, the ability to deliver 'Maturity at Entry Into Service' (EIS) is critical. The cost of failing to meet this obligation to the supplier, the airframer and the operator can be enormously expensive from both an economic and reputation perspective. Immaturity leads to the need for fleet hardware and software upgrades, delays, cancellations and dissatisfaction in both the flight crew and ground maintenance communities.

The following figures, 11.1 and 11.2 show Gant Charts illustrating some of the shortcomings of the component versus the system approach to the procurement process during the design and development phase of the program.

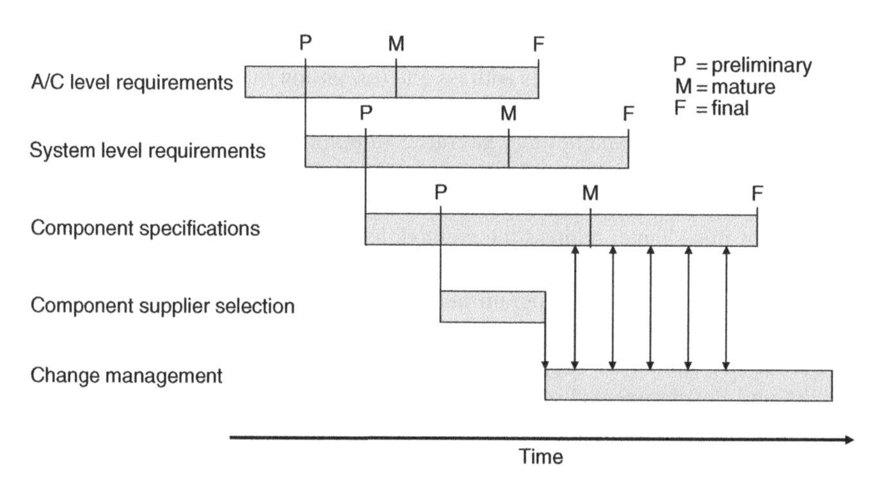

Figure 11.1 Component-based procurement.

As already stated, the main shortcomings of the component procurement approach is that the supplier community has no insight into the aircraft design objectives seeing only the specific design requirements and constraints provided in the specifications. There is no incentive for the supplier community to take an interest in or accept responsibility for aircraft-level issues that are not reflected within the specifications themselves.

As the aircraft and system designs evolve inevitable changes will occur, which can lead to serious program delays, additional program development costs and, in many cases, deterioration in the relationship between the supplier and airframe manufacturer. The procurement approach now in vogue with many aircraft manufacturers is to contract with competent suppliers capable of delivering and managing complete systems or major 'work packages' comprising major groups of products and equipment that operate as an integrated functional entity. In this new

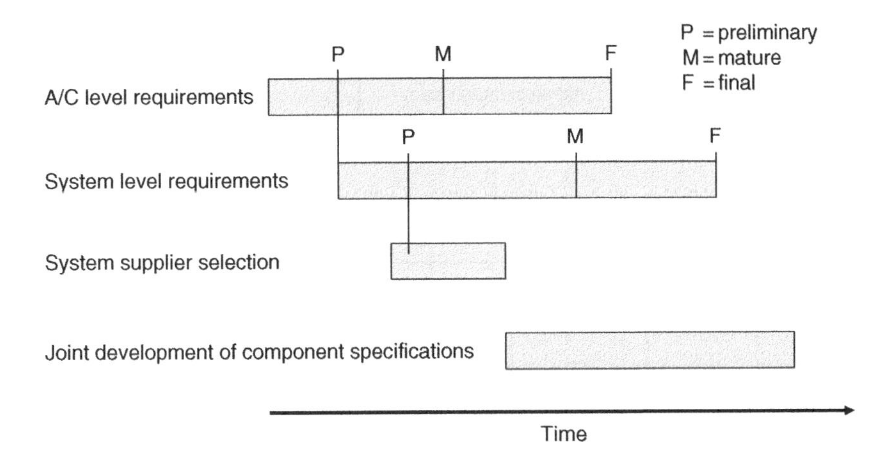

Figure 11.2 System-based procurement.

approach, the successful bidders are typically required to take responsibility for the functional performance of the equipment in the aircraft environment together with contractual constraints which include maintenance cost and weight guarantees, MTBUR and MTBF guarantees for major components and for the system as a whole.

The Gant chart of Figure 11.2 illustrates some of the benefits of this procurement methodology. Here the challenge is to select a supplier and establish contractual terms and conditions for the procurement of a complete system well before the system design has been finalized and for the equipment supplier and airframer to work together as an integrated team to develop, jointly, the final system solution and specifications for all of the components in the system.

Following supplier selection the fuel system equipment supplier, along with other major systems and structural suppliers, form integrated product development teams co-located typically at the aircraft manufacturer's facility.

Figure 11.3 shows a typical aircraft and system design environment showing how the on-site teams share access to the evolving aircraft Computer-Aided Design (CAD) database. This same database is utilized by the aircraft manufacturer's production team to evolve along with the aircraft design itself, the production tooling requirements. Component design evolution within the supplier community is coordinated, on a real time basis, via the internet with the operating units of the various system suppliers as shown in the figure.

This arrangement has the effect of bringing the supplier community into the aircraft design and development business as risk sharing partners with obvious benefits:

- The supplier teams are now actively participating with the airframer engineering team in the design of the aircraft.
- Direct participation in the design and development process by the supplier on-site teams provides a powerful motivation of the individuals involved to take an interest in the global issues associated with the aircraft design as it evolves.
- The added level of interest and knowledge of the supplier community regarding aircraft-level design issues results in a major benefit to the program in terms of cost, risk and maturity at entry into service.

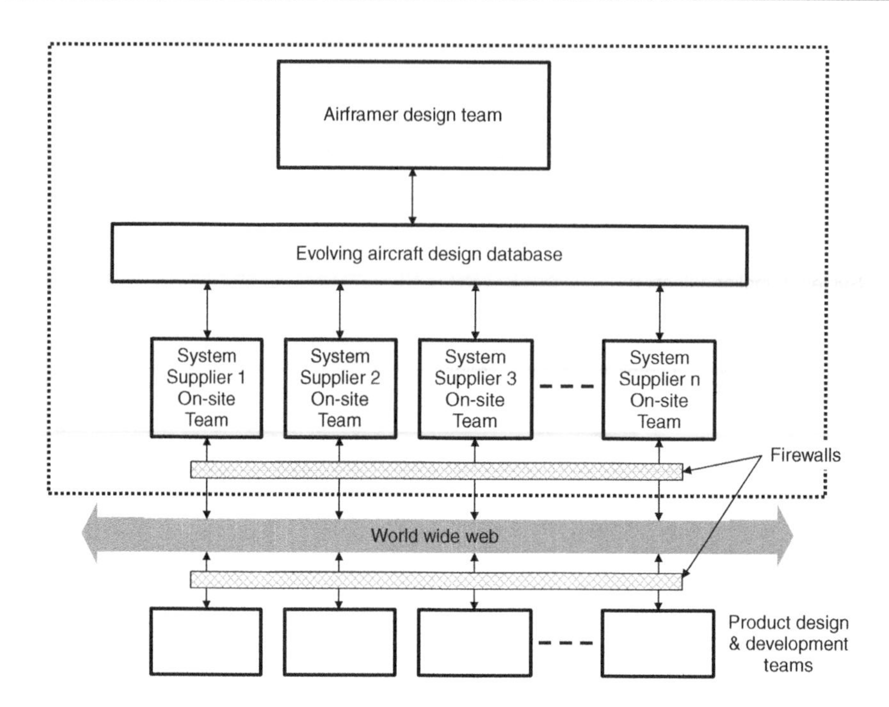

Figure 11.3 Typical system design environment.

It is important for the executives of both the airframer and supplier communities to 'Buy in' to this new integrated approach to aircraft development and to actively encourage their engineering teams to accept this new approach to aircraft design and development in order to fully realize the potential benefits that can significantly reduce cost and risk to the program.

The primary benefit that can be brought to bear from this new 'joint working' arrangement between the airframer and supplier engineering teams is the feeling of 'ownership' that develops among the participating on-site teams which often manifests itself in a recognition that we (all the on-site engineering teams) are designing an aircraft together and the prevailing attitude becomes 'What's best for the aircraft' in the decisions that are made in the early design phase.

This 'aircraft design team mindset' can realize substantial benefits to the program in terms of the level of effort applied by the team members and the quality of design solutions that evolve as a result of the positive attitude that develops among the supplier participants. The overall effect of this 'joint working/team building' phenomenon is to create an environment wherein the probability of achieving mature system functionality at entry into service is maximized.

An important caveat that must be recognized in this situation is the need to have the right people assigned to the integrated project development teams. When team member selection is done well the result is a well knit and effective team. This can be likened to 'Tissue typing' where rejection of those that do not fit can result in dysfunctional behavior that can seriously impair (or even destroy) the effectiveness of the unit.

The integrated design methodology outlined briefly above forms the basis for the design, development and certification procedures that are addressed here in this chapter. For additional reading on this topic the reader is referred to a paper presented to the International Council on Systems Engineering (INCOSE) symposium in Brighton UK in 1999 entitled 'Collaborative methods applied to the development and certification of complex aircraft systems' reference [21].

This paper discusses the new methodologies outlined above as they applied to the development of the Bombardier Global Express ™ and Airbus A340-500/600 aircraft fuel systems.

11.2 System Design and Development – a Disciplined Methodology

Complex integrated systems typical of modern transport aircraft fuel systems demand a disciplined approach to the overall design, development and certification process in order to ensure safe functionality of the system in service. A secondary benefit of using a formal, disciplined methodology is that it provides a working environment that supports the need for a high degree of functional maturity of the system, as installed in the aircraft, at entry into service. While safety will always be the primary issue associated with aircraft design, development and certification, the issue of maturity remains one of the most important requirements from the perspectives of all of the major stakeholders including the equipment suppliers, the aircraft manufacturer and the end user of the aircraft.

Within the past decade, one of the most practiced methodologies for complex aircraft systems development and certification is that described in the Society of Automotive Engineers (SAE) Aerospace Recommended Practice (ARP) 4754, reference [1]. This document is entitled 'Certification considerations for highly integrated or complex aircraft systems'.

The specific methodology described in this document was developed initially following a request from the FAA to the SAE to establish and document a system-level design and development procedure that aligns the criticality of function with the safety provisions necessary to satisfy the functional safety and certificability of the evolved design.

While the primary focus of the initial assignment was aimed at systems with substantial electronics and software content, it can also be applied to the development and certification of other systems; however; when applied to relatively simple systems, the formality of the development structure, processes and documentation may be reduced substantially.

The methodologies promoted in ARP 4754 are supported from a detailed safety process perspective by a second Aerospace Recommended Practice, namely ARP 4761, reference [2], entitled 'Guidelines and methods for the safety assessment process on civil airborne systems and equipment'.

Together these two advisory documents form the basis for the definition of an integrated process discipline illustrated by the safety assessment process model overview shown in Figure 11.4.

This figure a simplified version of the process described in ARP 4754 and illustrates the inter-connected activities that are involved with the design, development, implementation and certification of a generic system. The diagram is divided into the safety assessment process on the left and the system development process on the right.

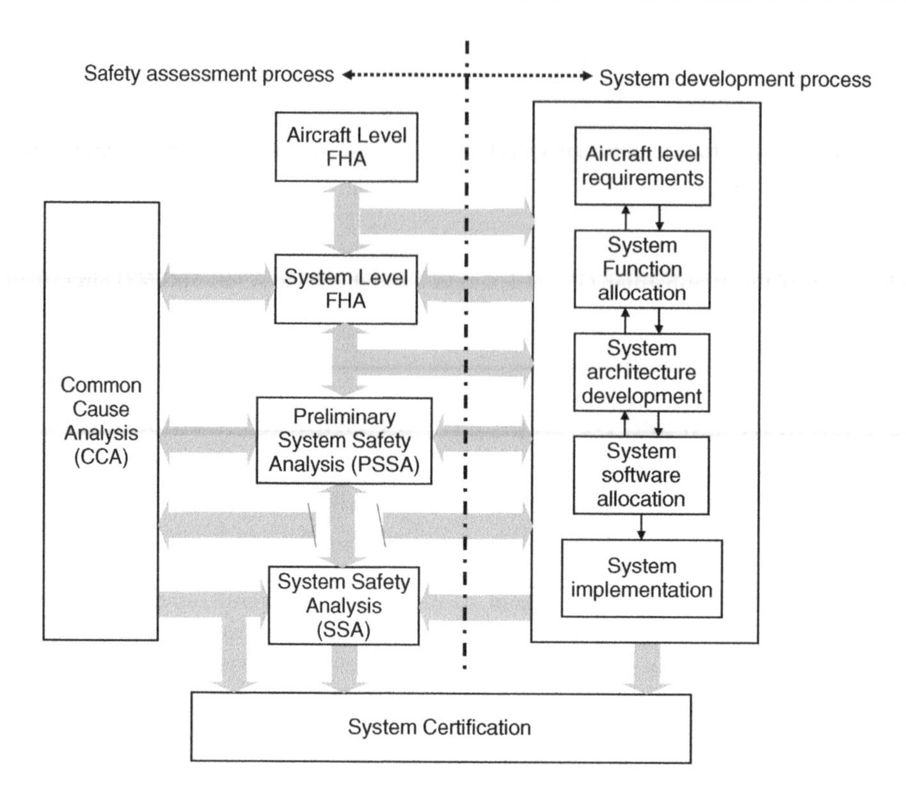

Figure 11.4 Safety assessment process model overview.

The Functional Hazard Assessment (FHA) is the starting point and is addressed initially at the aircraft level and subsequently at systems equipment levels to define the failure conditions and their associated effects from which safety requirements can be established. A key aspect of this activity is the allocation of failure classifications which are defined as follows:

A	Catastrophic	Probability of occurrence $<10^{-9}$ per flight hour
B	Hazardous/Severe	Probability of occurrence $<10^{-7}$ per flight hour
C	Major	Probability of occurrence $<10^{-5}$ per flight hour
D	Minor	Probability of occurrence – none defined
E	No safety effect	Probability of occurrence – none defined

In this methodology the letters A through E assign a 'Design assurance level' to the requirement or function which, in turn establishes the degree of rigor that is allocated to the design, validation and/or verification processes and associated documentation.

As the system design proceeds from the aircraft-level through the system and sub-system levels to the component level, the supporting safety assessment processes including the Preliminary Safety Assessment (PSSA) and Common Cause Analysis (CCA) activities are updated through many iteration cycles as the system design and architectural issues are finalized and ultimately documented in the System Safety Analysis (SSA). This end product of the safety assessment process becomes the primary supporting documentation that supports the application for system certification.

The brief outline presented here does only minimal justice to the process methods and practices covered in detail in the SAE Aerospace Recommended Practices (ARP's) referenced herein, and the reader is encouraged to review the actual ARP material to ensure a complete understanding of the issues that need to be addressed in a disciplined manner to ensure the introduction of a safe, certifiable and mature system into operational service.

The following paragraphs provide a different perspective of the design, development and certification process in order to give the reader a more pragmatic view of the issues involved in the deployment and management of the 'Joint working' integrated development teams that are commonly used today in the 'risk-sharing' contractual environment. Here the aircraft manufacturer's engineering team works closely with the selected system supplier's engineering team in the evolution of the design and the establishment and documentation of system safety-related decisions that result.

11.2.1 The 'V' Diagram

The most common illustration of the system design and development process is the 'V' Diagram which is shown in its simplest form in Figure 11.5 with top-level requirements established on the upper left based on operational requirements and airworthiness regulations. Descending down the left hand side the requirements are broken down first into a subsystems-level and then into a component-level.

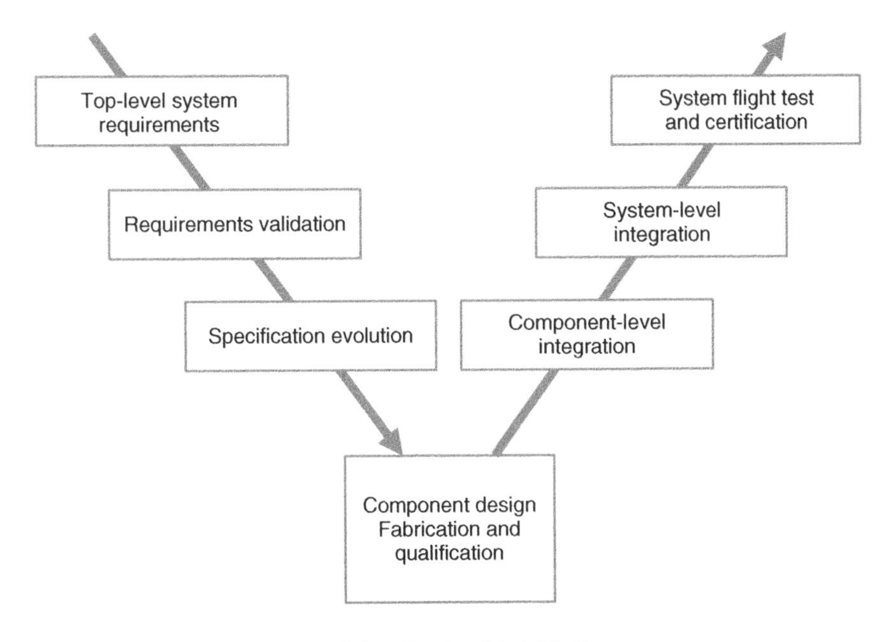

Figure 11.5 The simplified 'V' diagram.

At the base of the 'V', specifications are available to support the design, fabrication and qualification of the hardware (and software where applicable) associated with specific components.

Ascending up the right side of the 'V' the components are integrated in stages and their performance evaluated via verification testing to generate evidence that the requirements at the various levels of integration have been met. Eventually a fully integrated system is available for rig and/or ground and flight testing in the aircraft. Documentation of certification compliance is assembled during this process and forms the basis for the formal application for type certification approval.

The most significant aspects of the process (as also emphasized in ARP 4754) are as follows:

- At the earliest stage the functional requirements must be assessed from a safety criticality viewpoint.
- The left hand side of the 'V' focuses on the *validation* of requirements in order to ensure that the requirements are both correct and complete.
- The right hand side of the 'V' focuses on the *verification* of the requirements in order to ensure that the design solution meets the requirements as defined.

This differentiation between validation and verification is a key feature of the process and should not be confused with alternative methodologies which often combine the validation and verification process using the term 'V & V'.

The activities at the top left hand side of the 'V' address the fundamental issues associated with the vehicle application and its associated mission and as a result the primary focus here is in the design drivers that address functional safety and operational requirements that must satisfy the airworthiness regulations dictated by the civil (commercial) and/or military authorities. These issues are addressed in detail in Chapter 2.

11.2.2 Software Development

Modern fuel measurement and management systems rely heavily on software for their functionality and the design and development processes involved are usually driven by a combination of the customer requirements and, in the case of a commercial aircraft, RTCA document DO178B, reference [6]. For the experienced system supplier most of the key software processes will have been optimized over multiple programs of varying complexity and are thus well able to support the development of software in the aggressive program timescales that are typical of modern major aircraft programs

The high level requirements management and the extensive traceability required by DO178B should be fully supported by the use of commercially available but fully qualified tools.

The software life-cycle processes should also mirror the DO178B processes which are influenced and supported by a number of tools to support:

- Requirements Management
- Software Design Environment
- Configuration Management
- Code Coverage
- Emulators for the target processor (hardware/software integration support).

The following paragraphs summarize the process activities associated with the software design and development for the typical fuel measurement and management system.

Planning
The software planning process activities are primarily associated with the production of the D0178B planning documentation which defines the software development approach and specifies the necessary infrastructure requirements. This process leverages from the maturity of established practices and experience in the use of the software development infrastructure toolset. The following is an example of the software development and management plans and documents that ensue:

- Software Development Plan
- Software Verification Plan
- Software Development Standards
- Software Configuration Management Plan
- Software Quality Assurance Plan
- Software Certification Plan
- Software Life Cycle Environment Configuration Index

Requirements
The software requirements process develops the software high-level functional, performance, interface and safety related requirements. The key inputs to the process are the system requirements allocated to software components, the target platform interface requirements and the system architecture.

Derived Requirements
The importance of the close management of derived requirements cannot be over-emphasized. Software derived requirements are generated at both high and low levels during the requirements generation and design processes. At the lower, as well as higher levels of the specification, dedicated tools are used to capture and manage software derived requirements and their associated justification/rationale. The validation/verification of software derived requirements is of particular importance in achieving the goal of a mature system.

Infrastructure support for the establishment, maintenance and reporting of requirements traceability is viewed as a key consideration. Manually maintaining traceability is an error prone and highly labor intensive task particularly in complex systems where tool support is considered essential.

Design
The software design process is applied to the outputs from the high-level software requirements. The design activity is highly iterative and involves the development of low level requirements and the creation of the static and dynamic software architecture. One of the principle steps in the design activity is the development of the object architecture which will be employed to deliver the functionality specified in the high level requirements. The high level sequence diagrams are elaborated to specify a detailed sequence of processing steps in the context of the object architecture.

Coding

The software coding process involves the implementation of the low level requirements and the software architecture in a target compatible language. C language is frequently used for the software implementation on all processing platforms. A qualified verification utility should be used to automatically verify compliance with the coding standards.

Integration

The integration process encompasses software / software integration and hardware / software integration activities. Preliminary software integration activities are performed on a development platform. Advanced software and hardware / software integration activities utilize representative or final target hardware. The process activities include the linking of all components necessary for the application software to load and execute in the target environment.

Verification

A rigorous and disciplined approach to the verification process is critical. The process uses three techniques: reviews, analyses and testing. The objective of reviewing and analyzing life-cycle data is to detect and report errors introduced during the development processes (Requirements, Design and Coding). The objectives of testing are to demonstrate that the software meets its functional and performance requirements and that, with a high degree of confidence, errors that could lead to unacceptable failure conditions have been removed. It is recommended that the software development program employs an incremental adaptive approach to the verification of the software. The incremental aspect of the approach involves the progressive population of the Software Verification Results as a means of demonstrating certification compliance on an incremental basis. The adaptive approach is applied once a baseline has been established and involves the selective application of verification activities based on the impact of changes applied to a baseline

11.3 Program Management

Effective program management is critical to the success of any new aircraft development program since it plays important roles in many key aspects of the process including:

- managing and mitigating risks
- decision making
- schedule management
- budget management
- recovery planning following 'surprises'
- ensuring good communication throughout the team.

The soft skills associated with the assembly of an effective team that works well together can make an enormous contribution to the success of the program. This is particularly important in the situation where supplier teams are co-located at the aircraft manufacturer's facility and may therefore be required to live away from home for long periods.

It is equally important for the equipment supplier program managers to have a good working relationship with the aircraft manufacturer's staff including both the technical/engineering specialists and the aircraft program management people since a successful outcome to the design, development, certification and deployment process cannot be achieved without a commitment to full cooperation from everyone involved.

11.3.1 Supplier Team Organization

The organization structure described here is just one example based on the design and development of a typical fuel system where the all of fuel system equipment is procured from a single source. In this situation it is typical for the supplier to have a number of different operating units that focus on specific equipment groups for example:

- fuel gauging and avionics products
- fuel pumps
- fuel valves and related fluid-mechanical products.

In addition to the product-focused operating units it is necessary to utilize a systems engineering team to provide coordination, communication and interpretation of aircraft and system-related issues that evolve throughout the design and development phase of the program. Many of the systems support team may be located on site at the aircraft manufacturer's facility(s) from the initial concept study phase through the end of the flight test phase. Staffing levels will vary as the focus of activities move from conceptual/system-level to equipment specification definition to the product design, fabrication and qualification phase where the center of gravity of the effort moves to the component level. As the component effort moves into the integration phase where verification testing takes place on system test rigs, aircraft 'Iron bird' test rigs and eventually onto the aircraft itself, the focus of effort will move back to the aircraft manufacturer site(s).

Figure 11.6 shows a typical organization chart showing the reporting relationship between the on-site team and the various operating units as well as how the program manager relates to both the supplier operating unit teams and the aircraft manufacturer's team.

The fuel system program manager typically reports to a 'fuel systems integrator' who is the focal point for all technical communications between the supplier and the aircraft manufacturer. In addition the program manager via his responsibilities for schedule, budget and commercial issues must also be accountable to the aircraft manufacturer's program office.

In view of the broad responsibilities of the program manager covering both the aircraft/systems level, it is typical to assign an 'on-site' team leader to manage day-to-day events at the aircraft manufacturer's facility leaving the program manager the freedom to coordinate and communicate with the product operating units and their management teams. This becomes extremely important where technical decisions affecting product specifications must always try to recognize what is the best solution for the aircraft within the constraints of the overall budget.

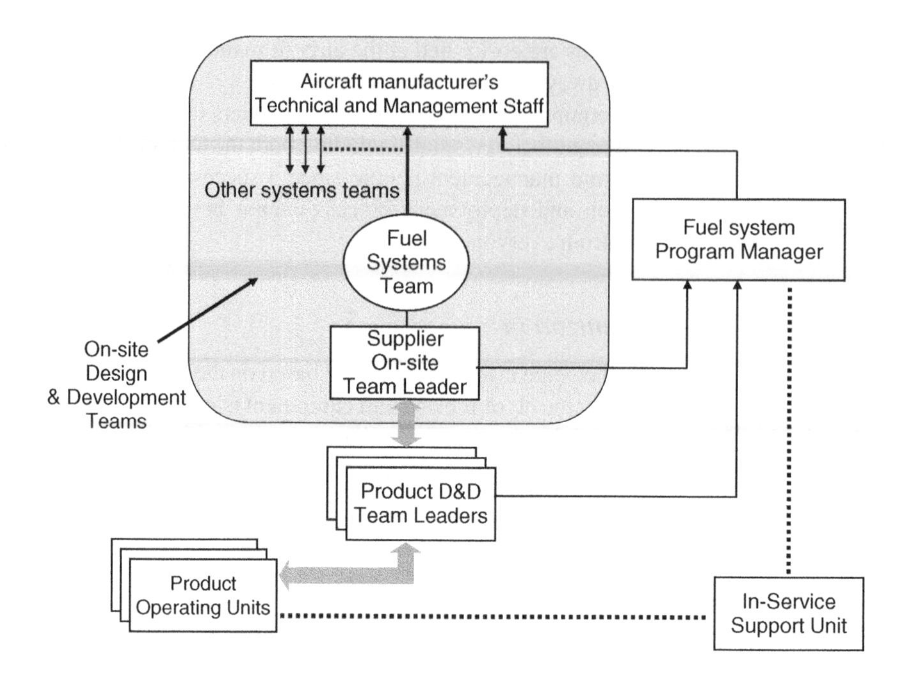

Figure 11.6 Typical program management structure.

An important aspect of the organization structure depicted in the Figure 11.6 relates to the need to involve the in-service product support organization from day one. In many organizations this operational function is provided by a centralized unit that becomes the primary contact with the commercial airline community or the military logistics organizations.

Involving the in-service experts as early in the program as possible can play a major role in maximizing the level of maturity of system functionality at entry into service.

11.3.2 Risk Management

Managing risk is an important aspect in the design, development and certification of any complex integrated system and there are a number of formal methods practiced today.

The key aspect of managing risk is the pre-emptive nature of the process, whereby potential risks are identified at the outset and mitigation plans put in place to continually evaluate and minimize their potential impact.

The first step in risk management is risk identification and this can involve a number of topics including:

Technical

Technical risks may be related to relatively advanced technology proposed as part of the system design solution or technical design challenges that may require special purpose test rigs to verify acceptability. In the area of electronics and software, issues such as the selection of the appropriate microprocessor and the timely availability of critical software development tools may pose serious risks to the program.

Schedule
In many program situations, the schedule defined for the aircraft may be aggressive relative to the design and development tasks that must be accomplished and this in itself can pose a significant risk to the success of the program. There also may be dependencies on the timely availability of special design support tools or equipment that could be a serious risk to achieving the planned schedule.

Inter-system Support
The typical fuel system has numerous dependencies upon the performance of interfacing systems including, for example, the propulsion, power generation, inertial navigation and flight management systems. Systems architectural issues such as the use of modular integrated electronics can pose a significant additional dimension of risk to the program.

The above are typical examples, however, each program will have its own unique risks that need to be identified as early in the program as possible. The risk management effort then involves the following:

- identification and tabulation of each specific risk;
- assessing the probability of occurrence of each risk from a qualitative judgment perspective;
- assessing the impact to the program of the effect of the occurrence of each risk in terms of cost schedule, maturity etc.;
- managing the execution of the mitigation plans.

Once the salient risks and mitigation plans have been identified, a weighting factor can be applied to each risk probability and impact-effect which, when multiplied together will clearly identify the priority of importance for each risk.

Figure 11.7 shows a risk analysis chart showing a number of hypothetical risks, their probability and impact assessments and weighting factors.

Risk description	Mitigation plan	Risk Probability H, M, L	Risk Impact H, M, L	Weighting		Total score priority ( )
				Prob 4,2,1	Imp 8,4,2	
Late availability of special development tools	Assign dedicated Support engineer To supplier	M	M	2×4		8 (3)
Effectivity of water management system	Build special Purpose test rig	H	M	4×8		16 (2)
Availability of selected microprocessor	Generate alternative design for back-up device	H	H	4×8		32 (1)

Figure 11.7 Risk analysis chart example.

In the example shown here, each risk is described along with a brief description of the mitigation plan. In the remaining columns, a rating defined as High, Medium of low (H,M or L) is assigned to each risk for both the probability of occurrence and the impact to the program. Weighting factors are assigned as shown for both the probability and impact ratings and a total score for each risk calculated as the product of weighting factors assigned for each risk. While these assignments are relatively subjective, it becomes relatively easy to prioritize each risk based on the total score. Once this has been achieved, the task is to maximize the effort of risk mitigation and oversight on the top priority items.

11.3.3 Management Activities

Effective day-to-day management of complex programs typified by the modern transport aircraft fuel system requires a focused, dedicated team operating closely with the aircraft manufacturer's team of design specialists. It is critical that the program planning process be a joint activity including the formalization of the overall process, risk management and maturity management plans (more about maturity management later). These joint efforts must cover the validation of requirements and the establishment of minor and major program milestones to the point where both the supplier and aircraft manufacturer communities feel that they each agree and own the plan.

An important part of the complete design, development and certification process is the holding of progress reviews both formal and informal and from design team level up to senior management level. These reviews represent a key communication tool for keeping everyone involved in the program, on both the aircraft manufacturer and system/equipment supplier sides, fully informed of all of the relevant issues and decisions that affect the program.

Program progress reviews involve the day-to-day reviews that take place from the engineering team level to the formal reviews that identify key program milestones involving both team-level and senior management participation by both the aircraft manufacturer and equipment/system supplier organizations.

Figure 11.8 shows the major program phases from the bid and proposal to the in-service support efforts.

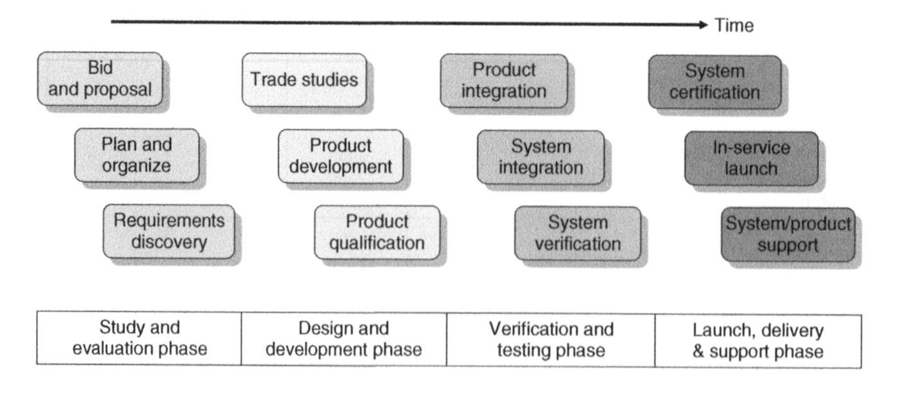

Figure 11.8 Major program phases.

Typically there are a number of formal reviews associated within each phase that serve as a 'stage gate' which must be completed before progressing to the next phase of activities.

The study and evaluation phase (sometimes referred to as the 'joint definition phase' since it involves joint participation by the system supplier and aircraft manufacturer) is a largely iterative process that is illustrated by Figure 11.9 which shows how the process begins with the top-level requirements and leads to the establishment of a system design description after several iterations.

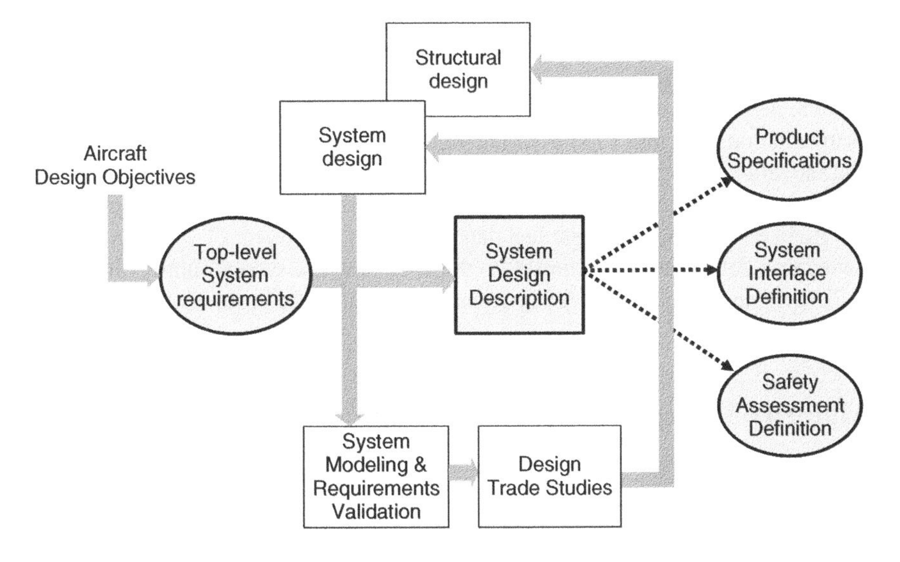

Figure 11.9 Study and evaluation phase.

This system design description includes a complete functional definition including the redundancy and signal partitioning strategy that is required to meet reliability, operational and safety goals. Separation of hardware and software functions is also established at this stage.

From the system design description the product specifications, interface requirements and system safety assessment can then be captured and documented.

Within the design and development phase two major reviews occur, namely the Preliminary Design Review (PDR) and the Critical Design Review (CDR). Each of these formal reviews are held at both the system level and at the equipment level for each product grouping, e.g. system, fuel gauging and management sub systems, fluid mechanical components. The latter group of products includes all of the fuel pumping and handling equipment as well as vent system products.

The detailed design (hardware and software) takes place in this phase of the program together with the identification of any special test equipment that will be required to support qualification and functional testing at the component and sub-system levels.

During this phase the focus of activity moves towards the system supplier and it is important here to maintain the concept of 'joint working' by having the aircraft manufacturer's engineering team members actively participate with the system supplier at his facility.

The verification and testing phase of the program is where integration testing of the various components begins to confirm that all of the design requirements have been met and appropriate test evidence is obtained to support the certification process.

As this phase of activity continues, the focus moves back to the aircraft manufacturer's domain as the level of integration is extended to include more of the other aircraft systems that the fuel system interacts with. This will include more complex test facilities including, perhaps an 'Iron bird' test facility capable of representing a simulated environment equivalent to a real aircraft.

When the aircraft becomes available, it is intended that any functional issues have been identified and fixed since identifying and solving problems on the aircraft is a very expensive and time-consuming activity.

The launch, delivery and support phase of the program involves the finalization of certification documentation and close support of the flight test program which is required to generate functional and reliability evidence in a timely manner. Here the system supplier may be required to provide on-site support at the flight test facility and actively participate in generation of flight test cards, flight de-briefing and flight test data analysis.

Following first deliveries of the aircraft to the operators, the system supplier is expected to provide engineering support as necessary to ensure that any field problems arising are quickly assessed and resolved.

11.4 Maturity Management

Delivery of mature functionality at entry into service (EIS) is one of the most important attributes of the system design, development and certification process.

From a fuel system perspective, the level of complexity and functional inter-connectivity with other aircraft systems makes the achievement of maturity at EIS particularly challenging.

The importance of maturity at EIS becomes evident when the penalties associated with immaturity are evaluated. Problems discovered after entry into service can result in:

- dispatch delays and/or cancellations
- inability to meet mission objectives
- increased workload for both the cockpit and ground support crews
- increased operating costs
- operator and end-user dissatisfaction
- expensive and time consuming equipment retrofits
- reduced operational safety
- loss of reputation within the industry for both the system provider and the aircraft manufacturer.

Effective maturity management requires a team effort from both the system supplier and the aircraft manufacturer. A successful maturity management plan emphasizes the need for a proactive approach that addresses problem avoidance through the linkage of past lessons learned and known program risks to the validation and verification process itself as indicated by the diagram of Figure 11.10.

The figure shows how the maturity management activities tie into the program management process represented here by the 'V' diagram. Lessons learned and identified risks generate

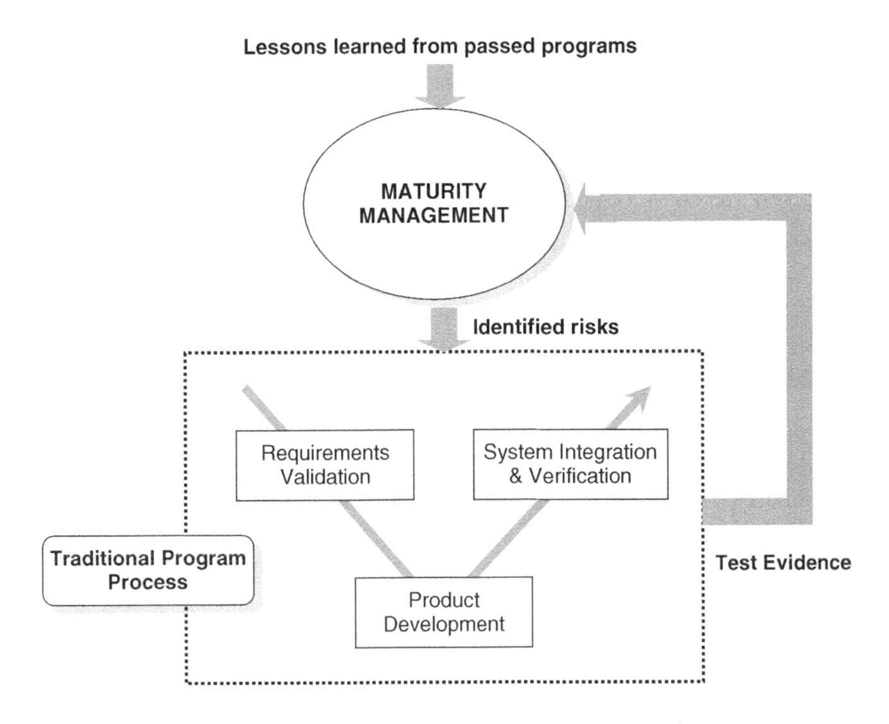

Figure 11.10 Maturity management process overview.

specific tasks that ultimately lead to the generation of test evidence in order to provide supporting proof that each line item has been accommodated within the system and/or equipment design. This test evidence can then be used to formally close-out each established maturity line item.

Maturity management can be considered to be similar to quality control. In each case it is desirable to 'design in' rather than 'inspect or verify out'. The latter approach is essentially reactive in nature, typically resulting in late changes, schedule pressure, patched software code and unnecessary complexity.

The proactive approach brought about via the application of a formal maturity management methodology focuses on problem avoidance applying anticipatory effort *before* problems arise.

Perhaps the most controversial issue regarding maturity management concerns the slippage of major requirement decisions that can have a significant negative impact on schedule. This is where it becomes crucial to have active participation in the maturity management process by the aircraft manufacturer. It is all too easy for the aircraft manufacturer's engineering team to hold off on making key requirements decisions early in the program such that the system supplier's schedule is put in jeopardy. If sufficient attention is not paid to the recognition and recovery from this type of event, schedule slippage, and most importantly the system test exposure time can be reduced to the point where system functional maturity prior to completion of integration and flight testing can be substantially impaired.

One approach to addressing this problem is to recognize that missed major milestones by either the system supplier or the aircraft manufacturer's teams will result in a continuing decline in potential maturity status until the milestone is achieved. By linking this maturity status factor to information regularly available to senior management of both the system supplier and aircraft manufacturer's senior management via some formal reporting mechanism, appropriate attention to the issue can be quickly brought to bear.

In order to have the opportunity to deliver functional maturity at EIS requires:

- a development schedule that provides the opportunity to expose the equipment (hardware and software) to the aircraft environment (simulated and/or real) for a sensible amount of time, i.e. several weeks for continuous evaluation and problem solving;
- key decisions affecting design requirements must be made in accordance with an agreed schedule to avoid late implementation of functional requirements either in hardware or software;
- a dedicated maturity manager reporting to an Executive Maturity Review Panel comprising senor management from both the system supplier and aircraft manufacturer communities. This Panel would have the responsibility to report on maturity trends on a regular basis (e.g. monthly) and with the authority to invoke corrective action activities with support from the executive managers of both the system supplier and aircraft manufacturer.

Maturity management should take place outside the normal day-to-day program management activities but should be reviewed regularly as a key part of a regular schedule of executive reviews which take place quarterly as a minimum. At this level the importance of maturity as a major program progress indicator is fully understood and therefore appropriate actions to keep maturity on plan should be forthcoming.

11.5 Installation Considerations

An important issue regarding fuel system design involves the installation design for the components and equipment. This is not usually an issue for the avionics equipment mounted in the avionics bay or in the pressurized area of the aircraft where environmental conditions are relatively benign. In the avionics bay, avionic module standards are used that clearly define unit installation requirements together with the associated environmental conditions that apply.

The more significant installation issues associated with fuel systems equipment are related to components, piping wiring harnesses etc that are located within or close to the fuel tanks themselves where the operational environment can require specific engineering expertise to avoid in-service problems that may arise as a result of the installation design.

The location of key components relative to the rotor burst zones is also a critical installation design issue and this was discussed previously in Chapter 2.

Since this is a 'systems' book a detailed treatment of equipment installation is considered to be out of the scope of the intent of this publication; however, a brief overview of some of the issues involved is covered in this section for completeness.

Figure 11.11 shows a Computer-Aided-Design (CAD) layout of the inside of a typical fuel tank showing the plumbing and equipment within the tank. As shown this installation is quite complex and demands careful attention to detail in order to ensure reliable functionality and easy maintenance.

Figure 11.11 Fuel tank plumbing and equipment layout (courtesy of Parker Aerospace).

The tank structure must be designed to handle inertia loads induced by the fuel under worst case accelerations. The structure must also be designed to allow migration of air along the upper surface. Fuel and water drainage along the lower tank surface must also be provided for. This is illustrated in Figure 11.12 which shows a sealed rib with stringers on the upper and lower surface.

The air vent holes are supplementary to the main vent system whose job is to ensure that tank pressure differentials never exceed some predetermined value usually between four and five psi. The addition of these vent holes ensures that the top portion of the tank allows free passage of air between rib bays.

The vent-hole size is typically a minimum of 0.1 inch and a sufficient number of holes are required to meet a total vent area requirement of typically >0.25 in^2.

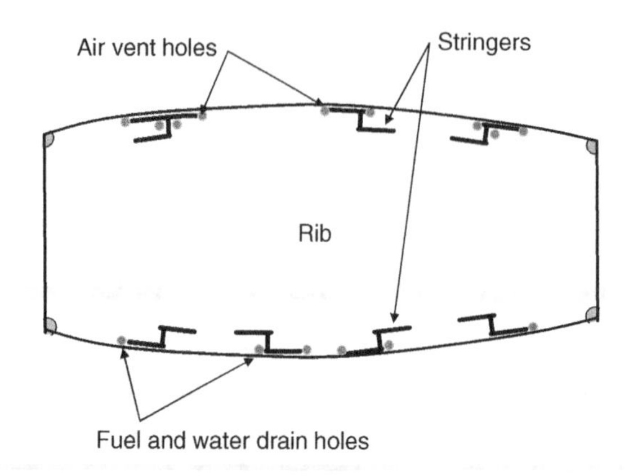

Figure 11.12 Air vent and drain holes.

A comprehensive drainage scheme is necessary to allow fuel and water to move unimpeded to the low point of the tank under gravity. This is necessary to minimize unusable fuel and to allow any water that settles out from the fuel to pool in an area where water drain valves are located. The equivalent minimum area requirement for fuel water drainage across each rib is typically 50 % larger than for the air vent requirement. Location of the drainage holes is selected to minimize the possibility of trapping water at the rib boundaries.

The piping design must take into account a number of key requirements:

- Tube sizing must ensure that fuel velocities are below 30 feet per second under worst case conditions.
- Clamping of rigid tubing should not result in installed stresses. Also induced stresses due to structural deflection in flight must be acceptably small.
- Relative movement between rigid tubes connected via flexible couplings should remain within the nominally acceptable tolerance of ± 3 degrees considering both installation tolerances and in flight deflection
- Routing of tubing should ensure that no unwanted siphoning can occur.

Location of tubing in relation to control valves should discourage accumulation of water at the valve that can freeze during flight with the possibility of rendering the valve inoperative.

Bonding

Bonding is a major issue concerning the installation of in-tank fuel system equipment from the perspective of ensuring fuel tank safety following a lightning strike. The purpose of controlling bonding standards is to ensure that lightning-induced currents can pass through in-tank equipment to the aircraft structure without generating significant voltage differential to cause arcing.

Standard design requirements exist that address equipment installation interfaces with aircraft structure. These requirements call out chemical treatments and surface finish standards for conductive components where they attach to structure in order to minimize the electrical resistance at these interfaces.

11.6 Modeling and Simulation

Modeling and simulation tools are used extensively throughout the design and development process. Modeling is necessary to provide valuable insight into the behavior of the highly non-linear aspects typical of fuel systems and their functions. They are also invaluable in the support of the system design, development and verification processes from the concept development phase to the completion of systems integration and aircraft certification.

The following is a summary of the problems that are typically addressed during the design and development process that rely heavily on the use of simulation and modeling to provide functional insight and thus to minimize the risk associated with engineering design and development.

- system requirements validation
- refuel panel ergonomics evaluation
- fluid network design and analysis including pump location optimization
- verification of refuel times
- worst case vent system sizing (emergency descent case)
- fuel gauging probe array optimization and measurement accuracy analysis
- unusable and ungaugable fuel analyses
- fuel management algorithm evaluation including BIT validation and verification
- aircraft environment simulation for system equipment verification, integration and certification.

A significant aspect of the use of dynamic models and simulations during the conceptual phase of the program is that models do not have to be 'real time' models. In fact it is desirable that models used to study system dynamic behavior run significantly faster than real time so that many missions, operational situations, failure modes etc. missions can be evaluated within a short space of time. The power of today's computers make it possible to develop complex, high fidelity models that can run 20 or 30 times faster than real time thus allowing the evaluation of many simulated test cases in a short space of time.

As implied from the above discourse, both static and dynamic models are used in the design, development and certification process depending upon the objectives of the evaluation task.

Presented below are specific fuel system modeling examples:

Tank Analysis Models

One of the biggest challenges in fuel system design is having to cope with the complex geometries of the fuel storage system. Tank geometry models are required to compute fuel surface locations for various fuel quantities, aircraft attitudes and g forces.

Fuel gauging probes location analysis tools are required that can optimize the probe array (numbers and locations of probes) for the various combinations of fuel quantity and aircraft flight conditions.

Most aerospace systems suppliers have proprietary programs as an essential part of their engineering portfolio. Such tools typically have the following capabilities:

- An interactive mode that allows the operator to set attitude, g forces, quantities etc.
- Capable of supporting multiple tanks and complex geometries

The outputs from these tools include valuable design support information such as:

- Fuel quantity gauging accuracy versus quantity and attitude
- Unusable fuel quantity versus fuel boost pump inlet location and aircraft attitude

Fluid Network and Fuel Management System Models

The geometric core of this tool can also be used to support a fluid network model for use in the evaluation of refuel performance in terms of refuel times, CG and balance issues and surge pressure. Similarly simulations derived from the tank analysis and fluid network models can be used to exercise the fuel management control logic and to evaluate the effects of system functional failures.

An example of a fuel system model used for validation of the design of the fuel management system of a twin engine transport aircraft is shown in Figure 11.13.

The upper schematic provides an overview of the aircraft functions that provide functional inputs to the fluid network model shown in more detail in the lower schematic.

The flight deck inputs from the fuel panel, engine throttles and APU on/off selection, together with the operational flight conditions determine the fuel flows demanded by each engine and the APU. The fuel management control logic selects the various pumps and valves and reacts to their status and fuel tank quantities.

This model can be used to study the system operational characteristics of the system during a typical flight mission through take-off, climb, cruise, descent and landing. This particular fuel system comprises four fuel tanks with pump and control facilities for feed transfer and lateral balancing. A simplified aircraft model defines operating conditions in terms of pitch attitude, altitude and Mach number. In addition simple engine models determine the fuel flow delivered to the engines as a function of throttle setting and flight condition.

The fuel system model operates within this environment where flows and pressures throughout the fluid mechanical system can be computed from pump and valve design data, line pressure drop estimates etc. Due to the highly non-linear features of the fluid network, tank geometries and engine fuel flow characteristics, together with the complexity of the aircraft environment, digital computation is the only practical method of model implementation available to the fuel system designer. The computing power of the modern desktop computer allows models such as the one described to be developed quickly and with reasonably high fidelity. As mentioned above, such models can provide simulated output substantially faster than real time thus allowing a thorough analysis of a typical system to be completed in a timely manner.

In the model shown in the figure various control algorithms for pump and valve operation to facilitate easy evaluation of various control schemes.

Real Time Models

During the verification phase of the program, represented by the right side of the 'V' diagram, as illustrated by Figure 11.14, the task of evaluating the functional performance of the system components as an integrated entity takes place.

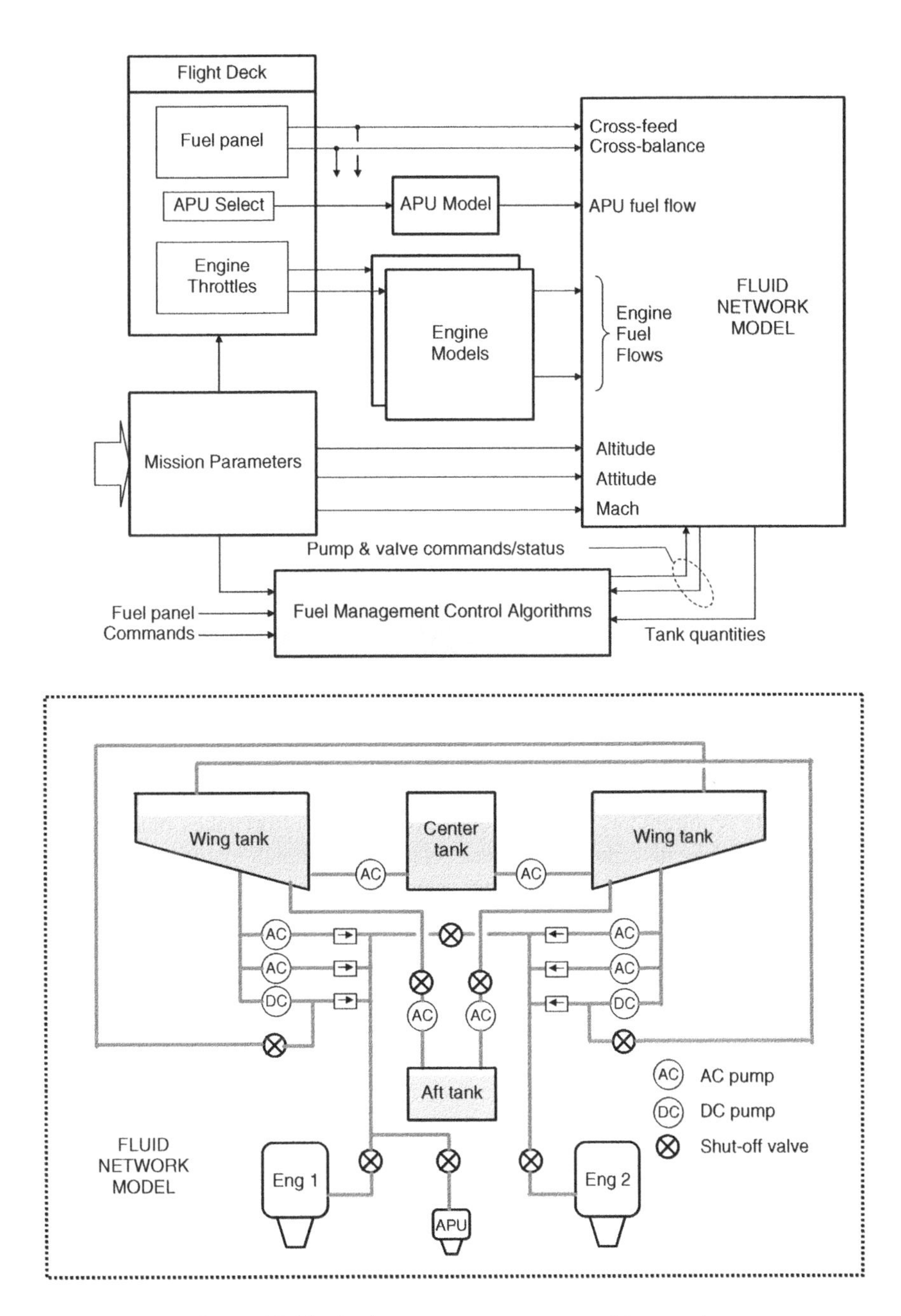

Figure 11.13 Fuel management system model example.

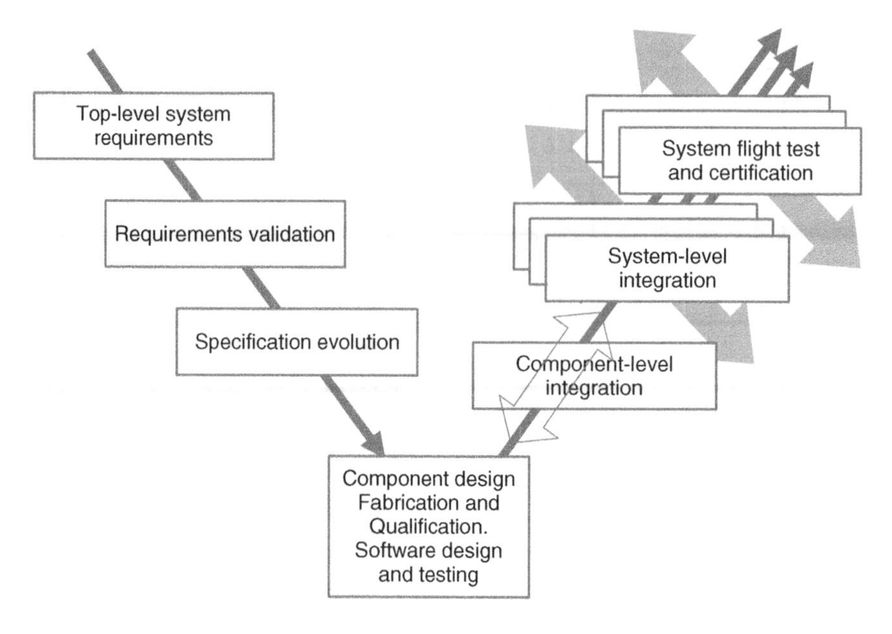

Figure 11.14 System integration phase.

Initially the components of the system are integrated and their performance evaluated as an integrated entity. The next phase involves an evaluation of the complete system functionality. Schedule requirements typically demand that this effort takes place well before the aircraft is available in order to enhance system functional maturity though extensive testing of the fuel measurement and management systems. To accomplish this, a real time model of the aircraft environment must be provided to test the system hardware and software against. In this way the system software responds as though it were installed in the aircraft. The system can therefore be evaluated during all of its functional aspects including:

- refuel performance
- simulated flight missions
- failure accommodation.

An important issue in the system integration test process is to consider the systems interactions of all other interfacing systems including, Electrical Power Management, Flight Management, Inertial Reference etc. This point is illustrated in Figure 11.14 by the arrows normal to the verification line of the 'V' diagram. This is equivalent to imagining several other 'V' diagrams existing simultaneously, in parallel to each other. In this environment, the system software can be certified through witnessed testing of the system during normal operation and in the presence of failures.

Figure 11.15 shows the typical test environment associated with the systems integration process showing the Special Test Equipment (STE) comprising the real time aircraft model, sensor interfaces, displays and mission management facilities. The objective of this arrangement is to have the avionics equipment under test respond as though it were installed in the aircraft.

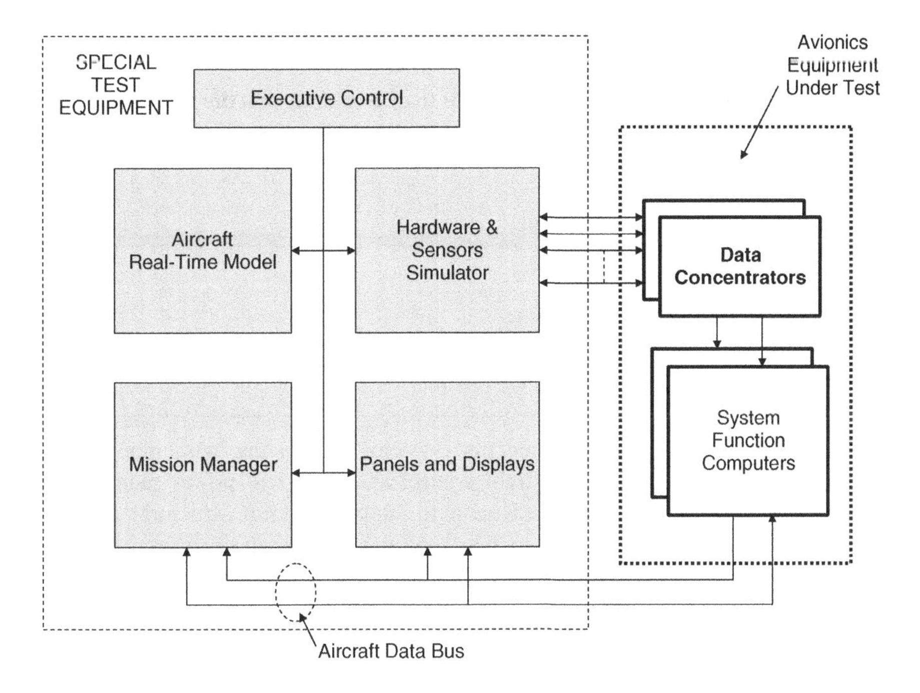

Figure 11.15 Systems integration testing environment.

11.7 Certification

11.7.1 Certification of Commercial Aircraft Fuel Systems

Before an aircraft can be introduced into commercial service it must receive a Type Certificate from the Certification Authorities which establishes that the required airworthiness standards have been met for that aircraft type. Each aircraft built to that design standard must then be granted a Certificate of Airworthiness.

The primary certification authority is usually with the country of the Original Equipment Manufacturer (OEM). In the case of large transport aircraft, therefore, this typically means the Federal Aviation Authority (FAA) for aircraft manufactured in the USA with requirements defined in Federal Airworthiness Regulations (FAR's) part 25 'Airworthiness Standards: Transport Category Aircraft' reference [15]. For aircraft manufactured in Europe, the European Aviation Safety Agency (EASA) is the primary certification authority under 'Certification Specification for Large Transport Aircraft CS 25 Certification Requirements'.

For fuel system design, the majority of applicable requirements are contained in the power plant section of the relevant airworthiness codes.

It should be noted that aircraft having gained a Type Certificate from either the FAA or the EASA, when exported to other countries will require acceptance against the national airworthiness standards of that country and therefore a validation exercise and/or compliance demonstration will be required. Therefore it is prudent to consider the airworthiness codes of all potential export countries as well as the primary authorities early in the design phase of the program.

A key element in achieving a Type Certificate is to demonstrate that the system design meets the minimum acceptable performance and safety standards. This demonstration process covers all areas including detailed reviews of:

- the system design and implementation
- design calculations
- simulation and model validation
- conclusion of system safety assessments
- rig and aircraft testing
- equipment and software qualification test results.

Modern fuel systems are complex integrated systems interfacing with many other aircraft systems such as fire protection, flight control and electrical power management. In addition, the fuel system may be called upon to support aircraft structural performance (wing bending relief and flutter) and to help optimize cruise performance (aircraft CG control). The certification process must therefore include the provision of data to these other interfacing systems so that they can demonstrate compliance with their associated requirements.

Before entering the design process too far it is vitally important to consider how compliance to the airworthiness requirements for the various features of the system will be demonstrated, and to get acceptance from the Airworthiness Authorities on the approach to be used. This is particularly important when considering rig and aircraft tests since their late inclusions can lead to new rigs being required and extra flight tests and instrumentation requirements, all of which are not only costly but also potentially cause the test program to be extended.

The certification process must therefore begin early in the system definition phase of the program through the generation of a preliminary certification plan which is up-dated as the system design is evolved. This plan addresses each of the airworthiness regulations and establishes a compliance method as either by similarity with existing designs, by analysis, by test or by a combination of methods. The importance of maintaining continuous communications with the airworthiness authorities from the outset in the certification planning and execution process cannot be over-emphasized.

11.7.2 Flight Test Considerations

The objective of the flight test campaign is to demonstrate the achievement of acceptable system performance over the complete range of conditions over which the aircraft is to be operated and with all of the fuel types that will be used.

It is generally not possible to perform flight tests on every fuel type with which the aircraft will be exposed. Therefore it is essential to establish a strategy of how the results of the flight test can be read across to the complete envelope of conditions and fuels, and to obtain acceptance of this strategy from the certification authorities.

Today, most commercial transport aircraft will be exposed to either Jet A or Jet A1 which are kerosene type fuels. Earlier systems were also cleared to use Jet B (similar to JP4). This is a wide cut fuel and although its use today in commercial applications is rarely encountered its use cannot be totally excluded.

Rather than design specifically for JP4 today, an approach which can be considered is to determine the operational restrictions to be observed when using this fuel and list them in the appropriate aircraft documentations. Although there are still many national fuel specifications around the world there is fortunately a convergence to the Jet A/A1 fuel type.

Design and testing for use with JP4, particularly when considering operation with boost pumps both operating and not operating, can give an important level of confidence in system capability since JP4 could be considered as an envelope-limiting fuel and that any other fuel likely to be encountered around the world would fall between Jet A/A1 and JP4, and therefore the amount of work and testing necessary to clear the system for other fuels could be reduced.

On past programs this has proved to be important in that the original fully instrumented flight test aircraft may not be available to perform testing with a new fuel type and therefore the cost of instrumenting another aircraft for fuel system testing (which may well be a customers delivery aircraft) is avoided.

The trend becoming popular today is to do complete certification testing with both JetA/A1 and TS1 fuels. TS1 is a kerosene type fuel produced against a Russian specification. This fuel has a vapor pressure, which is slightly higher than the Jet A/A1 specifications so hot fuel testing tends to be performed using this TS1 fuel.

When devising the flight test plan the fuel specification regarding the use of fuel additives should also be taken into account. If the additives are mandatory then there is usually no problem since all of the uplifted fuel will contain the additive. However in the case of optional additives consideration needs to be given to the possible effects the additive may have on the fuel and on the fuel system components. For instance the use of certain fuel system icing additives may affect the vapor pressure and even though the effects of such additives are likely to be small they should not be overlooked.

When developing the flight test plans it is important to keep in mind that a lot of certification evidence on system operation /performance can be collected without having to perform specific/dedicated tests. Significant amounts of relevant data will collected as a by-product from all of the aircraft tests performed throughout the test campaign.

There are however, some important areas which will require dedicated tests to be performed as follows:

Hot Weather Operation

In these flight-tests the fuel is required to be heated to a pre-defined minimum temperature and to initiate a rapid climb to altitude. The main purpose of this test is to ensure that during the climb as the tank air pressure reduces and more fuel vapors and air in solution evolve, the fuel boost pumps continue to operate satisfactorily by maintaining engine feed system pressure at a value which is above the pressure required to meet the specified engine inlet conditions.

Another critical part of the test is that to establish the limiting gravity feed altitude. In this test the fuel boost pumps are selected off and the aircraft climbed to an altitude at which an engine malfunction occurs.

During the climb at various altitudes the engine throttles are typically cycled from low power to high power settings to detect any malfunction tendencies. The altitude at which the engine operation is judged to be unsatisfactory is usually referred to as the gravity feed ceiling.

Negative g Operation

In this test it is required to demonstrate that the aircraft can operate satisfactory for a specific period (typically 8 seconds) when the aircraft experiences negative g conditions. During this test the aircraft is flown so that the fuel is under zero g conditions and therefore has a tendency to float away from the boost pump inlets, which potentially allows air to enter the system. How much air enters the system will depend upon the type of design precautions used in the design. Design precautions have included installation of pumps in collector cells which prevents the fuel moving/flowing away from the pump inlets. Additional features include the use of air release valves and specific boost pump design features which ensure that pump pressure is restored quickly and that any air which enters the feed system is discharged back into the feed tank.

11.7.3 Certification of Military Aircraft Fuel Systems

Certification of fuel systems and related equipment for military aircraft is very different from commercial aircraft fuel systems as normally there is no agency such as the FAA or other civil aircraft certification agency involved. Generally, military aircraft fuel system certification responsibility resides with the prime contractor with oversight from the military agency procuring the weapon system. A large part of fuel subsystem certification is performed on an iron bird (fuel system test rig) and/or as part of the flight test program. All of the fuel system equipment is typically subjected to a very extensive qualification test program as stipulated by the detailed equipment specification. Frequently fuel system equipment design and qualification requirements are based on the general fuel system requirements of MIL-F-8615, reference [22]. Table 11.1, which is extracted from MIL-F-8615, shows the tests normally required for fuel system equipment. The environmental test requirements are provided in MIL-STD-810, reference [23], which is a sub document of MIL-F-8615. The requirements of MIL-F-8615 and MIL-STD-810 are very often tailored within the equipment specification to meet the specific air vehicle requirements.

Results of the equipment qualification test program are submitted in a report to the responsible subsystem organization. Typically this qualification test report will be reviewed and approved by the responsible fuel subsystem engineer and his support staff. This qualification test report will then become a support document to the certification of the subsystem.

Table 11.1 Testing required for military aircraft fuel systems.

First Article Inspection Program								
Inspection		Requirement	Test	Test article				
				1		2		3
0	Examination	3.2, 3.1.4	4.5.2	x		x		x
1	Break-in Run	3.2	4.5.3	x		x		x
2	Calibration	3.7.1	4.5.5	x		x		x
3	Speed	3.7.7	4.5.7	x		x		x
4	Leakage	3.6.2	4.5.4	x		x		x
5	Electrical Insulation	3.6.5.6	4.5.16.2	x		x		x
6	Fuel Resistance	3.4	4.5.11			x		
7	Corrosion Resistance	3.4	4.5.14	x				
8	Endurance	3.7.5	4.5.7	x				
9	Contaminated Fuel	3.5	4.5.13	x				
10	Altitude	3.7.2	4.5.6			x		
11	Gravity	3.7.4	4.5.9			x	or	x
12	Acceleration	3.7.3	4.5.8			x	or	x
13	Vibration	3.6.3.3	4.5.18.4			x	or	x
14	Water	3.7.6	4.5.17	x	or	x		
15	Icing	3.7.6	4.5.17.3	x	or	x		
16	Dust	3.7.8	4.5.18.3	x	or	x		
17	Pressure Surge	3.7.9	4.5.10	x	or	x		
18	Mechanical Shock	3.6.3.2	4.5.15.1					x
19	Mechanical Load	3.6.3.1	4.5.15.2	x		x		
20	Overspeed	3.6.11	4.5.15.3					x
21	Electrical Actuators	3.6.5.3, 3.6.5.4	4.5.16.1					x
22	Explosion Proof	3.6.5.5	4.5.16.3					x
23	Electrical Compatibility	3.6.5.7	4.5.16.4	x	or	x		
24	Thermal Protectors	3.6.6	4.5.16.5					x
25	Humidity	3.4, 3.7.6	4.5.18.1	x				
26	Fungus Resistance	3.4.1	4.5.18.2	x				
27	Acoustical Noise	3.6.3.3	4.5.18.5			x	or	x
28	Thermal Shock	3.7.10	4.5.18.6			x	or	x
29	Bonding and Lightning	3.6.4	4.5.19			x	or	x
30	Disassembly	3.1	4.5.20	x		x		x

There is an exception to the non-involvement of commercial aircraft airworthiness authorities and this is regarding the conversion of commercial aircraft for military use. Air force tanker aircraft are a good example of this situation:

- the KC-135 version of the Boeing 707
- the KC-10 derivative of the DC-10
- the conversion of the Boeing 747 for duty as Air Force One.

In Europe a similar situation will apply to the new military transport the A400M which is being constructed by the Airbus, the commercial aircraft business segment of EADS.

In all of the above situations some level of approval from commercial airworthiness authorities (in this case the FAA in the USA and the EASA in Europe) is required.

11.8 Fuel System Icing Tests

As part of the fuel system certification process, system icing tests are required to verify resistance to icing that can occur during operation with excess water in the fuel. Federal Airworthiness Regulations (FARs) require that the fuel system for a turbine engine-powered aircraft must be capable of sustained operation throughout its flow and pressure range with fuel initially saturated with water at 80 °F and having an additional quantity of free water per gallon added and cooled to the most critical condition for icing likely to be encountered in operation.

The intent of this test is to verify that the system can provide uninterrupted flow and pressure to the engine inlet during the most adverse, yet realistic, operating conditions to ensure continuing sustained engine operation. These icing tests do not include engine functional requirements which are tested separately. Also, according to the specifics of the icing test requirements, ice blockage in subsystems, while not desirable from a system functional perspective, does not constitute an icing test failure as long as the flow and pressure to the engine inlet is not interrupted for the duration of the test.

Due to the lack of information regarding icing test procedures in the existing FARs, the SAE advisory document 'Aerospace Recommended Practice (ARP) 1401' has been used to define specific test requirements even though the original intent of this publication was to serve as a guideline. This has resulted in much confusion regarding the appropriate testing methods that should be used to determine a fuel system's ability to operate satisfactorily in severe icing conditions.

Perhaps the most significant contentious issue that has been debated over the years is the amount of free water per gallon of fuel system capacity that should be added above the water saturation limit to represent the 'Emergency operation' condition per ARP 1401. Of equal importance are the test times and fuel temperatures to be used in the test procedure.

This ARP is soon to be replaced by Aerospace Information Report (AIR) 790C under the cognizance of the SAE-5 Committee. This new document will define specific test procedures based on the extensive experience that has occurred following the conducting of icing tests by major systems subcontractors over the past twenty years or more.

Meanwhile the following commentary regarding icing test procedures and related issues, based on a consensus has emerged from the aerospace community on this subject, is offered to the readers of this book. It is anticipated that much of this information will be contained in the new AIR document when it is finally released. The primary issue associated with the test process relates to what is defined in ARP 1401, reference [24], as the 'Emergency operation' condition. The most significant requirements for evaluating a system's ability to operate under this emergency condition are:

- The quantity of free water to be added at the start of the test shall be 0.75 cc's per gallon of fuel in the system.
- The test duration which is recommended in the ARP as 30 minutes
- The operating temperatures to be maintained during the test. Three temperatures are defined as follows:

- +28 degrees Fahrenheit
- +13 degrees Fahrenheit
- Some lower temperature corresponding to the lowest temperature for the application under test (typically –20 to –40 degrees Fahrenheit)

It should be understood that an icing test is not the same as a low temperature test and it should also be understood that once the fluid temperature is below +13 °F the water/ice is transforming into its final solidified crystalline form and has the tendency to pass straight through boost pump screens and be digested by the engine. In the worst case situation, these ice crystals may plug holes and orifices. For an engine feed system, experience has shown that the critical temperatures occur between the first two specified test temperatures, i.e. between +28 °F and +13 °F. In this temperature range the water begins to form into an icy slush ice which is not yet not totally solidified yet, but very sticky. This sticky slush will adhere to almost anything it comes into contact with. The most significant areas of interest are the effect of this icy slush on engine feed pump inlet screens and/or feed ejector pump orifices. It is important to recognize that dynamic changes in operating conditions can impact the fuel system icing performance. For example, transient throttle changes causing a sudden increase in engine fuel flow may dislodge ice accumulation on screens and other locations within the engine feed system. It is therefore prudent to conclude any icing test procedure with a throttle transient that is representative of a 'Go-around' maneuver to ensure that there are no subtle icing problems that could manifest themselves during such critical events.

11.8.1 Icing Test Rigs

Icing test rigs can be full-scale or partially full-scale test rigs fabricated and tested in lieu of flight testing. To certify a fuel system to the FAA icing test requirements the test rig must duplicate the actual aircraft fuel system as near as practical and conform to certain aircraft fuel system drawings with respect to component locations and coordinates; however, test rigs do not need to be an exact duplicate of the aircraft fuel tank internal structure and sound engineering judgment is important here to establish the optimum blend of simplicity (hence lower cost) and functional representation. Certification Authority inspection of the test rig is typically required.

In general, rigs should be constructed to duplicate the flow path that the water will take along the lower Inner Mold Line (IML). Holes through the bottom of the ribs and drain holes in stringers must duplicate the aircraft as near as practical.

The principle here is to allow as much of the water/ice to migrate to the engine feed pump (or test component) as would be expected in the aircraft.

In some cases actual aircraft structure has been used for icing testing. This is possible for small aircraft or helicopters where it may be more economical than the fabrication of a special purpose rig. A benefit of this approach is an assurance that fuel tank internal structure is fully representative of the aircraft.

11.8.2 Fuel Conditioning

The configuration of the fuel chilling system is very important for accurate and repeatable icing tests. The chilling system (see Figure 11.16) should not have any low points where water can settle. The circulating circuit from the conditioning tank through the fuel chiller should

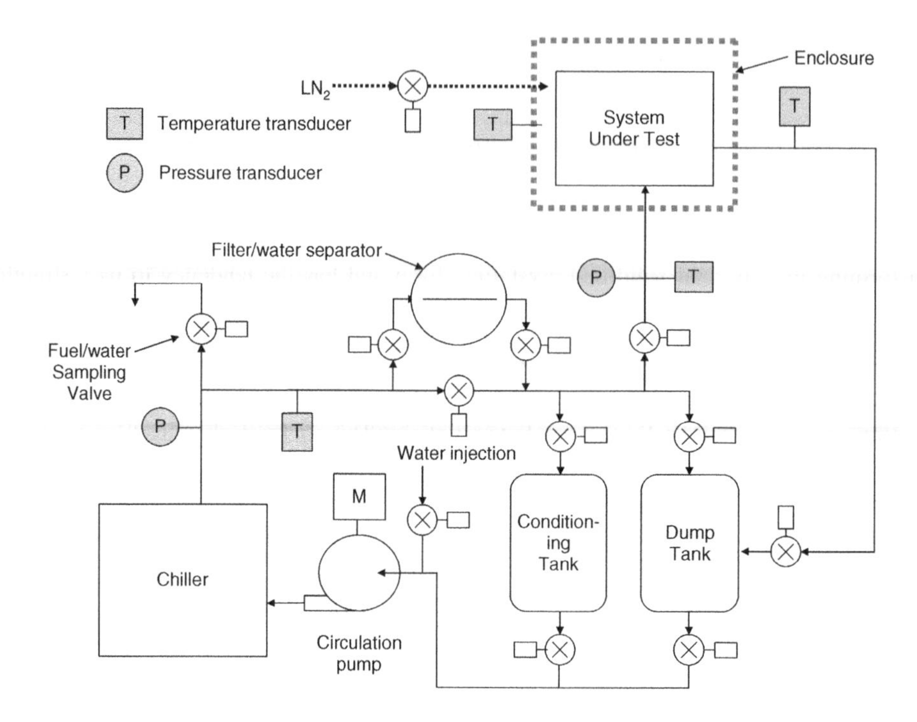

Figure 11.16 Icing test rig schematic.

terminate at a spray bar along the bottom of the tank to ensure that water does not settle on the bottom of the tank. It is important that the chiller circulation pump should be of sufficient size to keep the fuel/water agitated.

12

Fuel System Design Examples

This chapter presents a number of specific fuel system examples covering a broad range of applications from business and regional aircraft applications to the only supersonic transport aircraft that has been operated successfully in regular passenger service, i.e. the Concorde. This fuel system example provides the reader with a valuable insight into some unique fuel system design issues that may be extremely beneficial to the prospective fuel system designer by providing an in-depth understanding of some of the key operational constraints that are involved in the design and certification of such an interesting application.

While it was the intent to present here a military aircraft fuel system application, security issues have prevented a timely approval of this example which continues to be worked for presentation in subsequent editions of this book. Nevertheless, the examples described herein do provide the reader with a variety of real world applications that serve to give the reader a good grounding in the basic principles of aircraft fuel systems design that can be applied to all fuel system design applications.

The objective of this chapter is to show how many of the lessons presented throughout this book have been applied in practice and also to take the opportunity to comment, where appropriate, from both a positive and negative perspective on the solutions that evolved and the problems encountered during development and certification. This point is not meant as a criticism of the system that was ultimately certified as part of the aircraft's Type Certificate since there are typically many real world constraints covering diverse issues including business, schedule and technology risk that must often take precedence over what may be otherwise considered as the optimum design approach. Also, as we all know hindsight is the easiest technology to apply in these situations.

The examples covered in this chapter are as follows:

- ***The Bombardier Global Express Business Jet***
  This aircraft is a top-of-the-line business jet with very long-range capable of cruising at altitudes as high as 51,000 ft. The aircraft design, which received its Type Certificate in

Aircraft Fuel Systems R. Langton, C. Clark, M. Hewitt, L. Richards

1997, had a novel electrical system comprising variable frequency ac power generation and a computerized power management and distribution system that had a significant impact the design of the fuel system.

- ***The Embraer 170/190 Regional Jet***
  This regional jet was first introduced into service in 2005 and represents the current standard of high technology/low operating cost commuter aircraft designed to meet the growing demands for small capacity, medium-range jet-powered aircraft.
- ***The Boeing 777 Wide Bodied Airliner***
  This extremely successful transport aircraft first entered service in 1995. Some five versions with variations in capacity and range are currently in production. The fuel system comprises the first use of ultrasonic quantity gauging.
- ***The Airbus 380-800 Very Large Commercial Transport***
  This aircraft is the largest commercial transport aircraft configured with a true double-deck passenger cabin spanning the full length of the pressurized cabin. From a fuel system perspective there are many technical challenges that have been met including the need to load more than 250 tonnes of fuel in less than 50 minutes.
- ***The Concorde Supersonic Transport***
  This aircraft was operated by British Airways and Air France for some 25 years primarily the trans-Atlantic routes before being taken out of service in 2004. The fuel system design challenges for this application are considerable including demanding operating environmental conditions (temperatures and altitudes) and the need to provide flight critical balancing of the aircraft during transition between subsonic and supersonic flight regimes.

In the presentation of the above fuel systems which follows, the key design and development challenges and solutions are covered in order to reinforce the lessons covered throughout the various chapters of this book.

The comments and functional interpretations of these aircraft examples are the opinions of the authors and do not in any way represent the views of the aircraft manufacturers involved.

12.1 The Bombardier Global Express™

The Bombardier Global Express™ is a twin engine ultra long-range business jet (see Figure 12.1) designed to fly 8–19 passengers, with extensive on-board amenities in many configurations with an operating range of up to 6700 nautical miles (dependent upon payload and cruise speed). Cruise Mach numbers for the Global Express™ range from 0.80 to 0.88 with an initial cruise altitude 43,000 ft achievable from a maximum take-off weight departure in less than 30 minutes. A final cruise altitude of 51,000 ft is available with this aircraft. Operational and market requirements led to a fuel system specification that was capable of complying with ETOPS regulations that could be legislated in the future as applicable to twin-engine business jets of this class. The fuel system design that evolved provides a high degree of functional integrity from a fuel handling perspective with a fuel measurement and management system that ensures both high gauging accuracy and fully automated functionality.

Figure 12.1 The Global Express™ Business Jet on its Maiden Flight in 1997 (courtesy of Mike Nolte).

12.1.1 Fuel Storage

The fuel storage system comprises four fuel tanks; two wing tanks, a center tank located between the wings and an aft fuselage tank aft of the rear pressure bulkhead. The center tank comprises two compartments; a main center tank and a forward auxiliary tank. These tanks are connected via large diameter tubes and therefore, from a fuel system functional perspective can be regarded as a single tank.

The fuel tank arrangement is shown in Figure 12.2. Note that the wing tanks are divided into four semi-sealed compartments connected via baffle check valves that allow fuel migration inboard while restricting outboard flow. This concept, which was discussed previously in Chapter 3, minimizes aircraft CG variations with changes in aircraft attitude and fuel quantity. The inboard compartments of each wing tank are designated as the engine feed tanks since all of the fuel pumps associated with both feed and transfer are located within these compartments.

The vent system design adopted for the Global Express is somewhat unconventional in that the surge box is located within the fuselage rather than outboard of the wing tanks. With the surge box being located well above the fuel tanks the need for float actuated vent valves is eliminated; however, airworthiness regulations require that the surge box and vent lines within the pressurized area have double walls with an inter-space drain. Also easy access to the vent system components within the cabin must be provided for maintenance and periodic inspection.

A schematic overview of the vent system is shown in Figure 12.3.

The climb vents and center tank vents connect directly to the bottom of their respective surge box and a vent line across the top of the cabin connects both surge boxes. This upper vent line also connects to the aft tank and to the dive vents located outboard on each wing tank. Air scoops are located on the underside of each wing and connect to the bottom of

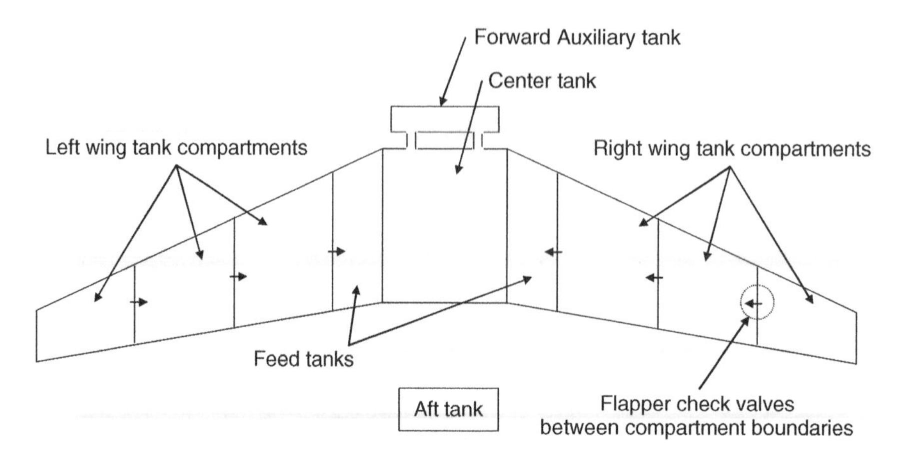

Figure 12.2 Fuel storage arrangement.

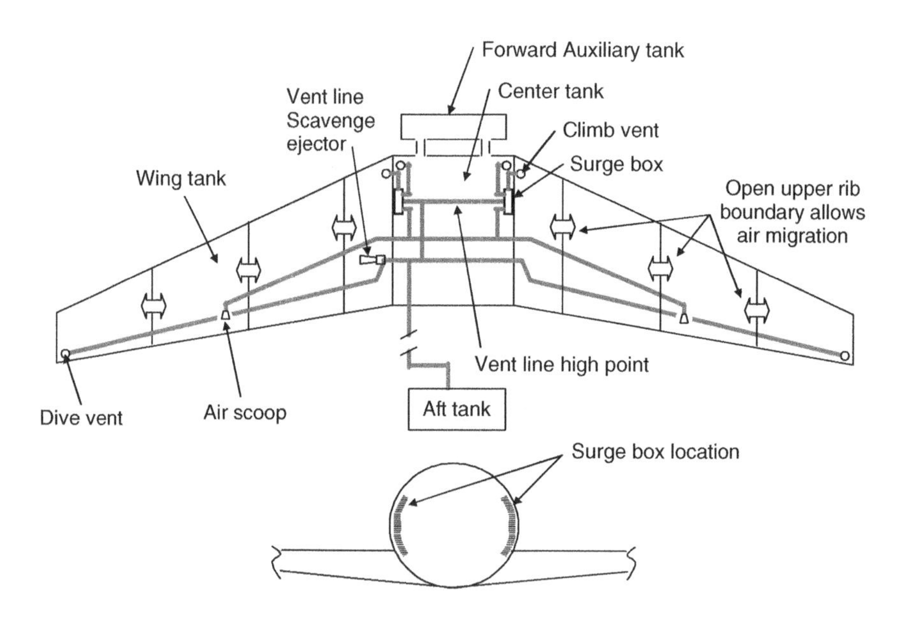

Figure 12.3 Vent system overview.

each surge box. As indicated in the figure, the upper boundaries of each wing tank compartment have open sections to allow free migration of air to either the climb or dive vent outlets.

To keep fuel from accumulating within the vent system, vent ejectors located in the feed tanks continually scavenge the main vent line both during the refuel process, using the refuel pressure as the motive pressure source, and during flight, using feed line pressure for motive power.

12.1.2 Fluid Mechanical System Design

The fluid-mechanical features of the Global Express™ fuel system are shown in the schematic diagram of Figure 12.4. As shown, the system employs motor-driven pumps for both engine (and APU) feed and transfer from both center and aft auxiliary tanks. The ac pumps use induction motors powered by a variable frequency 115 volt supply. The frequency range from engine idle to maximum power is approximately 350–700 Hz and, therefore, pump speed varies with engine rotational speed by a factor of about two to one.

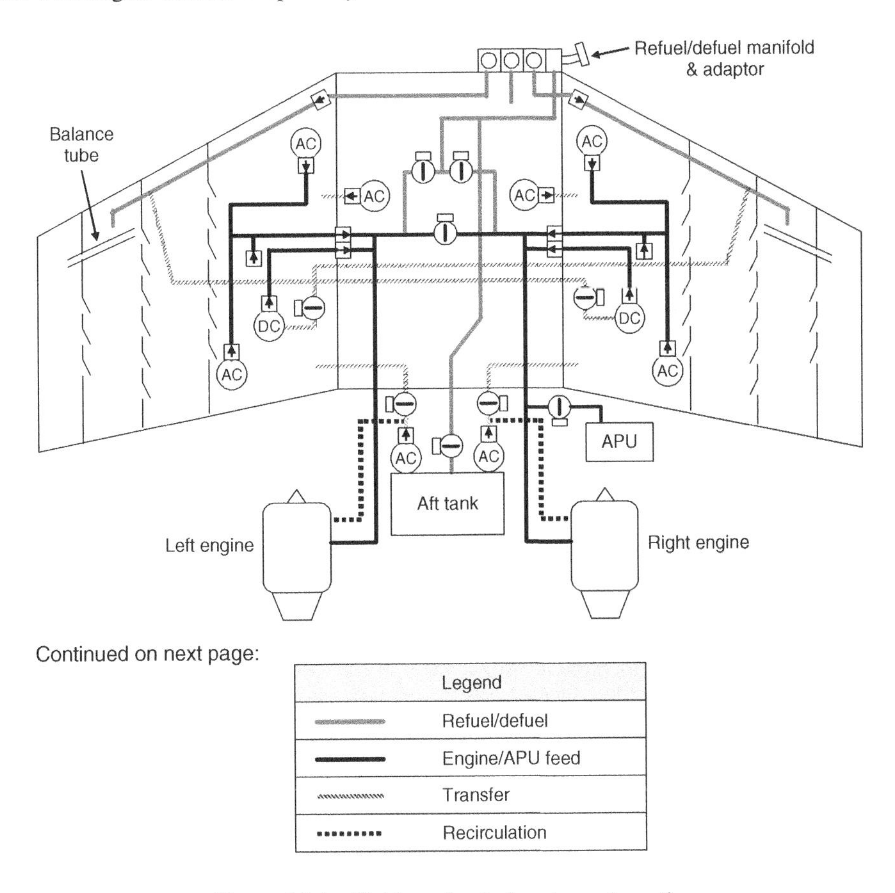

Figure 12.4 Fluid-mechanical system schematic.

The dc pumps are used for APU starting, lateral balance and as a back-up to the feed system in case of loss of ac power.

Each of the fluid mechanical functions is described in detail in the following paragraphs.

12.1.2.1 Refuel and Defuel Operation

The fuel lines associated with the refuel and defuel system are shown gray in the above figure. Feed lines are shown in solid black in the figure.

The Global Express™ uses a refuel/defuel distribution manifold similar to that used on the Boeing 737 series of aircraft. (See Chapter 6, Figure 6.15.) This manifold comprises three shut-off valve modules and a standard refuel adapter. Each shut-off valve can be actuated manually or electrically under the control of the fuel quantity gauging system. As discussed previously, the manifold distribution approach keeps surge pressures from the refuel piping within the aircraft. In this application, however, the decision to add the aft fuel tank was made after the basic three-tank system had been established and therefore the aft tank refuel shut-off valve was located at the base of the aft fuel tank.

The refuel distribution system is extremely simple. Each wing tank is refueled into the second outboard-most compartment and the flapper check valves at each compartment boundary allow the uplifted fuel to migrate inboard to the feed tank and second compartment. Balance tubes between the outboard and refuel compartments allow uplifted fuel to pass to the outboard compartment once the refuel compartment is close to full. Orifices located at the output of the distribution manifold are sized to allow equal flow rates into each wing tank.

12.1.2.2 Engine and APU Feed

Each engine is fed from two variable frequency ac-powered pumps located forward and aft in the feed tanks.

With this power supply system the pump sizing requirement becomes the emergency descent condition where the engine power setting is at flight idle (corresponding to a low pump speed) and the aircraft pitch attitude is nose down. Thus with the aft engine location the fuel feed pumps design requirement is to deliver flight-idle fuel flow to the engines with a minimum of 5psi above the fuel vapor pressure taking into account the feed line losses and the head difference between the feed pump outlet and the engine inlet. (This point was made in Chapter 4 with Figure 4.5.)

In the event of an ac pump failure, the auxiliary dc pump for that side of the aircraft is automatically selected on.

The APU is fed from the right hand feed line and all three feed lines have their own dedicated LP shut-off valve.

A suction feed check valve allows the engines to continue to operate in the unlikely event of total loss of feed pump pressure up to an altitude limit of approximately 24,000 ft.

A cross-feed valve selectable by the flight crew connects both feed lines to accommodate the single engine out condition.

12.1.2.3 Fuel Transfer

Refer to the patterned lines in Figure 12.4.

The fuel burn sequence requires that center tank fuel is consumed first. This is accomplished via two ac transfer pumps located in the center tank. These two pumps keep the wing tanks topped up as fuel is consumed until the center tank is empty. At some predetermined wing tank quantity any fuel in the aft tank is then transferred to the feed tanks using the aft transfer pumps and their associated transfer shut-off valves.

In addition to fuel burn sequencing, a lateral balance system is provided using the dc auxiliary pumps together with lateral transfer shut-off valves. The lateral balance system is designed to maintain any lateral imbalance at less than 500 lb. Control of the lateral balance system can be either automatically or by selection by the crew. If any of the dc pumps is selected on following a feed pump failure the automatic lateral transfer system is disabled.

12.1.2.4 Recirculation System

The fuel recirculation system allows warm fuel from the engine to be recirculated to each wing tank in the event that wing tank fuel temperatures approach operational minimums. This system was added after type certification when it was found that during very long-range operations the outer compartment of the wing would reach limiting temperatures more frequently than had been expected. By connecting the engine recirculation line upstream of the aft transfer valve this modification was relatively simple and has been proven effective in service. Selection of the recirculation system is via the flight crew.

12.1.3 Fuel Measurement and Management

The Global Express employs a common avionics computer defined as the Fuel Management and Quantity Gauging Computer (FMQGS) to perform all of the fuel management and quantity gauging functions. A top level schematic of the FMQGC and its functions is shown in Figure 12.5.

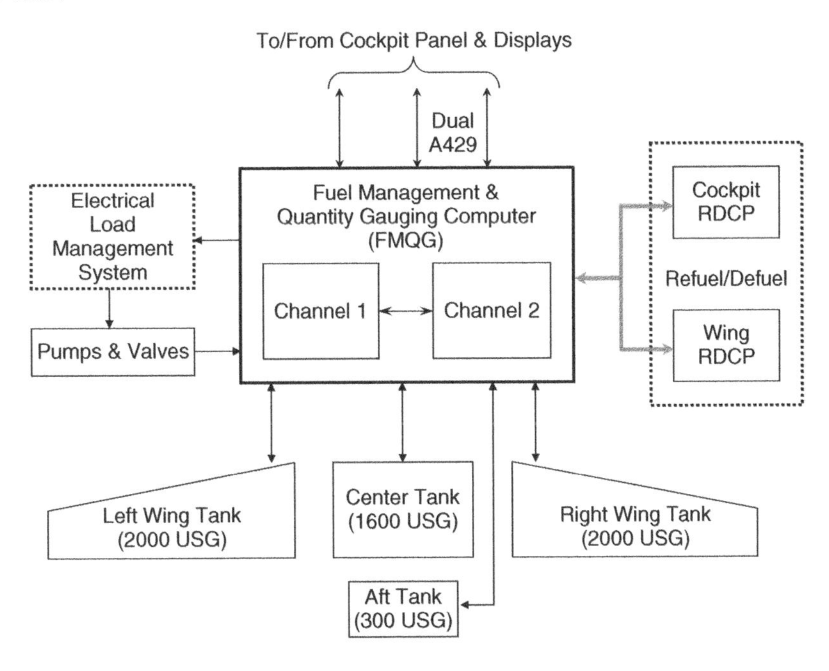

Figure 12.5 Fuel Management and gauging schematic.

The ac capacitance gauging system is designed to meet or exceed the requirements for a Class III system as defined by MIL-G-26988C which is +/− 1% of indicated quantity +/− 0.5% of full contents. The complement of in-tank sensors includes 34 probes, 6 probe/compensator combination units, three temperature sensors and two high level sensors.

The high level sensors are, in fact, miniature capacitance probes which provide verifiable capacitance values for each state which is a big help in fault detection. An important aspect of this design is the introduction of the Fuel Properties Measurement Unit (FPMU) which is located in the left feed tank and contains a densitometer, a compensator and a temperature sensor. This unit is connected to the refuel gallery so that the up-lifted fuel is infused into the unit displacing any residual fuel from the last mission. Upon completion of the refuel process, the density, permittivity and temperature of the up-lifted fuel are acquired and stored by the Fuel Management and Quantity Gauging Computer (FMQGC) providing an accurate characterization of the fuel on board.

As indicated in the figure, the management and gauging system architecture uses a dual channel arrangement with extensive Built-In-Test. The probes and compensators are interleaved so that total loss of either channel will not lose gauging information even though some degradation in accuracy may occur.

The FMQGC fuel management functions include:

- Control and status assessment of all system pumps, valves and in-tanks sensors
- Fuel transfer – burn sequencing and lateral balance
- Flight crew and ground crew interface
- Automatic/manual refuel/defuel operation
- System fault detection and annunciation via the EICAS

12.1.4 Flight Deck Equipment

On the flight deck, an overhead panel (see Figure 12.6) contains selector switches for wing-to-wing transfer, aft transfer and crossfeed. The ac boost pumps, normally selected on during the engine start process can be inhibited individually via this panel. Selection of the recirculation system is also associated made using this panel.

The EICAS display includes a fuel system synoptic as well as advisory, caution and warning messages.

A refuel control panel and display is also provided on the flight deck so that the crew can control the refuel process from that location. The same refuel panel is installed at the refuel station on the right side of the aircraft close to the wing root.

12.1.5 Operational Considerations

Because of the extra long range and hence long mission times, the aircraft systems of the Global Express include a substantial degree of equipment redundancy to ensure that there is no loss of any of the primary functions following all single and many multiple failure situations. For example, each engine has two ac power generators together with an APU with its own generator that can be operated in-flight.

The power management system is a state-of-the-art microprocessor-based system that communicates with the fuel system Fuel Management and Quantity Gauging Computer (FMQGC) via an ARINC 429 data bus to provide health status information related to the power bus availability associated with key fuel systems equipment. For example, should the ac power bus availability degrade to the point where a single subsequent failure would loose power to one or more boost pumps, the auxiliary dc pump is automatically powered on. This feature is

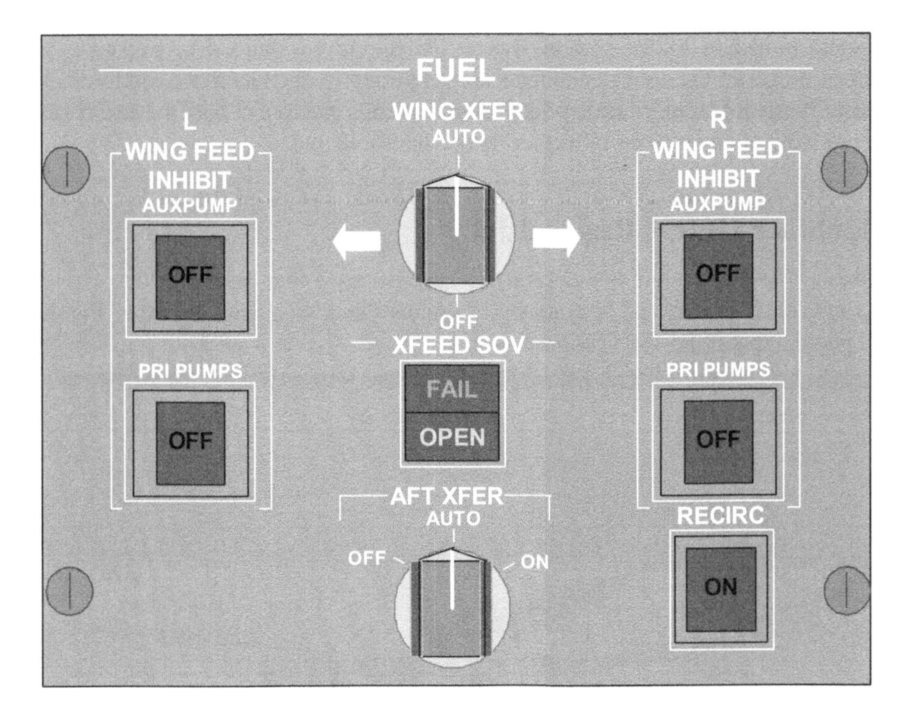

Figure 12.6 Fuel panel.

critical for an aircraft that can cruise at altitudes up to 51,000 ft where loss of feed pressure could result in an engine flame-out.

An interesting aspect of this issue is the fact that during normal operation with engine rotational speeds at the cruise condition or higher, there is a substantial boost pump excess capacity as a result of the pump sizing condition described above. This excess capacity could be used to provide motive power to transfer ejector pumps as an alternative to the motor-driven solution that was adopted. One reason that this solution was not adopted in the Global Express™ is the fact that the decision to incorporate variable frequency ac power was made late in the development program when much of the fuel system design work was already completed and any delay to the development schedule associated with system redesign was considered to be an unacceptable risk to the program. In hindsight the use of ejectors for fuel transfer may have been a more effective solution than the system that was eventually certified for the following reasons:

- The electrical power budget for the fuel system would be reduced due to the elimination of the motor-driven transfer pumps.
- Direct maintenance costs would be reduced because ejector pumps have no moving parts and hence have a higher operational reliability than motor-operated pumps.
- Ejector pumps have a better pump-down capability than the typical motor-operated pump and their introduction would therefore potentially reduce unusable fuel.

Nevertheless, hindsight is always twenty-twenty and the current solution has proved to be fully satisfactory, however, the above comments are supported by the fact that a modification was introduced shortly after entry into service to add scavenge ejectors to both the center tank and the inboard wing compartments to further reduce unusable fuel.

12.2 Embraer 170/190 Regional Jet

The Embraer 170/190 is a family of aircraft designed to accommodate from 70 to 110 passengers in single class seating and is available in standard and long-range versions. Figure 12.7 shows a photograph of the 190 version.

Figure 12.7 The Embraer 190 Regional Jet (courtesy of Mark Kryst).

This conventional commercial transport which entered service with in 2005 uses a simple two tank fuel storage design with state-of-the-art avionics and fluid mechanical equipment that emphasizes simplicity and reliability.

12.2.1 Fuel Storage and Venting

Figure 12.8 shows the fuel storage arrangement which comprises two integral wing tanks with three compartments and a collector cell in each wing. Total fuel capacity for the standard version is approximately 21,000 lb. The collector cells which are situated inboard and aft house all of the feed system equipment.

A surge tank at the outboard section of each wing is the vent source for the storage system.

As indicated by the arrows in the figure, flapper check and baffle check valves are installed at the boundaries of each compartment allowing fuel to migrate inboard and into the collector cell. The number of valves required between the various fuel tank compartments is determined by the maximum refuel rate into the outer compartment where flow through the flapper/baffle check valves to the inboard compartments must be sufficient to ensure that the outer compartment fills last (see Section 12.2.2 for a description of the refuel system).

The vent system, illustrated schematically in Figure 12.9 is very simple. Climb vents connect the forward upper corners of the left and right main wing and wing stub compartments to the

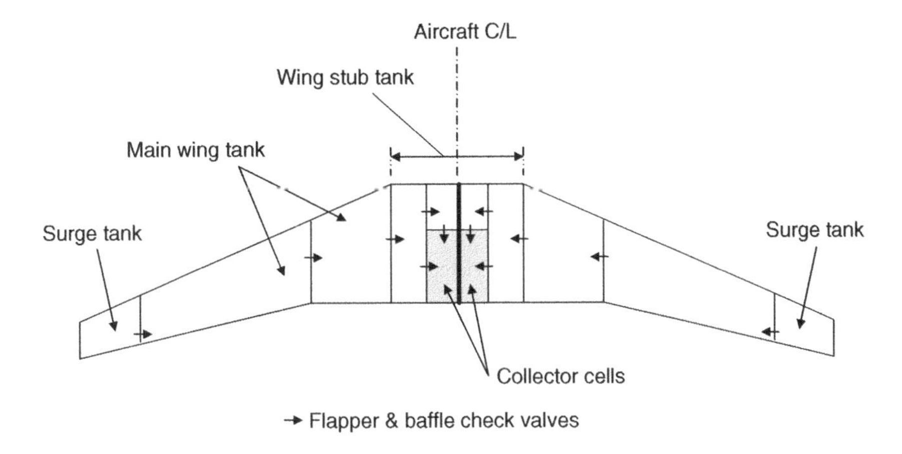

Figure 12.8 Fuel storage arrangement.

surge tank via the wing stub compartment. Float-actuated vent valves connect the outer wing compartments to the surge tanks for venting during cruise and descent.

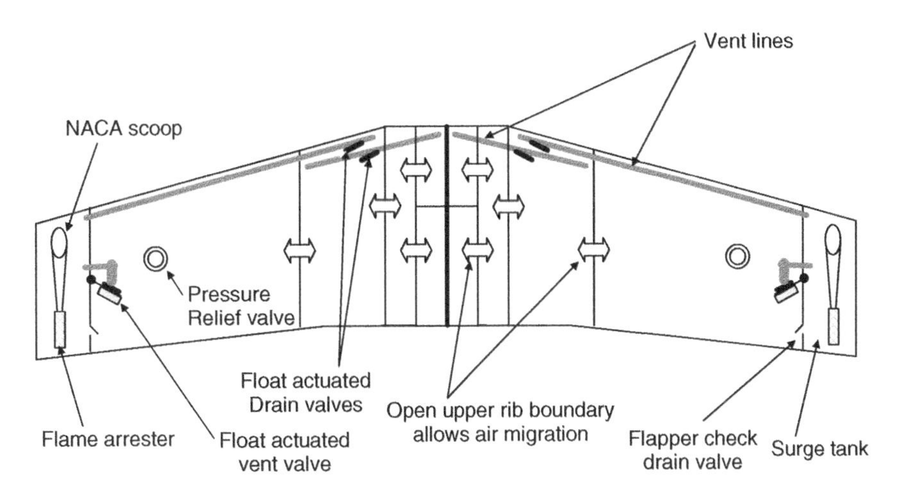

Figure 12.9 Vent system schematic.

The NACA scoops recover some of the airstream total pressure and connect to each surge tank via flame-arrestors.

There is sufficient open area at the upper boundaries of the semi-sealed ribs to allow air migration between the various compartments during normal operation while the main vent lines are sized to ensure that the pressure difference between the ullage and the outside air never exceeds 3 psi during a worst case emergency descent. Additional structural protection is provided via a high capacity relief valve mounted to the upper skin of each main wing tank to protect against a failed open refuel valve.

12.2.2 The Refuel and Defuel System

The refuel/defuel system is shown schematically in Figure 12.10. The standard refuel adapter is located on the right wing leading edge and connects to a single refuel gallery that discharges the uplifted fuel into the outer compartment of each wing via two refuel shut-off valves.

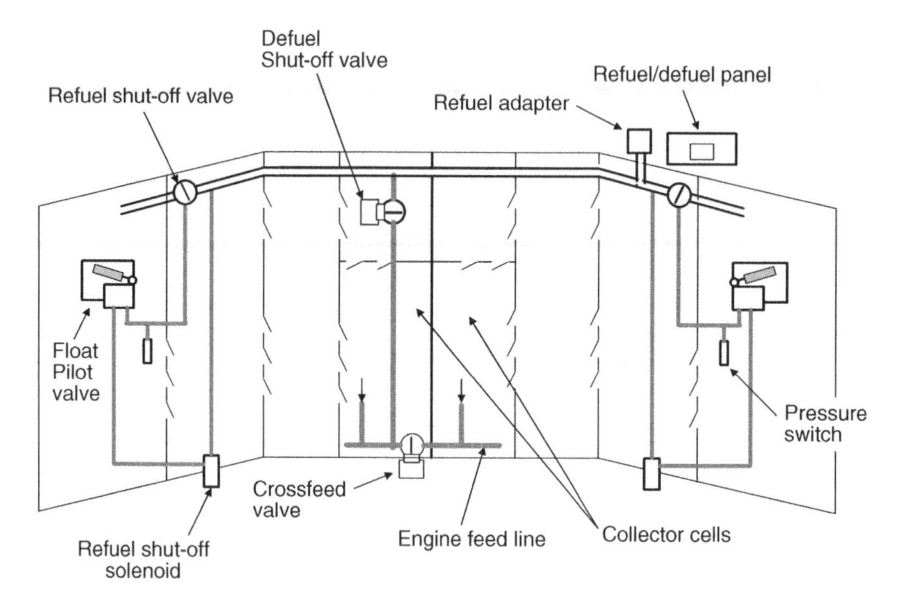

Figure 12.10 Refuel/defuel system schematic.

The refuel shut-off valves are hydraulically actuated by refuel pressure which is selected onto the actuation mechanism of the refuel shut-off valve by either:

- the high level float pilot valve, or
- the solenoid valve which causes direct lifting of the float.

The float pilot valves prevent over-fueling of the aircraft. The solenoid valves can be selected by the refuel operator to verify closure of the refuel valves (i.e. the pre-check function).

A pressure switch in the refuel shut-off valve actuation line provides an indication of the refuel valve status at the refuel panel. This pressure switch is actually installed outside the fuel tank not as shown in the schematic for intrinsic safety reasons.

Figure 12.11 shows the layout of the refuel/defuel panel defining the various switches, indicator lights and displays. As shown, the aircraft can be refueled either manually or automatically by making the appropriate 'Refuel selection' in the upper right hand corner of the panel. Gravity refueling is also available via adapters installed on the upper surface of each main wing tank.

During manual refueling, the operator controls the refuel shut-off valves directly and by observing the display can close the valves when the required quantity is reached. If automatic refueling is selected, the operator must pre-select the desired quantity on the display before selecting 'Auto' refuel and the system will automatically close the refuel valves by

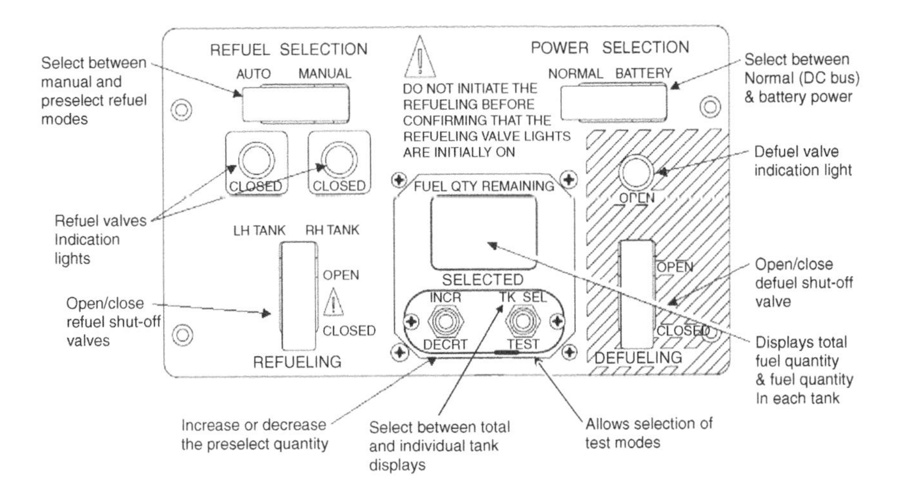

Figure 12.11 Refuel/defuel panel.

energizing the refuel shut-off solenoids based on the gauging system information, when the desired quantity has been uplifted.

Defueling the aircraft is accomplished by opening the defuel shut off valve which connects the left collector cell with the refuel gallery. Opening the crossfeed valve provides access to fuel on the right side of the aircraft. Suction defueling, pressure defueling (via the motor-driven feed pumps) or a combination of both is available with this system. (The defuel valve is actually installed on the aft spar with the motor and electrics outside the tank and not as shown in the schematic.)

The defuel valve also provides the ability to do a wing-to-wing transfer. This function is only available on the ground primarily for maintenance purposes and involves closing the 'Low' wing tank refuel valve, opening the crossfeed valve and switching on the opposite ac feed pump.

12.2.3 In-flight Operation

Figure 12.12 shows a simplified schematic of the fluid-mechanical system for in flight operation. As shown, the system uses a dedicated motive flow pump on each engine to provide motive pressure to their respective feed and scavenge pumps. With this arrangement, the feed system is completely self-sustaining without the need for electrical power once the engines are running.

A single dc pump located in the right hand collector cell allows the APU to be started via aircraft battery power. An ac pump on each side is available as back-up for the feed ejector pumps. Pressure switches in each feed line monitor ejector feed pump performance and initiate automatic switch-over to the back-up motor-driven pumps if feed pressure goes below a pre-set minimum value.

The ac pumps can also be used to correct any fuel imbalance that may develop by opening the crossfeed valve and powering the high side pump. The ac pumps are sized to be able to overcome the feed ejectors thus allowing crossfeed from the high side wing tank.

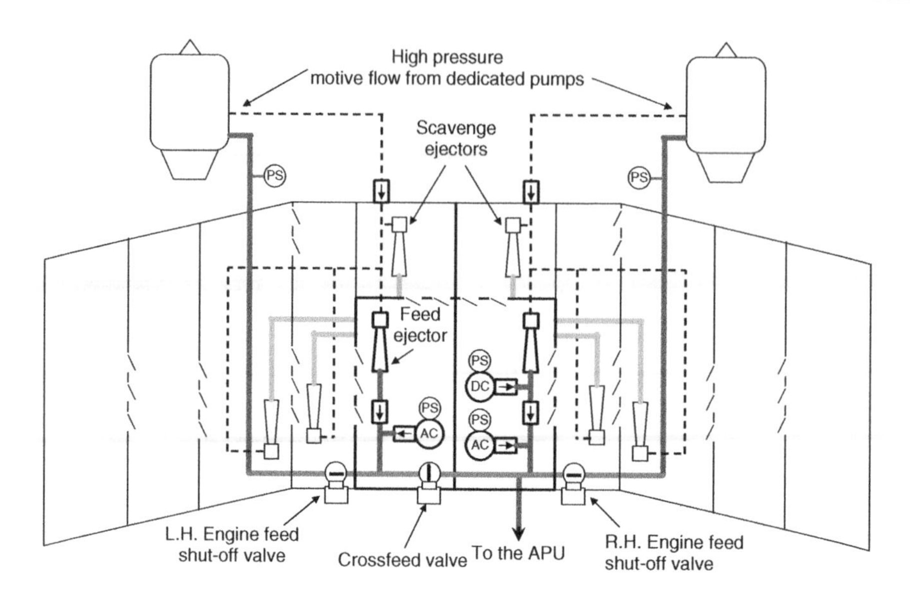

Figure 12.12 Feed and scavenge system schematic.

The scavenge ejectors minimize unusable fuel by scavenging the low section of each outboard compartment plus the forward section of the wing stub tank thus maintaining a full collector cell until the main wing and wing stub tanks have been depleted.

12.2.4 System Architecture

The Embraer 170 uses a state-of-the-art modular avionics suite wherein Modular Avionic Units (MAUs) with common hardware implement many different software functions. In the case of the fuel system, however, fuel quantity gauging information is required to support automatic refueling on aircraft battery power and therefore, the Fuel Conditioning Unit (FCU), which processes all of the gauging and temperature sensors was implemented using a separate dedicated, avionic unit.

Figure 12.13 is a high level schematic of the fuel system architecture and shows how the system interfaces with the modular avionics, the flight deck, the refueling station, the power distribution system and the fuel network components.

As shown, the MAU's communicate with the power distribution system and the flight deck via the Avionics Standard Communications Bus (ASCB). The Fuel Conditioning Unit transmits gauging information to the MAUs via an ARINC 429 standard data bus. The refuel panel and FCU communicate via a separate proprietary data bus.

The overhead fuel panel on the flight deck is shown in Figure 12.14 and comprises three pump selectors and a crossfeed valve selector.

The overhead panel pump and crossfeed valve selector states are wired directly into the MAUs as follows:

AC pump 1	to MAU1
AC pump 2	to MAU2
DC pump	to MAU1
Crossfeed	to MAU2

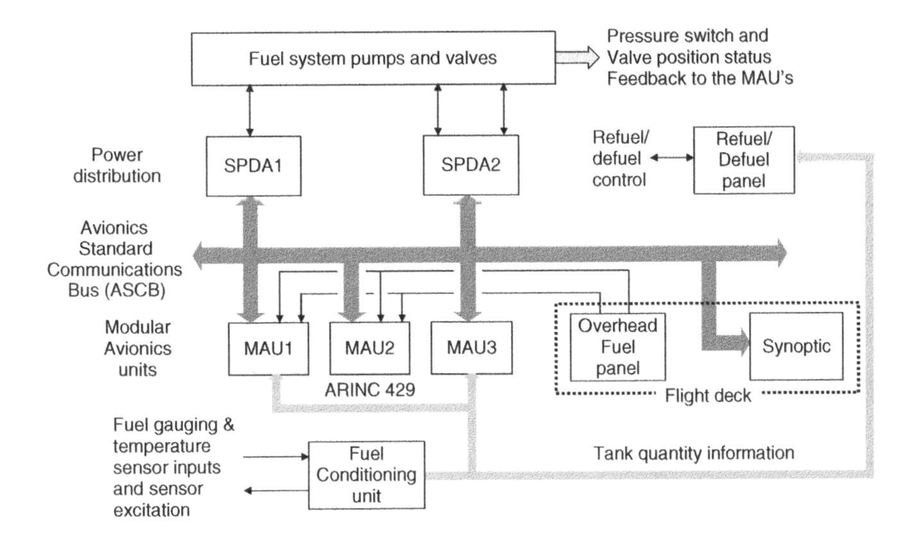

Figure 12.13 Fuel system architecture overview.

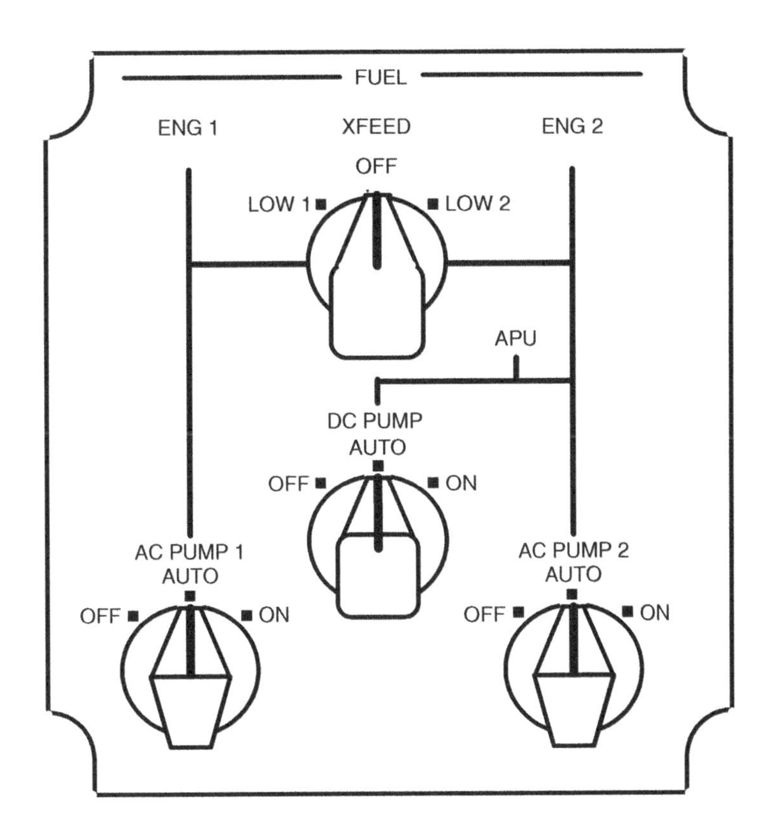

Figure 12.14 Overhead panel.

A grayscale interpretation of the fuel system synoptic displayed on the EICAS multi-function display is shown in Figure 12.15. The actual display uses color to enhance readability, for example the quantity information is presented in green amber and red; in both analog and digital formats. In the example shown in the figure the left wing tank (tank 1) has only 2010 lbs and is displayed in amber. The right wing (tank 2) has 4210 lbs and is presented in green.

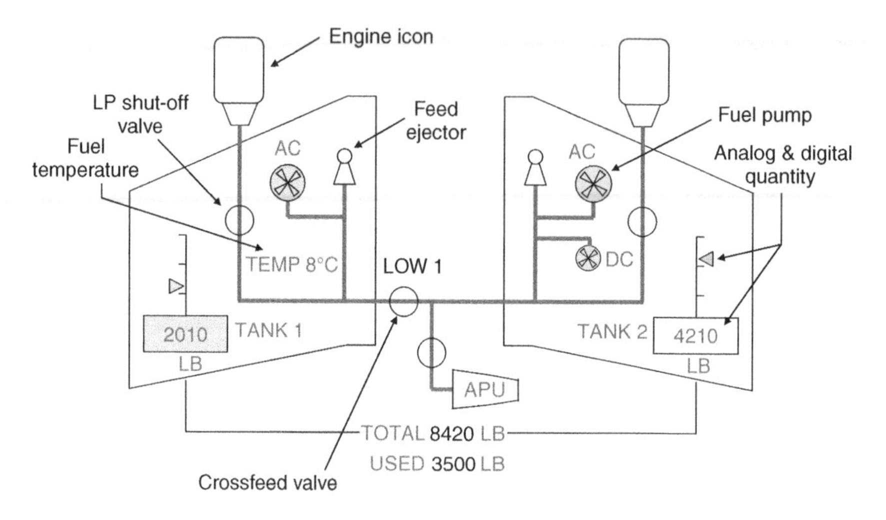

Figure 12.15 Fuel synoptic display.

Advisories, cautions and warnings use white, amber and red in the text displays. Here the low fuel state of tank 1 is displayed in white as an advisory.

Fuel lines are in green when fuel is flowing and otherwise in red.

12.2.5 Fuel Quantity Gauging

The Embraer 170 fuel gauging system uses ac capacitance gauging to achieve a nominal accuracy of ±2 % of full scale during normal flight conditions. This accuracy capability improves to ±1 % of full scale during ground refuel when attitude variation is minimal.

The selected architecture for the gauging system is a 'Brick wall' design comprising a dual channel, microprocessor approach where each channel is dedicated to its own (left or right) fuel tank.

Figure 12.16 shows a schematic of the gauging system showing that the refuel repeater/indicator is powered and supplied with quantity information from the right-hand gauging system.

By arranging each of the two channels to communicate with each other, the total fuel system gauging status is available from each of the two channels during normal fault-free operation to provide quantity gauging information to the aircraft via the Modular Avionic Units (MAUs) and to the refuel panel and repeater indicators.

Each wing tank comprises twelve capacitance probes and one compensator. A fuel bulk temperature sensor is located in the right wing tank.

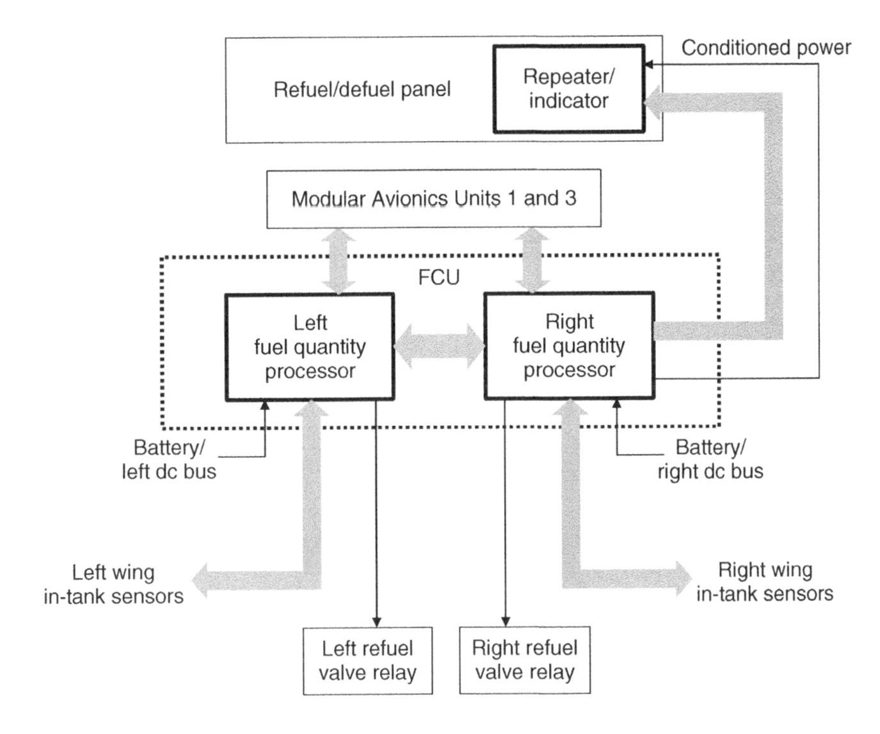

Figure 12.16 Fuel gauging system architecture.

Secondary gauging is provided by magnetic level indicators; three in each wing tank.

Low level indication is provided using capacitance sensors similar to tank probes but much smaller and located low down in each tank. The advantage of using capacitance devices for the low level detection function is that these sensors can utilize the same signal conditioning interface used by the primary gauging system and also they provide a measurable signal in both the true and false states which greatly enhances the fault detection and accommodation process.

In view of the requirement for independence of the low level warning system from the primary gauging system, the low level sensors are cross-coupled as shown in Figure 12.17.

This system architecture provides the ability to dispatch the aircraft following any single gauging failure (including loss of either FCU channel) since with the two tank system, symmetry of quantity between the two tanks can be safely assumed during normal operation and the low level indication system remains effective on the failed side of the aircraft.

12.2.6 In-service Maturity

Since its introduction into commercial airline service in 2004, the Embraer 170 fuel system has demonstrated a high degree of maturity. The only significant modification to the fuel system resulting from flight testing was the addition of a relief valve to the motive flow pump outlet since it was found during flight test that the feed ejector outlet pressure at high engine rotational speeds was able to overcome the opposite wing auxiliary pump when attempting to transfer fuel.

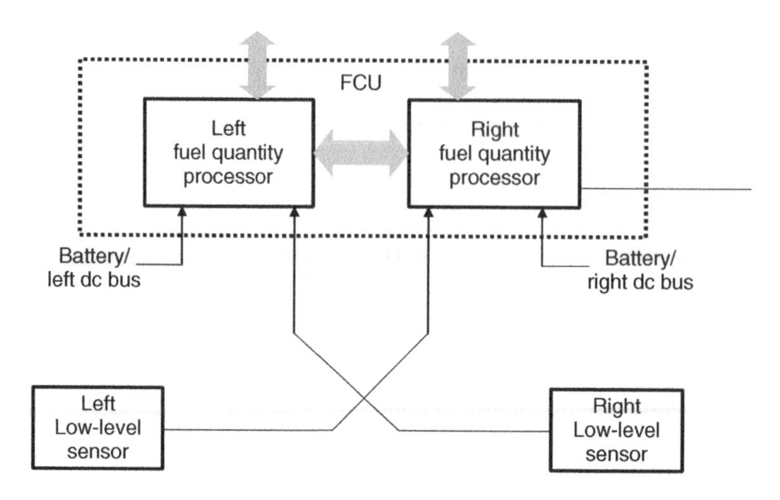

Figure 12.17 FCU architecture showing low level system implementation.

12.3 The Boeing 777 Wide-Bodied Airliner

The Boeing 777 (see the photograph in Figure 12.18) is a wide-bodied twin engine transport aircraft that continues to be one of the most successful commercial aircraft in service today. The initial production version, the 777-200, entered service with United Airlines in June 1995 with 180 minute ETOPS approval at the outset. This was a unique accomplishment in airline history and a credit to the Boeing Commercial Aircraft Company and to Pratt and Whitney,

Figure 12.18 The Boeing 777-200 airliner (courtesy of Arpingstone).

whose PW4084 engine powered this initial version, for successfully completing an extremely aggressive flight test program accumulating more than 3000 flight test hours in about 9 months. Certification of the aircraft with both the GE90 and the Rolls-Royce Trent 800 engines followed soon after.

Several extended range versions of the initial production aircraft are currently available designated the 777-200ER, 777-200LR, 777-300 and 777-300ER providing a wide selection of operational capability aimed at satisfying the disparate needs of most of the worlds airlines.

With a wingspan of almost 200 ft the B777 was originally designed to have the outboard-most 20 ft of the wing fold upwards to provide easier access to terminal gates. This design feature defined the outboard dimensional limit of the wing fuel tank. The wing fold concept was later disbanded as unnecessary and was never incorporated in production.

The B777 fuel system uses an integrated architecture based on both the ARINC 429 and ARINC 629 data buses for inter-system communication as shown on the system schematic overview of Figure 12.19. This diagram illustrates the refuel, gauging and management functions under control of the Fuel Quantity Processor Unit (FQPU). During the refuel process power is provided to the FQPU and Integrated Refuel Panel via the Electrical Load Management System (ELMS) Standby Power Management Panel.

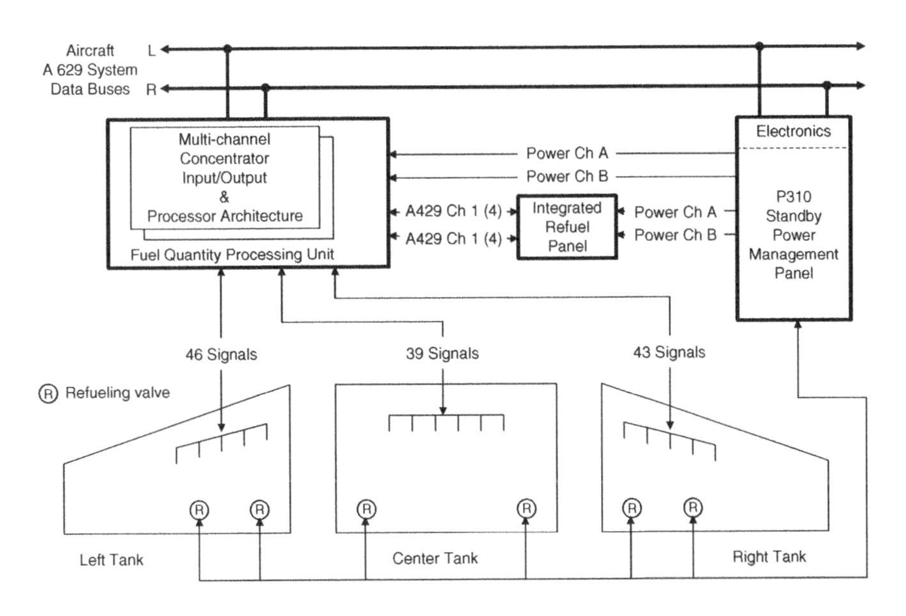

Figure 12.19 Fuel gauging and management system overview.

The following paragraphs provide a detailed description of the B777 fuel system covering the storage, fluid-mechanical, measurement and management functions.

12.3.1 Fuel Storage

The fuel storage arrangement for the B777 is very similar in concept to Boeing's previous twin engine transports, comprising two integral wing tanks and an integral center tank.

As shown in the tank layout of Figure 12.20, the center tank is configured as two compartments in the inboard section of each wing connected by large diameter tubes that pass below the floor of the cargo compartment. Thus the two main sections can be regarded as a single center tank.

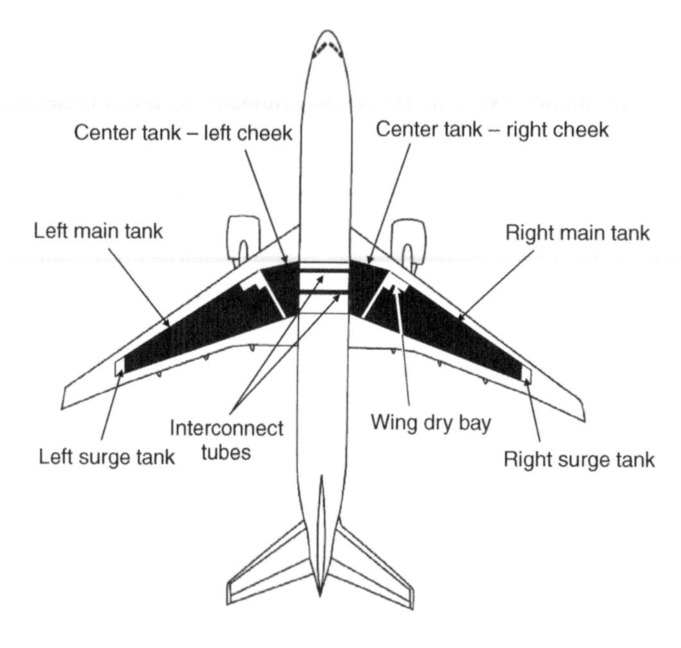

Figure 12.20 Boeing 777-200 fuel tank arrangement (courtesy of Arpingstone).

An interesting issue related to the design of this center tank is worth mentioning here. The interconnecting tubes may be considered as providing gravity cross-balancing between the two inboard sections of the center tank. This approach to lateral balance control has been used in other aircraft fuel systems where the main wing tanks are similarly interconnected by large diameter tubes with a cross-balance control valve that allows the crew to open thus allowing fuel from the heavy wing to flow to the opposite wing to provide lateral balance. In the authors view, this arrangement can be dangerous because of the potential for loss of lateral control that can occur. To illustrate this point, consider Figure 12.21 which shows the B777 arrangement in both steady level flight and in a sideslip.

As shown in the figure, in the presence of a sideslip, which usually only occurs transiently in normal operation but can exist in steady state when flying with one engine out, there is a rolling moment that is induced as a result of the change in lateral CG as the fuel transfers under the effects of gravity from the high side to the low side of the center tank. In the B777 application this moment is minimal because the center tank fuel is located close to the aircraft centerline. Clearly, if the main wing tanks were interconnected in this way the potential for very large and hence dangerous rolling moments could develop. For this reason, the use of a gravity cross-balancing system is not recommended under any circumstances.

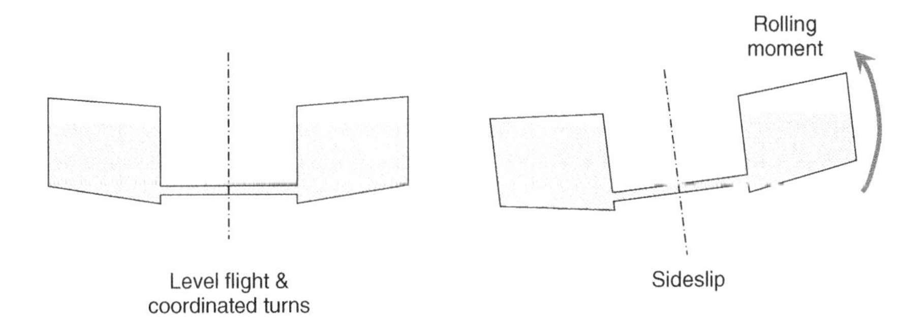

Figure 12.21 Effects of sideslip on center tank CG.

12.3.1.1 Vent System

The fuel tanks of the B777 are vented into surge tanks located outboard of each main wing tank. Each surge tank has a NACA scoop to recover outside air dynamic pressure and a flame arrestor to protect the fuel system from the possibility of a direct lightning strike igniting fuel vapors and propagating the resulting combustion into the fuel tanks. The surge tank also contains a re-settable relief valve that opens when a pre-determined pressure differential occurs in either direction between the outside air and the surge tank.

The tank vent lines are formed using vent channels formed using 'Hat' section stringers instead on conventional piping. Figure 12.22 shows a simplified rendition of the vent system arrangement for the B777 showing the main vent lines and outlets in each main tank compartment. As shown, float operated drain valves, located at the low point of the vent lines, are used to drain any fuel that may have entered the vent lines back into the center tank. At the outboard locations, float actuated vent valves are also used to close off the vent system when fuel is present thus preventing fuel from entering the surge tank. Should any fuel get into the surge tank (e.g. during ground taxiing maneuvers) a check valve between the surge tank and the main wing tank allows fuel to drain back into the main tank after take-off when the outboard wing tip is high.

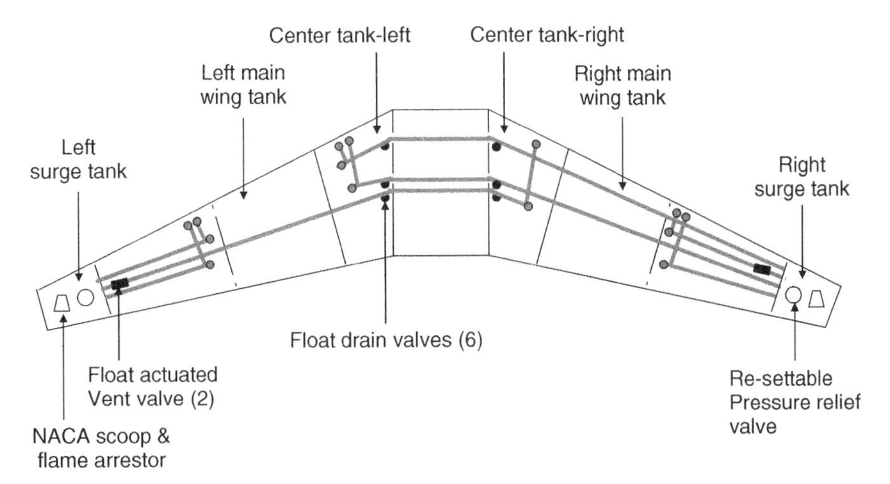

Figure 12.22 Vent system overview.

This vent system employs a relatively sophisticated pressure relief system to protect the structure from over-pressure situations. This device operates in either direction and can be reset on the ground. A cruder solution is to utilize a burst disk that protects the structure but requires significant maintenance down-time following an over-pressure event compared to the relief valve approach. Thus a dispatch versus maintainability trade-off is a choice for the system designer.

12.3.1.2 Water Management

Coping with water in fuel is a major challenge for the operator community with the high utilization rates demanded for economic success and this issue is particularly difficult in long-range operations where long, cold soaks at altitude followed by a descent into a hot and humid atmosphere can result in large quantities of water condensate mixing with the fuel. Since settling times are rarely adequate for effective use of water drain valves, alternative measures need to be taken. In the B777 two techniques are employed to manage water:

1. Small ejector pumps continually scavenge the fuel tanks in an effort to mix any residual water to form small droplets and to deposit the pump discharge into the each feed pump inlet. Motive power for these ejector pumps is taken from the feed pump discharge.
2. Ultrasonic water detectors are installed in the low points of each main tank and in the left and right sections of the center tank to detect the presence of free water. These sensors display advisory messages on the central maintenance display to alert the ground crew that there is water in the tanks.

12.3.2 Fluid-Mechanical System

The fluid-mechanical system of the B777 is illustrated by the schematic of Figure 12.23. This diagram is simplified for clarity by leaving out the APU feed system which is described later.

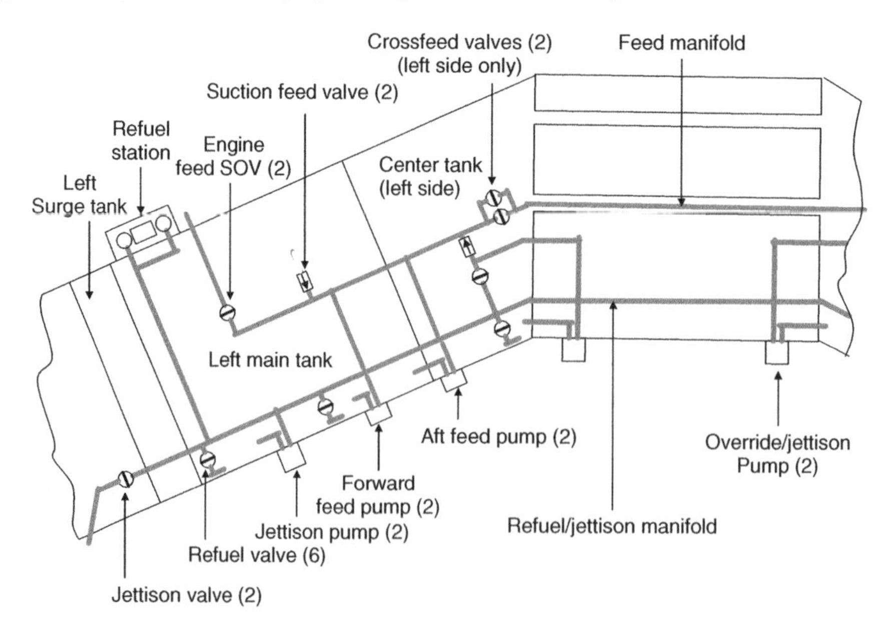

Figure 12.23 Fluid mechanical schematic.

All of the pumps and control valves are located on or close to the rear spar, well away from the engine rotating parts and therefore well outside the rotor burst zones.

Each of the main functions is described in the paragraphs that follow.

12.3.2.1 The Pressure Refueling System

The pressure refueling station is located on the leading edge of the left wing where there are two standard D1 adaptors alongside an integrated refuel and display panel. Since there are only three separate tanks on the B777, the refueling system is extremely simple.

Figure 12.24 shows the refueling system in schematic form.

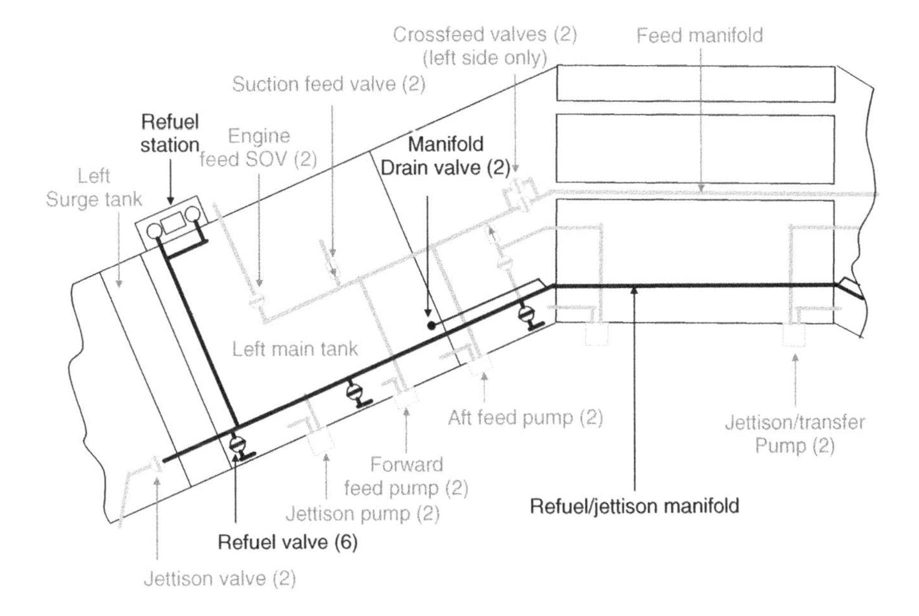

Figure 12.24 Pressure refueling system.

The refuel gallery runs along the aft side of all three tanks and also serves as the jettison gallery. There are six refuel shut-off valves in this aircraft (two per tank). The refuel gallery is drained at its low point to minimize unusable fuel. The manifold drain valves are located in each of the main wing tanks.

Control switches on the integrated refuel panel open and close the refuel shut-off valves. The panel provides auto-refueling by pre-setting the total fuel load required. Weight distribution between the three tanks is managed automatically and the appropriate refuel valves are selected closed when the correct weight in each tank is met.

The refuel valves will also close when the maximum volume for that tank is reached. The system test switch on the refuel panel provides the pre-check function. When auto-refueling is in process, pushing the test switch will cause the refuel valves to close and open automatically as auto-refueling is resumed.

Figure 12.25 shows the integrated refuel panel layout.

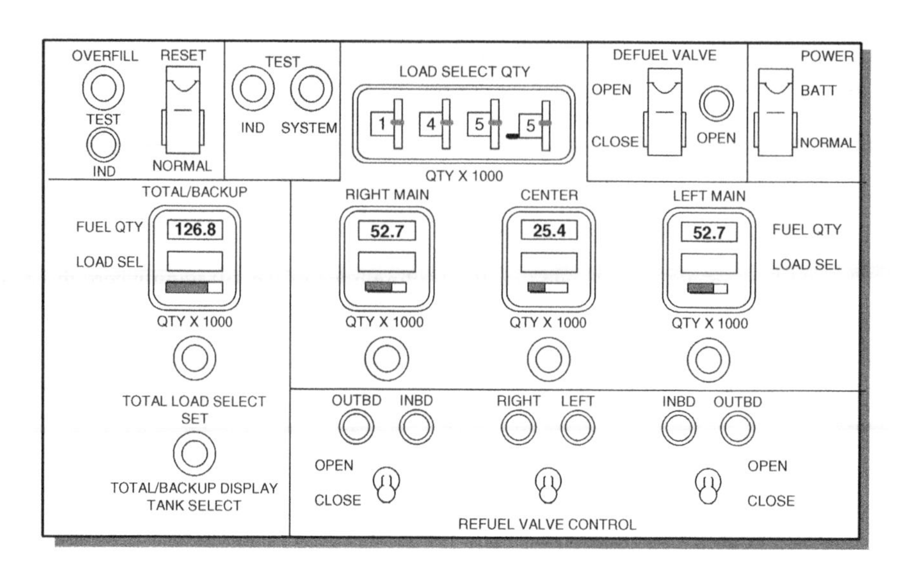

Figure 12.25 Integrated refuel panel.

An important operational issue with this refuel system is that power is required, either from the ground handling bus or from the aircraft battery, to operate the refuel valves. A manual back-up capability is not provided. Since there are two refuel valves per tank, however, single valve failures will not result in a dispatch delay.

12.3.2.2 Engine Feed System

The B777 feed system uses an override pumping system to provide the correct fuel burn sequencing, i.e. to burn center tank fuel first. This practice is a common feature of Boeing fuel system designs.

Although this arrangement requires significant upsizing of the center tank override pumps, this fact is somewhat compensated for by using the same pumps for the jettison function. Figure 12.26 shows the engine feed system in schematic form.

All feed pumps are spar mounted with snorkel lines. In the nomenclature used for the main feed pumps the term 'Forward' and 'Aft' refer to the location of the snorkel inlets. The override/jettison pumps are located in the fuselage with the inlets located in the left and right sections of the center tank respectively. The override pump outlets are connected to the engine feed manifold via check valves to protect the integrity of the feed line. With fuel in the center tank the higher outlet pressure from the override pumps overpowers the main tank feed pumps which operate essentially deadheaded (except for the small scavenge pump motive flow). Once the center tank is empty and the override pumps switched off, the main feed pumps automatically take over the engine feed task.

In the unlikely loss of both engine feed pumps the engine can operate in suction feed mode at altitudes up to about 25,000 ft.

Thermal relief valves (not shown in the schematic) protect the feed manifold from over pressure by relieving any excess pressure into their respective main wing tanks.

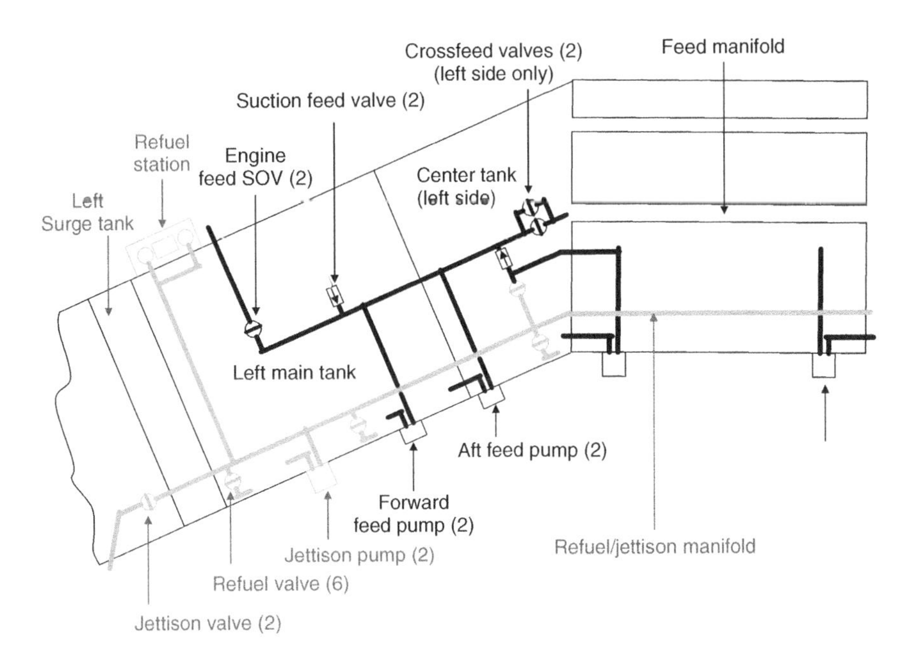

Figure 12.26 Engine feed system.

Crossfeed is provided via two separate isolation valves connected in parallel and located on the left side of the aircraft. This redundant solution ensures continued availability of function following any single failure which may be critical in ETOPS operations.

12.3.2.3 Jettison and Defuel Systems

The fuel jettison system allows the crew to dump fuel overboard following an emergency in order to reduce the aircraft weight to some value at or below the maximum landing weight. The main wing tanks have dedicated jettison pumps while the center tank override pumps are used to support the jettison function when selected by the crew.

Figure 12.27 shows the jettison system schematically (together with the defuel system to be described later).

In view of the functional integrity required for the jettison system, a two-step selection process is involved. The crew must first arm the system before selecting the system via a guarded switch. Both left and right jettison valves can be selected independently. Following selection the jettison system will automatically stop when the maximum landing weight is reached or, if required, a fuel remaining target can be selected by the crew. Installation of the main tank jettison pump is arranged so that the pump inlet becomes uncovered at some predetermined safe minimum quantity.

The defuel system is also shown on the above figure which also uses the refuel/jettison manifold. Opening the defuel valve via the integrated refuel panel connects the feed manifold to the refuel/jettison gallery. This valve is shown dotted in the figure because while it is functionally correct in the diagram, it is, in fact located on the right side of the aircraft.

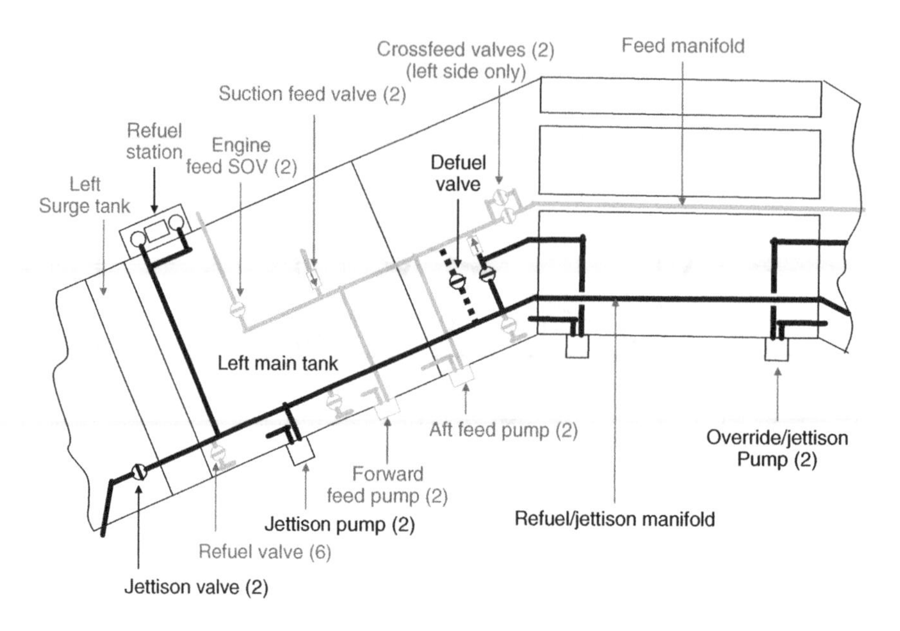

Figure 12.27 Jettison and defuel systems.

Defueling can be accomplished using the feed pumps or by applying suction to the refuel nozzles. It is desirable to disable the jettison valve during defuel to prevent inadvertent spillage.

12.3.2.4 APU Feed

The APU fuel feed system is illustrated by the schematic of Figure 12.28.

A dedicated dc powered feed pump, mounted on the rear spar pumps left main tank fuel through the APU fuel shut-off valve to the APU at the rear of the aircraft. Fuel lines through pressurized areas are double walled with overboard drains suitably protected with flame arrestors.

An isolation valve allows the APU feed line to be pressurized via the main engine feed line in the event of an APU pump failure.

12.3.3 Fuel Measurement and Management

12.3.3.1 FQIS System Overview

The Boeing 777 provided the first platform for the universal application of ultrasonic gauging to a commercial airplane fuel quantity indicating system. Boeing had until this point always used ac capacitance gauging systems for their commercial airplanes. The ultrasonic technology was seen as a way to further improve gauging reliability by eliminating the need for in-tank harnessing shield continuity essential to ac capacitance gauging systems. This ultrasonic gauging solution utilizes a fault- tolerant and redundant system architecture which blends this new approach to gauging while retaining many traditional tried and tested Boeing gauging practices. The capacities of the three tank configuration of the Boeing 777-200 are shown in

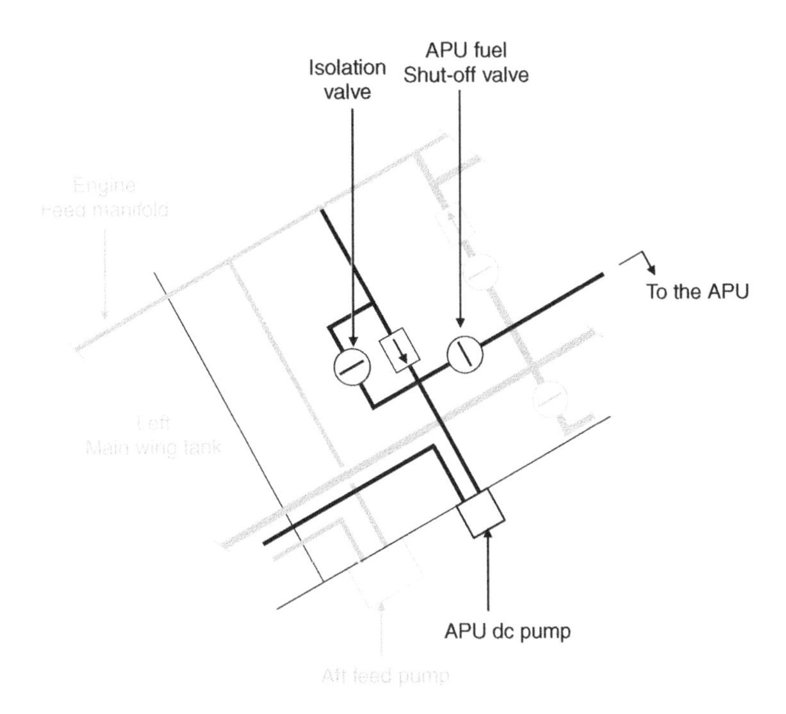

Figure 12.28 APU feed system.

Table 12.1. As already described, the center tank comprises left and right sections sometimes referred to as 'cheek' tanks located on each side of the fuselage interconnected by two large diameter pipes.

Table 12.1 Boeing 777–200 fuel tank quantities

	Boeing 777–200 tank quantities			
	USGalls	Liters	lbs*	kgs*
Main tanks	9,300 ea	35,200 ea	62, 870 ea	28,515 ea
Center tank	12,400	46,940	83,825	38,020
Totals	31,000	117,340	209,565	95,050

* Based on a nominal fuel density of 6.76 lb/USGall, 0.81 Kg/L

The three-tank Fuel Quantity Indicating System (FQIS) comprises a number of ultrasonic probes, velocimeters and water detectors supported by a vibrating cylinder densitometer within each tank.

The left tank unit array also includes a temperature sensor. The tank units are interconnected by internal tank harnessing that features twisted pair wiring terminated with quick disconnect technology both at the tank units and at the tank wall. External tank harnessing interconnects the tank unit arrays to the central Fuel Quantity Processing Unit (FQPU) which is the heart of the FQIS.

This unit, which is housed in an air-cooled ARINC 600 5MCU located in the equipment bay, provides the following functions:

- determines the quantity of fuel within each fuel tank
- determines the total airplane fuel quantity
- communicates all quantities to the Airplane Information Management System (AIMS)
- communicates all quantities to the Integrated Refuel Panel
- commands refuel valves to open or close during the auto-refuel process
- provides FQIS health monitoring
- communicates all FQIS fault data to the Central Maintenance Computer System (CMCS).

The FQIS architecture is illustrated in Figure 12.29 showing the main elements within the FQPU. This processor is configured with 2 identical but independent channels, each comprising an input/output circuit and an ARINC 629 and 429 card.

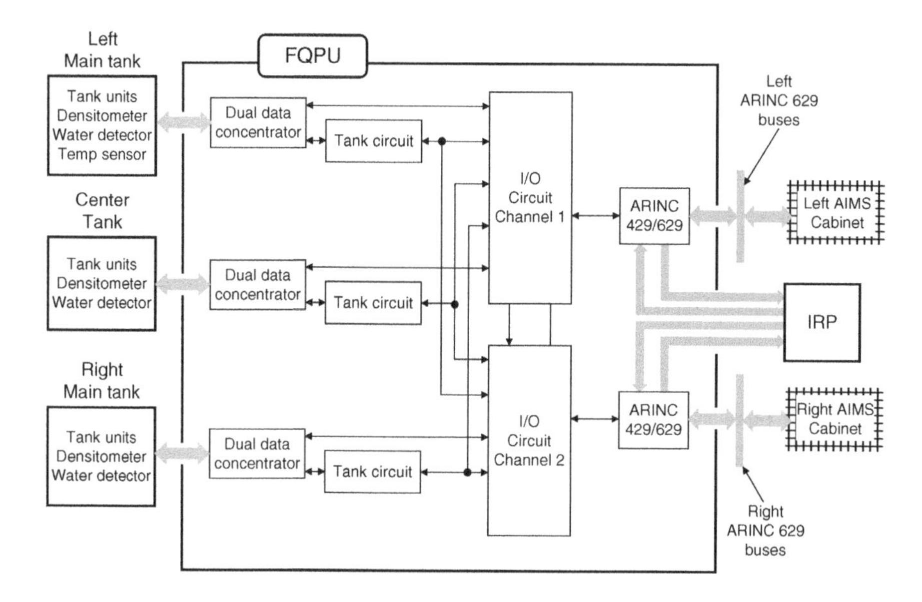

Figure 12.29 FQIS system architecture.

The Boeing 'Brick wall' policy of independence of the gauging of each tank is primarily implemented using 3 dual channel data concentrators and 3 tank circuits. A detailed treatment of ultrasonic gauging technology is discussed in Chapter 7.

Each of the channels within a data concentrator interfaces with all the tank units of a tank, a technique not readily achievable in ac capacitance systems. Each of the three calculated quantities is then input to each of the two independent channels for communication by the associated ARINC 629 and 429 busses. The FQPU has a highly fault tolerant architecture in that even if a data concentrator or tank circuit fails, gauging of the three tanks is maintained. The alternate data concentrator channel of the problematic tank may be directly used with the input/output circuit to compute fuel quantity as the latter circuit is also equipped to perform the tank circuit function should the necessity arise.

12.3.3.2 In-tank Gauging Sensor Implementation

Fuel Height Measurement Probes (Tank Units)

The left and right main tanks of the Boeing 777-200 are each gauged by 17 ultrasonic fuel height measuring probes and three ultrasonic velocimeter-type probes. The left and right cheeks of the center tank are each gauged by five ultrasonic fuel height measuring probes and 1 ultrasonic velocimeter-type probe.

Each ultrasonic probe measures the height of fuel at the probe location by sending a sound pulse from the bottom of the tank to the fuel surface and back as is described in Chapter 7.

Each ultrasonic fuel height measuring probe (see Figure 12.30) comprises a stillwell with an ultrasonic transmitter/receiver located at the bottom. The inside of the stillwell is specially coated to provide optimum acoustic properties. To avoid any processing confusion between the surface and any target reflections, speed of sound measurement is performed in a separate velocimeter.

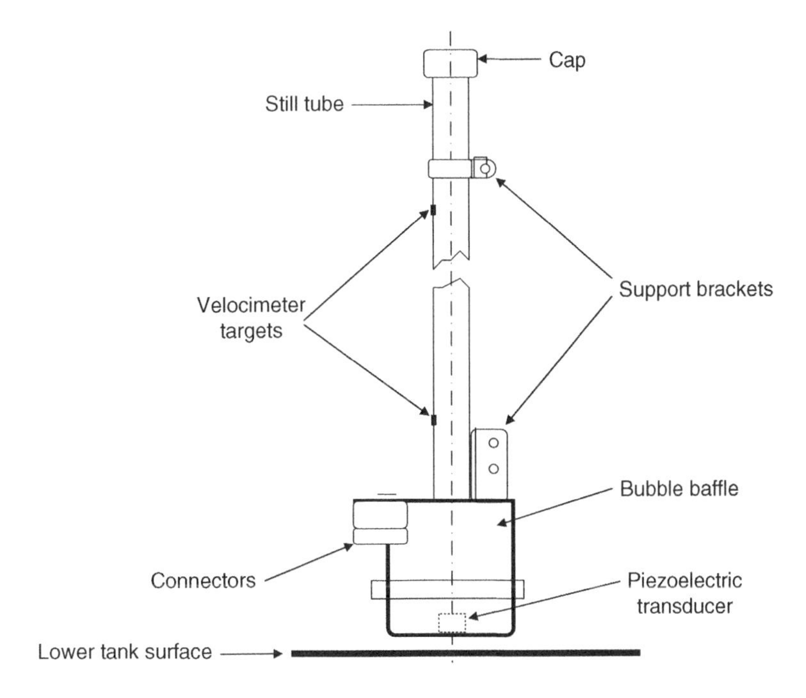

Figure 12.30 Ultrasonic fuel height measurement probe.

The transmitter/receiver assembly features a bubble shroud to prevent the ingress of large bubbles into the stillwell, such as those likely to be generated during the refuel process which can cause premature acoustic reflections and under-gauging. The transmitter/ receiver assembly contains a piezoelectric ceramic crystal transducer that has a three-resistor network mounted directly on, and electrically in parallel with, the crystal. This is necessary to provide safe discharge of any energy created by any abnormally large mechanical or thermal shock effects.

Electrical interconnection is provided by a 'quick disconnect' connector block located on the side of the bubble shroud. Because of the electrically pulsed nature of the input and output signal, the interconnection with the processor is inherently HIRF resistant and may be made using twisted-pair wiring. However, the externally-located, multiple probe, out-of-tank cabling between the tank wall connectors and the fuselage is double shielded to provide electro-magnetic protection against lightning strike.

Velocimeters

The fuel height measurement probes do not include the means necessary for the system to measure and compensate for variations in the speed of sound; this is accomplished separately by the velocimeters.

The velocimeters are identical in design to their fuel height measuring counterparts but with the addition of several targets located at regular intervals along the stillwell to ensure that the speed of sound can be obtained at the varying fuel levels within each tank. Velocimeter units are not directly used for fuel height measurement.

Densitometers

The fuel quantity measuring hardware is completed by the inclusion of three vibrating cylinder densitometers (Figure 12.31) to facilitate the conversion of fuel volume information into fuel mass. The principles of the vibrating cylinder densitometer are described in detail in Chapter 7. In the B777 application, a densitometer is installed in the inboard compartment of each wing tank. A third densitometer is installed is located in the left cheek of the center tank. As with the ultrasonic tank units, the densitometers feature quick disconnect terminal blocks.

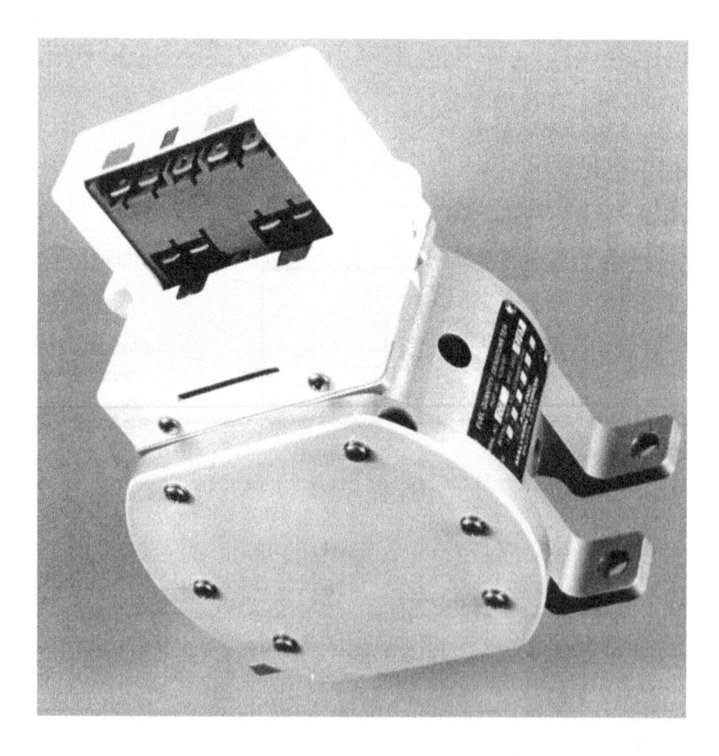

Figure 12.31 Fuel densitometer (courtesy of GE Aviation formerly Smiths Aerospace).

12.3.3.3 Water Detection

As already mentioned in the fluid-mechanical section, the B777 is fitted with ultrasonic water detectors. Four of these devices are installed in the system and provide a maintenance page message of the presence of 'pooled' water in the sump area of the affected tank(s). The water detector is similar in design to an ultrasonic fuel height measuring probe with the exception that it is designed for mounting in opposite fashion to a probe with the transmitter/receiver assembly uppermost. Also the stillwell features vertical slots to ensure full ingress of the fuel/water interface for detection. Since the transmitter/receiver is uppermost, the ultrasonic signal is projected downwards to the fuel/water interface. With increasing free water content, the interface rises to cause a proportionately reducing sonic pulse round trip travel time that is detected by the processor to generate the maintenance panel message.

12.3.3.4 Fuel Management

Due to the simplicity of the fuel storage and override center-to-wing transfer system, the fuel management task on this aircraft involves only the following tasks which are under the control of the FQPC:

- control of the auto-refuel process by closing the refuel valves when the correct tank quantities have been reached;
- communicating system health status to the AIMS for display on the EICAS.

Other fuel management tasks including control of the Jettison System, and de-selecting the override pumps following depletion of the center tank are under the direct control of the flight crew.

The resulting system is both simple and highly redundant with minimal opportunity to be the cause of a dispatch delay.

12.4 The Airbus A380 Wide-Bodied Airliner

The Airbus A380-800 which is the world's largest commercial transport aircraft is a double deck wide-bodied aircraft with a maximum take-off weight of 560 tonnes and a range of 8200 nautical miles. This aircraft entered service with Singapore Airlines in October 2007 configured with a three-class cabin having a total capacity of 550 passengers.

This version of the A380 has a maximum fuel capacity of 81,890 US gallons (equivalent to approximately 250 tonnes at 70 degrees F). All fuel is stored in integral wing tanks.

These tanks are clearly delineated in Figure 12.32 by the darker shade of the wing surface between the front and rear spars.

The fuel system design for this aircraft is complex and contains many features necessary to meet a number of demanding requirements which include:

- Superior dispatchability and functional availability. This is critical for such a large aircraft where the impact of delays and cancellations can be prohibitive in both coast and reputation.
- High functional integrity of the measurement, management and fuel handling systems in order to support critical aircraft operational modes including active CG control and wing load alleviation.

- Excellent quantity gauging accuracy with minimum degradation in the presence of equipment failures.

Figure 12.32 The Airbus A380-800 (courtesy of www.diecastairplane.com).

The functions provided by the fuel system include aircraft refueling and defueling, fuel feed to the engines and APU, fuel jettison, fuel measurement and management. The latter function includes many complex control functions including, refuel distribution & control, fuel transfer and fuel burn sequencing, wing load alleviation, lateral balance and active longitudinal CG control.

Each of the fuel system functions of this remarkable aircraft is described in the following paragraphs.

12.4.1 Fuel Storage

The fuel storage arrangement is shown in Figure 12.33. As shown there are five tanks in each wing and a trim tank in the horizontal stabilizer. The wing tanks are vented to a vent tank outboard in each wing. In addition there is a mid-span surge tank which is needed to accommodate the ground refuel condition when the fuel and engine weight results in an air bubble location in the region of the surge tank.

The trim tank is vented via a conventional outboard vent tank located on the right side of the trim tank.

During the refuel process the outer wing tank quantity is limited to prevent excessive wing bending moment due to fuel and engine weight. After the aircraft leaves the ground and WOW becomes 'False', the fuel transfer pumps move fuel from the mid and inner tanks to the outer

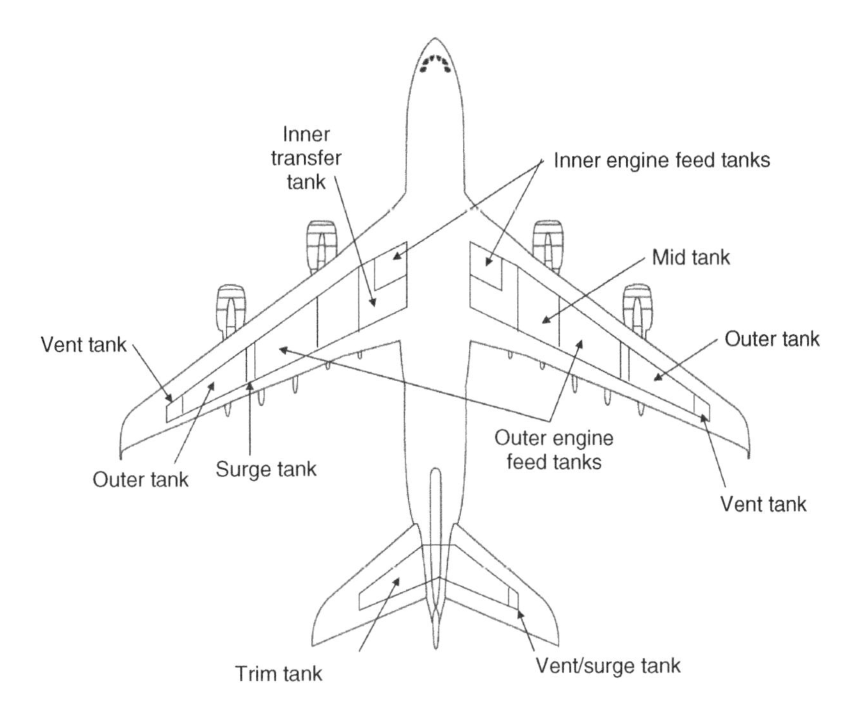

Figure 12.33 A380-800 fuel tank arrangement.

tank for wing load alleviation purposes. This outer tank typically remains full until the top of descent is reached at the end of the cruise phase. By then the aircraft weight is substantially reduced through fuel consumption and wing bending relief is no longer necessary.

Note that the A380-800 aircraft does not have a center fuel tank; however, provision has been made within the design to add additional fuel capacity using the center wing box for longer range versions of this aircraft.

12.4.2 Fluid-Mechanical System

The fluid-mechanical system of the A380-800 is very complex in view of the many functions and work-arounds provided and the large number of auxiliary fuel tanks involved. In fact there are 21 fuel pumps and 46 control valves in the fluid mechanical system. This does not include scavenge ejectors, air release and thermal relief valves which have been omitted from the schematics that follow, for clarity.

In order to provide a clearer depiction of the main system functions, the schematic diagrams for the specific function being addressed are depicted in bold while the remaining equipment is shown in grayscale.

The fluid-mechanical system implementation supports the following fuel system and auxiliary functions:

- refueling (and defueling)
- engine and APU feed

- fuel transfer in support of fuel burn sequencing
- fuel transfer in support of CG control and wing load alleviation
- fuel jettison
- hydraulic system cooling.

Each of the above system functions is addressed below from the fluid mechanical perspective. The issues associated with management, control and fault accommodation are addressed later in this section.

12.4.2.1 Refuel and Defuel

The pressure refueling system is supported by two wing-leading-edge refuel stations each providing access to the aircraft refuel circuit via two standard nozzle connections. The integrated refuel and display panel is located in the fuselage between the wings. As in all modern transport aircraft, the refuel process is controlled by the fuel management system to ensure that aircraft balance is maintained throughout the process from both lateral and longitudinal perspectives. A manual back-up capability is available where the ground crew can control the process by selection of refuel valves from the integrated refuel panel (IRP).

The fuel system refuel schematic is shown on Figure 12.34. This diagram shows the left-hand wing fluid mechanical arrangement which is repeated on the right-hand side of the aircraft.

The two main galleries and their discharge valves and associated diffusers are available to support the refuel function which provides a great deal of flexibility and the ability to accommodate multiple equipment failures. Defuel of the complete aircraft or individual tanks can be achieved using either the internal aircraft pumps and externally applied suction. This process is typically controlled manually by the operator using the IRP Refuel/Defuel switches for the targeted tanks.

The internal pumps are activated from the flight deck OverHead Panel (OHP). Complete gauging capability is provided during the defueling operation; however fuel characterization and integrity checks are inhibited.

12.4.2.2 Engine and APU Feed

The engine feed system is of traditional Airbus design comprising collector cells within each feed tank that are maintained full via large ejector pumps that draw fuel from the inboard forward section of each feed tank. These ejectors receive their motive flow. from the feed pump discharge and are sized to meet maximum engine take-off flow. During normal operation, the ejector pumps maintain a small positive pressure within the collector cell; excess flow being spilled back into the feed tank. Two motor-driven boost pumps located within each collector cell provide boost pressure to each engine. One pump operates as the primary pump while the second pump provides a back-up capability and is switched on automatically following loss of primary boost pump pressure. Small scavenge ejectors also driven from engine feed pressure are installed in the collector cells to induce any free water from the cell bottom and discharge it in small droplet form at the inlet to the boost pumps where it is combined with the feed flow and burnt by the engine.

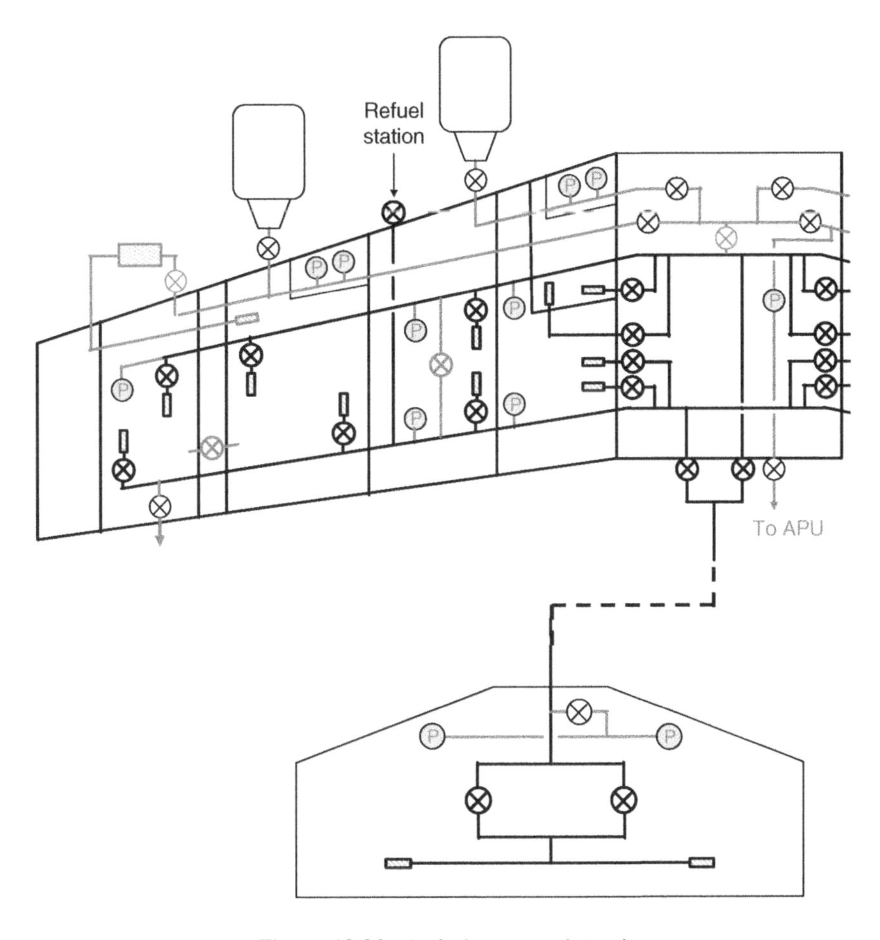

Figure 12.34 Refuel system schematic.

Figure 12.35 shows a schematic of the feed system for the left hand side of the aircraft. Note that the collector cell main and water scavenge ejectors are not shown for clarity.

As shown in the figure, four crossfeed valves allow fuel crossfeed to any of the four engines following a shut down of any of the four engines thus ensuring that feed tank fuel from a shut-down engine is made available to the other remaining engines.

The APU is fed from the right outer feed tank and has its own dedicated dc motor-operated fuel boost pump for initial start-up when the outer engine feed pump is not running. Once the engines are running, the dc pump can be switched off. A separate low pressure APU isolation valve (not shown) is provided close to the APU.

12.4.2.3 Fuel Transfer

The transfer system uses the forward gallery for all normal transfers. Figure 12.36 shows the transfer system schematic. In the presence of equipment failures, access to the aft gallery is available as a work-around.

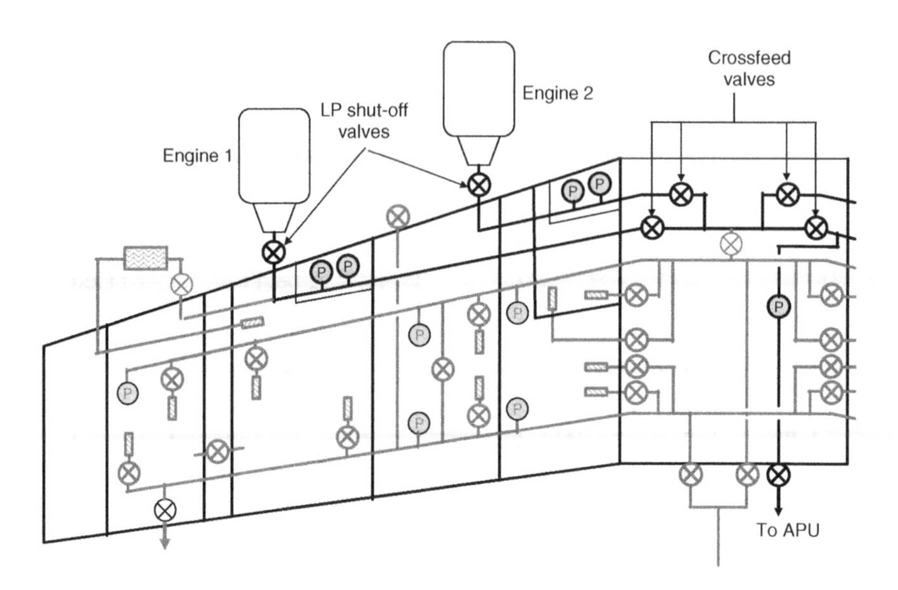

Figure 12.35 Engine and APU feed system.

The various fuel transfer functions which are many and, in some cases, complex include the following:

Fuel Burn Sequencing
During flight, the fuel management system transfers fuel from the inner and mid fuel tanks to the feed tanks so that the feed tanks remain essentially full until the inner and mid tanks have been depleted. The outer wing and trim tanks are typically the last of the auxiliary tanks to be transferred. During this process which involves a specific predetermined schedule, lateral balance of the aircraft is automatically maintained.

Wing Load Alleviation
To minimize wing bending stresses during flight, fuel is transferred from the inner and mid tanks to the outer tanks immediately following take-off and these outer tanks typically remain full until the end of the cruise phase.

Active Longitudinal CG Control
The initial quantity of fuel in the trim tank is dependent upon aircraft loading and is determined as part of the refuel distribution process. As fuel is consumed, the aircraft CG will move aft until the optimum cruise condition is reached. Any further CG shift outside a predetermined tolerance, will result in a fuel transfer between the trim tank and the wing tanks. So long as fuel remains in the inner and mid tanks, any forward transfers will be into the inner tanks. Once the inner and mid tanks are empty, forward transfers are into the feed tanks.

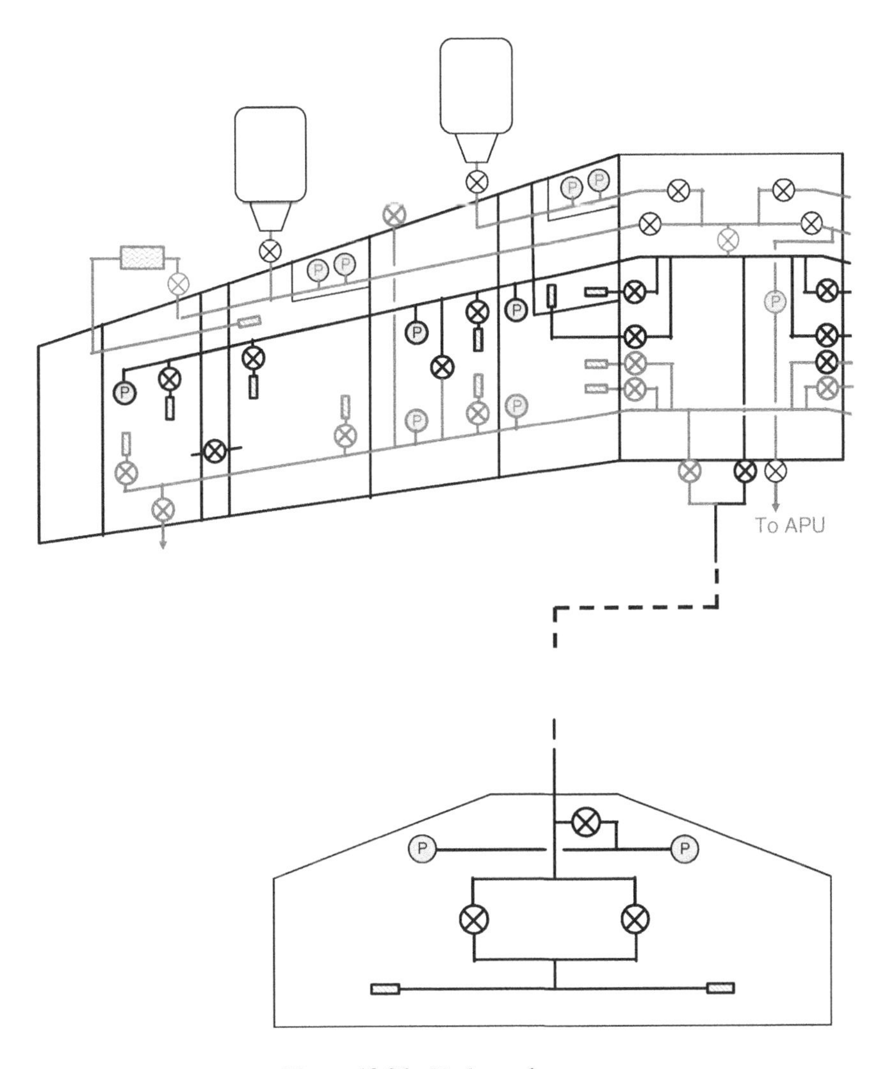

Figure 12.36 Fuel transfer system.

Gravity Transfer
Transfer of fuel from the outer tanks or from the trim tank in the presence of pump failures can be accomplished using gravity. Trim to wing transfer by this method is limited to a specific range of aircraft pitch attitudes.

Transfer from the outer tanks using gravity is achieved by opening the transfer valve located in the mid-wing surge tank.

12.4.2.4 Fuel Jettison

Fuel jettison is selected by the flight crew via guarded switches to arm and select the system. The crew can select a specific fuel load (Dump to Gross Weight) at which the jettison function is cancelled. Otherwise the jettison continues until the maximum landing weight is achieved.

Figure 12.37 shows a schematic of the jettison transfer system which uses the aft gallery exclusively. The jettison pumps are located in the inner and mid tanks as shown. During jettison, the trim tank is emptied into the inner tanks. Jettison valves connect the aft gallery to the fuel dump mast in each wing.

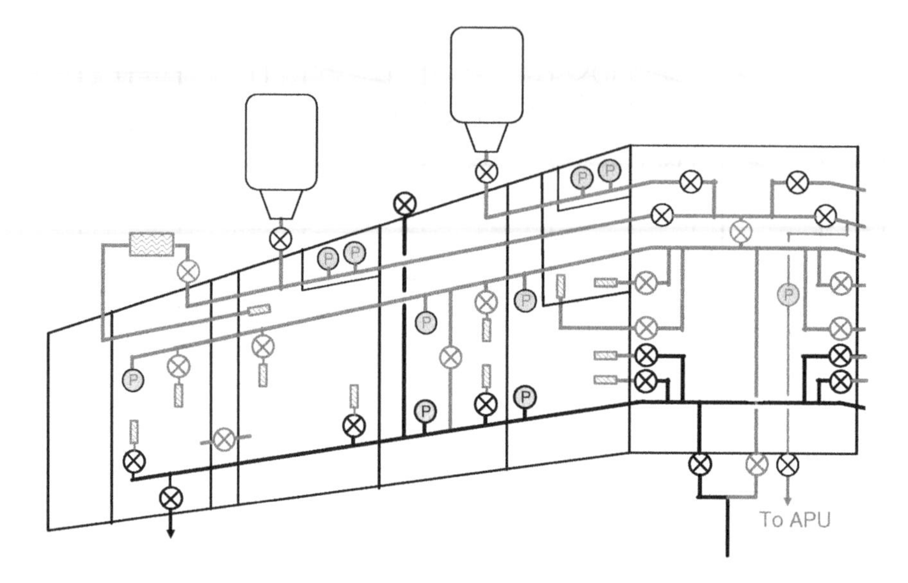

Figure 12.37 Jettison system schematic.

12.4.2.5 Hydraulic System Cooling

The fuel system is used to cool the aircraft hydraulic systems using feed flow from the outer engines as shown in Figure 12.38. The integrity of the engine feed system function must be addressed carefully when using this approach. Discharging hot fuel back into the feed tank

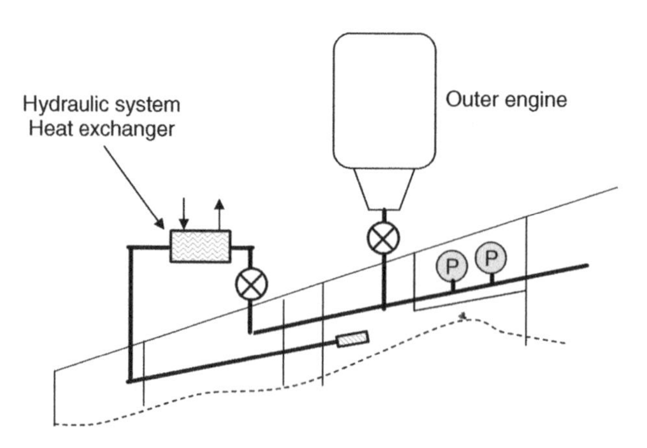

Figure 12.38 Hydraulic system cooling.

must be done carefully so that the accuracy of the fuel quantity gauging systems is not impacted as a result of local fuel heating near fuel quantity probes.

12.4.3 Fuel Measurement and Management System (FMMS)

12.4.3.1 FMMS Architecture

As mentioned in Chapter 7, the A380 features an Integrated Modular Avionics (IMA) suite comprising a number of Central Processor Input/Output Modules (CPIOM) units interconnected by an Avionics Full DupleX (AFDX) switched ethernet digital data bus to both operate and communicate between the numerous aircraft systems that include the Fuel Measurement and Management System.

Figure 12.39 is an overview of the fuel system architecture showing the control and monitoring information flow.

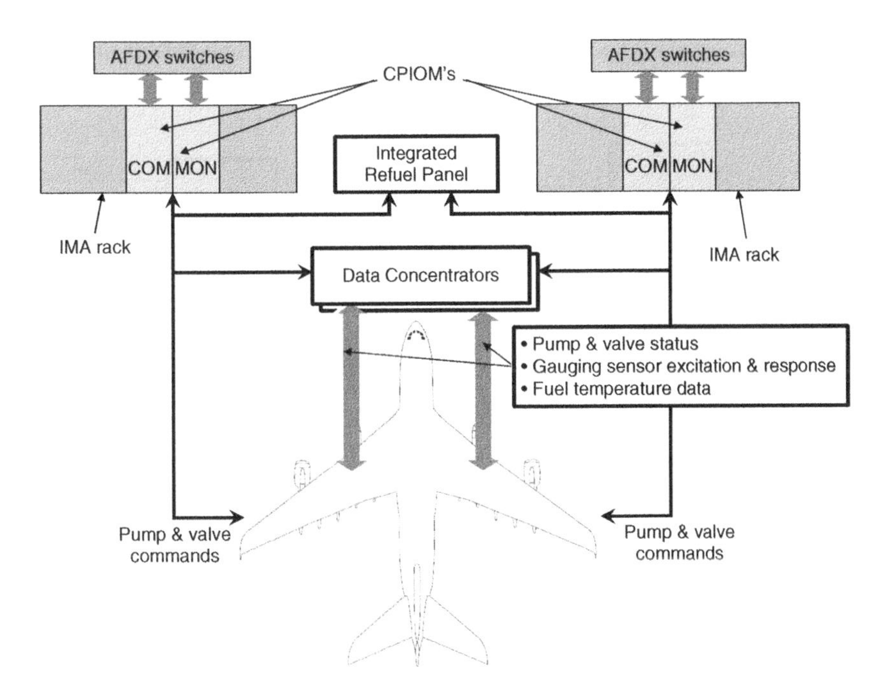

Figure 12.39 Fuel Measurement and Management System (FMMS) architecture overview.

Each CPIOM is a simplex computing unit having an operating system capable of running multiple partition resident software applications supporting a standard input/output interface. The functionality of the software is the responsibility of the Fuel Measurement and Management System supplier.

The principle of the IMA concept is to reduce life cycle costs through the use of standard avionics modules with common hardware. However, because of the disparate input/output and integrity needs of the various systems using IMA avionics, it has been necessary to develop several different CPIOM standards for the A380 aircraft. The interconnecting AFDX, also referred to the Aircraft Data Communications Network (ADCN), is an ethernet-based network adapted to aeronautical constraints. It allows all the CPIOMs to simultaneously transmit and receive data at 100 Megabits per second.

The CPIOMs are arranged in pairs to form computing lanes. Each lane is configured with one CPIOM designated as the 'Command' (COM) channel while the other CPIOM designated as the 'Monitor' (MON) channel. Each of the two fuel system computing lanes is capable of performing all of the system functions with one of the two lanes designated as the 'Primary' lane controlling the system while the other lane operates as a 'Standby'. The functional health of each lane is continually assessed by the BITE software within each MON channel and should the health of the primary lane deteriorate to a level below that of the standby lane, control of the system is switched over to the standby lane.

Each lane interfaces with the two Fuel Quantity Data Concentrators (FQDCs) which interface with the in-tank equipment.

FMMS Avionics

Figure 12.40 shows the avionics architecture employed by the A380 showing how the four CPIOMs interconnect with the data concentrators (FQDC's) and the Integrated Refuel Panel (IRP).

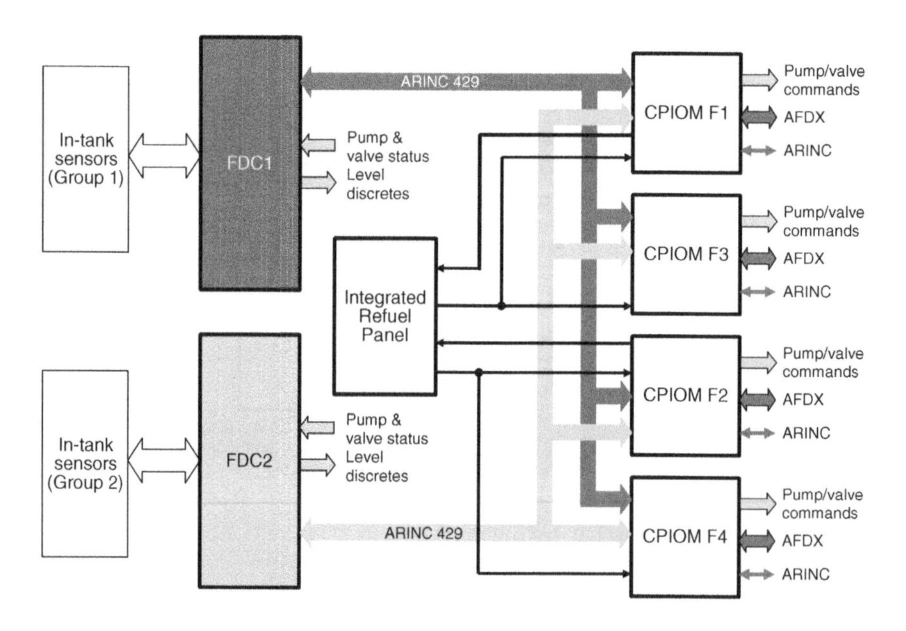

Figure 12.40 FMMS avionics architecture (courtesy of Parker Aerospace).

Fuel Quantity Data Concentrators (FQDCs)

The two fuel FQDCs function as data acquisition units. They acquire and process tank located component data, compute alternative fuel quantity data, acquire pump and valve feedback signals, and generate back-up fuel level warnings This data is transmitted to the CPIOMs by redundant high speed ARINC 429 data bus. Additionally, the unit performs both tank located component built-in-test (BIT) as well as extensive internal BIT.

Each FQDC is configured with three independent 'brick walled' processing channels comprising two Tank Signal Processors (TSPA and TSPB) and an Alternate fuel Gauging Processor (AGP)and Discrete Input (DIN).

Each TSP provides processed in-Tank component data for:

- tank capacitance probes
- densitometers
- temperature sensors.

The AGP provides:

- a second source of FQI data computation output for comparison purposes (see fuel quantity measurement below);
- Back-up Low Level and Overflow warnings;
- pump and valve feedback discrete status.

All of this data is transmitted over ARINC 429 to the CPIOMs.

CPIOM Software Partitioning

Each pair of fuel system CPIOM's within the IMA suite execute FMMS software with the COM and MON functions partitioned as shown in Figure 12.41. The fuel system supplier is responsible for the functionality of the embedded software in the CPIOM's.

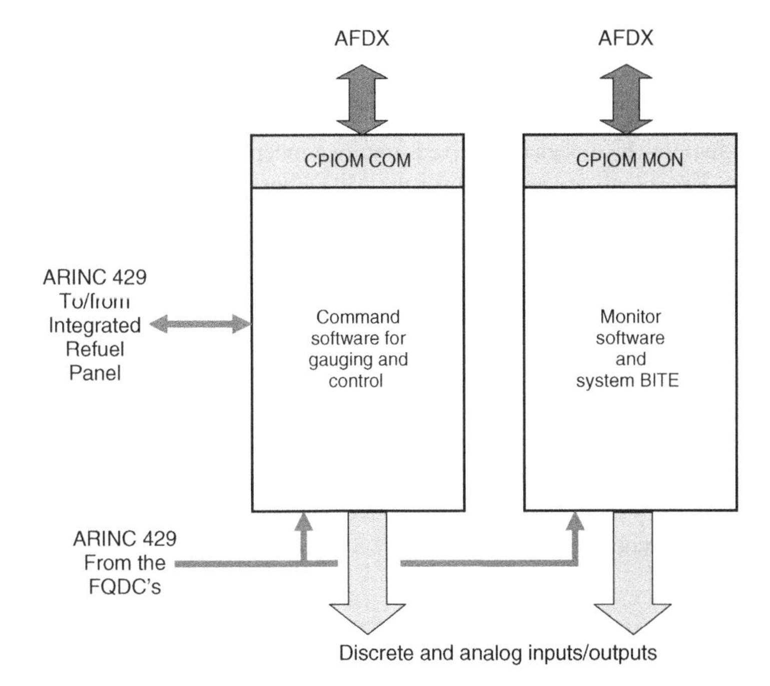

Figure 12.41 CPIOM software partitioning.

FMMS Functionality

The FMMS performs the following core functions:

(i) Fuel Quantity Measurement and Indication

The A380 gauging system is an ac capacitance system designed to accommodate multiple failures without accuracy degradation. This system measures, processes and monitors the fuel quantity in each tank with an accuracy of better than ± 1%.

A self-heal algorithm is incorporated to deal with various failures and performs an accuracy prediction. The system sends fuel quantity data to various aircraft systems including the ECAM, CDS and CMF.

The fuel measurement system provides a data display integrity of 10^{-9} per hour which represents the probability of the system displaying to the flight crew an erroneous but believable fuel quantity. This is accomplished by employing an Alternative fuel Gauging Processor (AGP) to compute fuel quantity using dissimilar algorithms from the main gauging system. This 'Third' fuel gauging channel is incorporated within the FQDCs and any significant difference between the main gauging system and the alternative fuel gauging processor is used to annunciate an integrity failure.

(ii) Fuel Temperature Measurement and Indication

Fuel temperature is measured, processed and monitored within each tank in a fault tolerant manner. Sufficient sensor redundancy is provided meet the catastrophic failure requirements, high integrity fuel temperature warning is provided for each feed tank and a low fuel temperature warning is provided for the trim and outer tanks. High accuracy fuel temperature measurement is utilized by the fuel gauging algorithm.

(iii) Fuel Level Determination and Indication

High integrity fuel low level signals, derived from two independent sources, are provided for each feed tank. The signals are relayed over the AFDX bus with a backup discrete provided from the FQDC in the event of loss of the IMA or AFDX. Fuel high level indication for each tank is produced using the fuel probes as level sensors. Indication is provided at the IRP and on the AFDX signal for the aircraft systems. Fuel overflow indication, derived from two independent sources, is provided for each surge tank. The signals are relayed over the AFDX bus with backup discretes provided from the FQDC in the event of loss of the IMA or AFDX.

(iv) CG Measurement

In order to meet the hazardous failure conditions, CG is determined by two independent methods. Using zero fuel and zero fuel aircraft CG, fuel and aircraft CG are computed, respectively. Also the accuracy of CG measurement is predicted. CG information is initialized when commanded by the aircraft interface.

(v) CG Management

The FMMS computes the aircraft CG targets and performs forward and aft transfers by transferring fuel to and from the trim tank to maintain aircraft CG within a predefined target. A CG limit warning is provided to the crew.

(vi) Fuel Transfer Control

The various transfer modes including fuel burn sequencing, wing load alleviation, etc. were described in the fluid mechanical section (see Section 12.4.2.3). Control of the transfer functions is accomplished automatically by the FMMS utilizing tank quantity information and controlling the pumps and valves appropriately. In order to provide adequate functional integrity for this task, the flight crew has manual back-up capability via the Over-Head Panel (OHP) shown in Figure 12.42. Here control of individual pumps and valves is provided to allow the crew to accommodate any failures if the automatic system.

(vii) Refuel Control

The FQMS provides both automatic and manual refuel control from an integrated refuel panel (IRP). Automatic refuel may also be controlled from the flight deck. The required fuel mass quantity is preselected by the operator. The individual tank quantities are targeted to maintain longitudinal CG and lateral balance to predefined limits during refuel operation. Each tank fuel quantity target prediction is performed to maintain post-refuel distribution. Automatic operation is aborted whenever refuel limits are not maintained. A complete status of refuel is provided on the IRP status message display.

(viii) Defuel Control

This may be achieved using the internal aircraft pumps and externally applied suction. It is controlled manually by the operator using the IRP Refuel/Defuel switches for the targeted tanks.

The internal pumps are activated from the flight deck Over-Head Panel (OHP). Complete gauging capability is provided during the defueling operation; however fuel characterization and integrity checks are inhibited.

(ix) Jettison Control

This is push-button selected by the crew in the cockpit. The crew select a quantity of fuel to be jettisoned via a Jettison Fuel to Gross Weight (JFWG) selection. The system stops jettison when the selected amount of fuel has been discharged or jettisons fuel down to predetermined quantity. The system activates the jettison control valve by applying high integrity series control from the Command and Monitor CPIOM outputs. The system provides complete gauging capability during the jettison operation.

(x) Transfer Status, Warning and Caution Indication

Comprehensive information on pump, valve and transfer status is provided by the system. Pump status identifies 'on', 'off', 'abnormal on' and 'abnormal off' conditions. Valve status identifies 'open', 'shut', 'failed open' and 'failed shut' conditions. Transfer status identifies transfer active', 'transfer failed' and 'transfer inhibited' conditions.

(xi) System BITE

The system features enhanced built in test (BIT). It performs power up BIT and a safety test to verify all safety requirements. A continuous cyclic test is performed to monitor all fuel system equipment to reduce failure exposure time. Extensive fault analysis of the tank located sensors (open, short, contaminated) of the fuel measurement system enables failures to be both isolated and identified for type of failure. Failures between the LRUs, CPIOMs and harnesses are isolated. The system interfaces with the aircraft Central Maintenance Function (CMF).

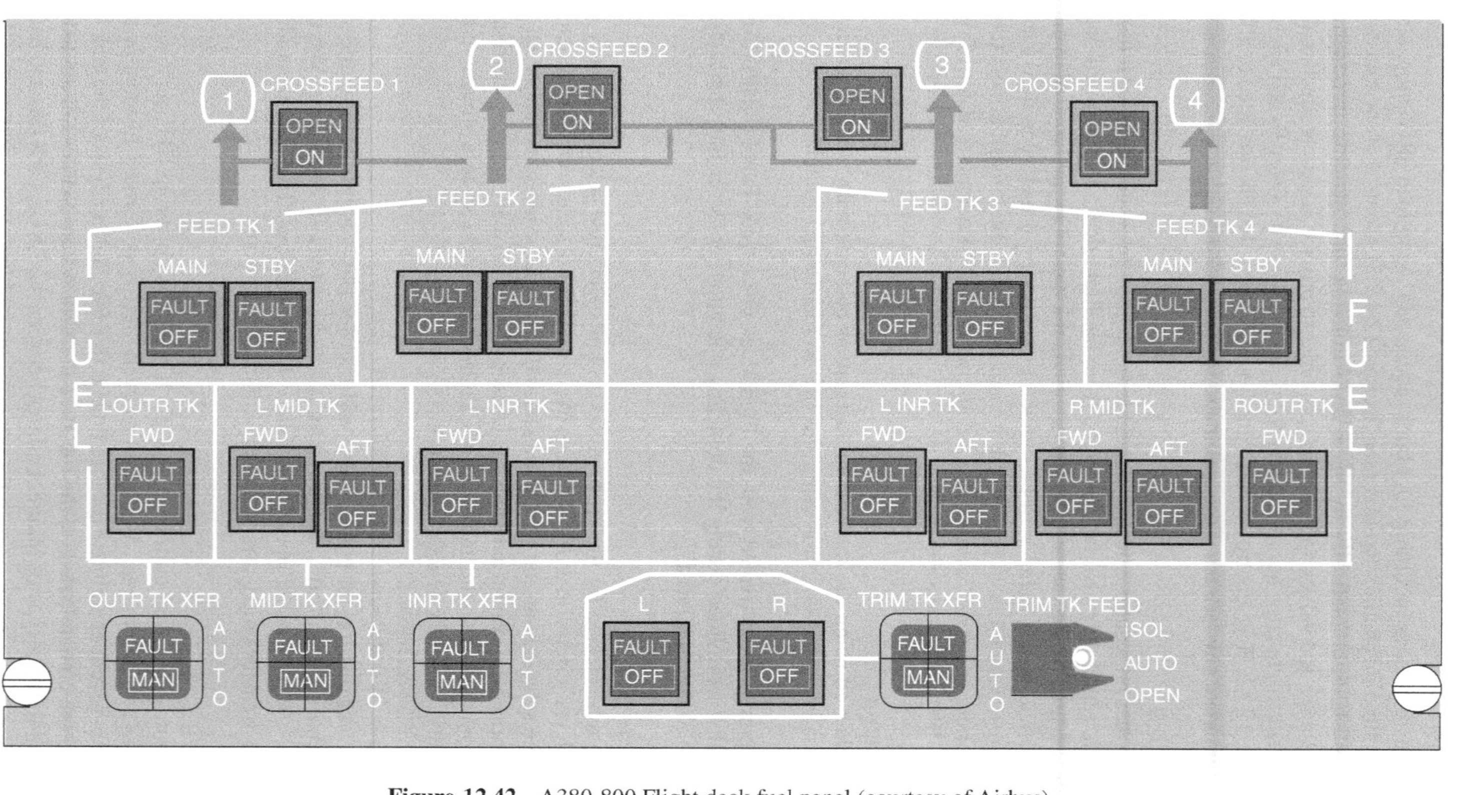

Figure 12.42 A380-800 Flight deck fuel panel (courtesy of Airbus).

It is interesting for the reader to compare the relative fuel system complexities of the Boeing and Airbus long-range transport aircraft. Boeing fuel systems are typically simpler with fewer tanks and a minimum of automated fuel management. The Airbus designs tend to be more complex with automated fuel management control modes including wing load alleviation and active CG control.

12.5 The Anglo-French Concorde

The Concorde Program initiated in the early 1960s was a joint Program between the French Aerospace Company Aerospatiale and the British Aircraft Corporation with the objective of introducing supersonic passenger service on the lucrative and rapidly growing trans-Atlantic routes between Europe and the Americas. The aircraft received its Type Certificate in 1972 but, due to largely political obstructions in the USA related primarily to the higher airport noise associated with its operation, it did not enter regular passenger service until 1975. The aircraft was retired from service in 2004 due to operational cost issues but not after becoming a popular form of transportation for the high end of the trans-Atlantic traveling public. With only some 16 aircraft in regular airline service shared between British Airways and Air France this program was a relatively disastrous venture from an economic perspective; however, its legacy was to be a catalyst for the formation of the Airbus Company which has today become an extremely successful builder of commercial transport aircraft with about a 50 % share of the world's market for passenger aircraft of 120 seats or greater.

The Concorde, see the photograph of Figure 12.43, is a four-engined delta-winged aircraft with beautiful lines in reverence to its very high speed cruise operational requirement and, with a passenger capacity of close to 100 people in a first-class environment. Compared with most of today's transatlantic wide-bodied aircraft, the Concorde is a relatively small aircraft with a length of 204 ft and a wing span of 85 ft.

Figure 12.43 The Concorde supersonic transport (courtesy of John Allan).

12.5.1 Fuel System Operational and Thermal Design Issues

The Concorde fuel system design incorporates the needs of both the supersonic and subsonic operational regimes. The resulting fuel system design, therefore, comprises all of the features that are typical of aircraft designed solely for subsonic operation and in addition provides the necessary additional attributes that support safe, reliable functionality during the supersonic flight aspects of its operation in service.

Figure 12.44 shows the operating envelope that was used to define the fuel system design requirements for Concorde in terms of altitude and ambient temperatures showing anticipated hot and cold day temperature extremes.

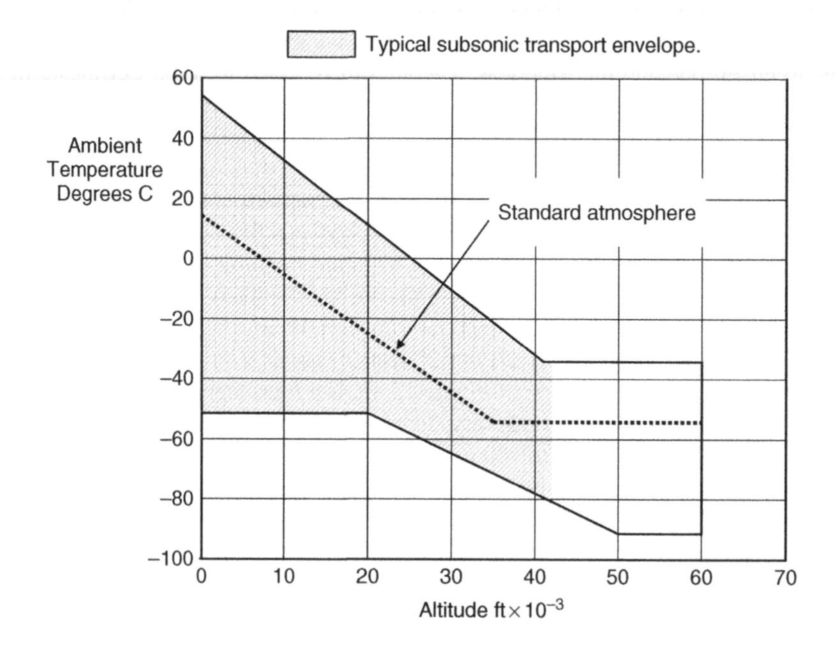

Figure 12.44 Concorde operating envelope.

This figure also indicates the operating envelope typical of today's subsonic transports showing clearly the extended area for Concorde and since this aircraft was required to cruise for long periods close to Mach 2.0, the free stream recovery temperatures are much higher than for the typical subsonic transport as demonstrated by the following equation for constant energy (adiabatic) fluid flow:

$$T_R = \left(1 + F_R\left[\left(\frac{\gamma - 1}{2}\right)M^2\right]\right)T_A$$

where:

T_R is the recovery temperature in absolute temperature units
F_R is a recovery efficiency factor
γ is the ratio of specific heats = 1.4 for air

M is the flight Mach number, and
T_A is the local ambient temperature in absolute temperature units

In the above equation, the recovery efficiency factor represents the fact that full recovery of the free stream energy will be less than 100 % and a factor of 0.9 is commonly used for design purposes.

If we calculate the recovery temperature for a local ambient temperature of -60 degrees C (213 degrees K) which is quite typical for the cruise flight conditions of the Concorde, the recovery temperature for a flight Mach number of 2.0 is:

$$T_R\left(1+0.18\left(2.0^2\right)\right)213 = 366.4^\circ K \text{ This comes to } +93.4^\circ C(+200^\circ F)$$

This means that the skin temperatures of the Concorde during cruise flight can approximate the temperature of boiling water for typical operational atmospheric conditions.

As a matter of comparison, the same equation applied to the traditional subsonic transport with a cruise Mach number of 0.85 yields a recovery temperature for the same local ambient of –32.3 degrees C (-26.1 degrees F).

From the above example it becomes clear that the wide range of operating conditions associated with the Concorde application makes the fuel system design a challenge in terms of the types of materials which can be used and the design features that must be incorporated to ensure that the required system performance is achieved. Perhaps the most significant issue to be considered is associated primarily with the supersonic regime, and this is the need to be able to reject heat from other aircraft systems such as air conditioning, hydraulic systems as well as some engine systems into the fuel system.

In order to maximize the available stored fuel heat sink the fuel tank design configuration must attempt to minimize the surface area in contact with the air stream (thus minimizing air stream heating) while maximizing internal volume. The tank storage configuration is therefore substantially different from the traditional subsonic transport and is shown in Figure 12.45 which shows that the fuel is stored in 13 tanks comprising four main groups, as indicated in the figure.

12.5.2 Refuel System

Figure 12.46 shows the refuel system in schematic form. The two main fuel galleries connecting the forward and aft trim tank groups are utilized by the refuel system to distribute the uplifted fuel to all of the fuel storage tanks. The refuel control units in conjunction with the nearby refuel panel allow the fuel tanks to be loaded either separately or simultaneously while ensuring that critical aircraft CG limits are not exceeded during the refuel process.

12.5.3 Fuel Transfer and Jettison

Referring to Figure 12.46 the left-hand and right-hand transfer tanks store the bulk of the fuel in both the wings and part of the center fuselage. These tanks are defined as the LH and RH groups of the transfer storage system as shown. A third group comprises the trim transfer tanks which store fuel in two forward wing tanks and in a single aft tank in the rear fuselage of the

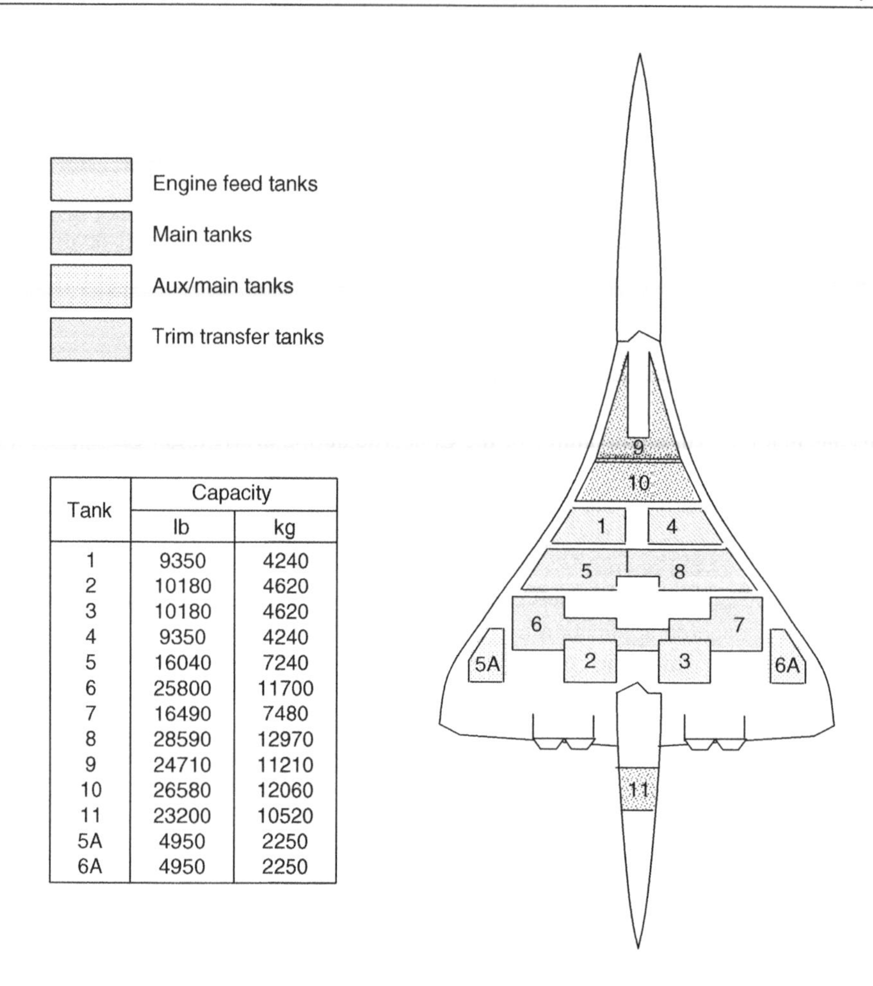

Tank	Capacity	
	lb	kg
1	9350	4240
2	10180	4620
3	10180	4620
4	9350	4240
5	16040	7240
6	25800	11700
7	16490	7480
8	28590	12970
9	24710	11210
10	26580	12060
11	23200	10520
5A	4950	2250
6A	4950	2250

Figure 12.45 Fuel tanks configuration and capacities (courtesy of Airbus Industrie).

aircraft. A feed/collector tank dedicated to each of the four engines defines the fourth tanks group.

Figure 12.47 shows, schematically, the main transfer, trim transfer and jettison systems.

The trim transfer tanks are located at the forward and aft extremes of the storage system and can therefore accommodate the large shift in center of pressure that occurs during transition from subsonic flight to supersonic flight by transferring fuel from the forward trim tanks to the aft trim tank. The main transfer group is available to provide incremental control of the aircraft longitudinal CG throughout the cruise phase although in practice this was found to be largely unnecessary since the CG movement during the cruise phase is very small due to the favorable location of the tanks. The ability of the fuel system to maintain accurate longitudinal CG control is critical in minimizing cruise drag and hence maximizing the operational range of the aircraft.

An important design issue associated with the trim transfer system is to accommodate the critical safety case of all engines out during supersonic cruise to provide a sufficiently fast

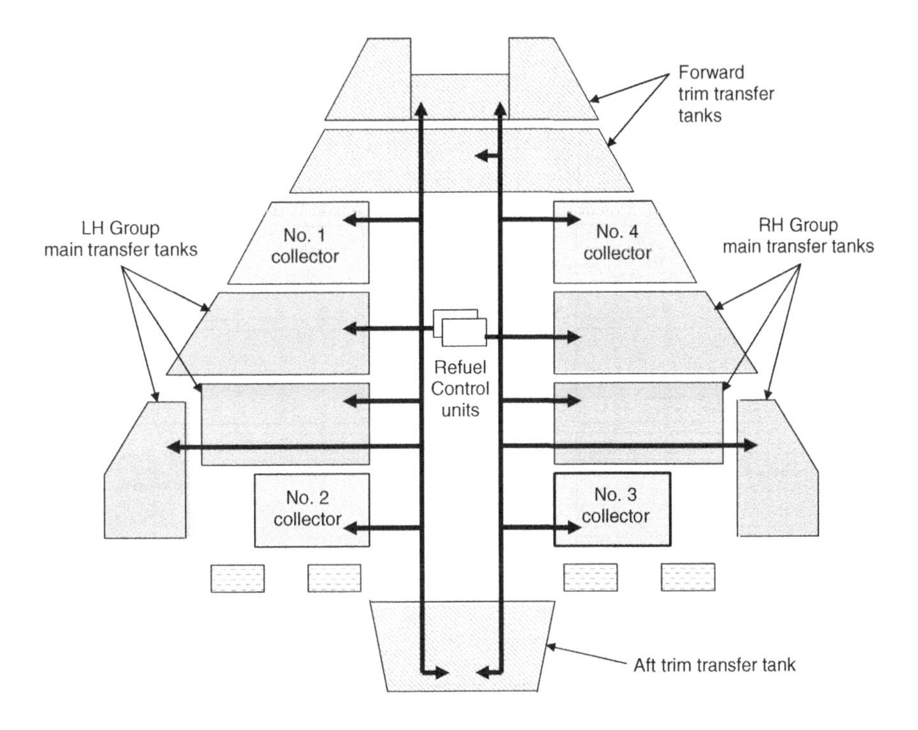

Figure 12.46 Fuel tank configuration and refuel system.

forward transfer in order to maintain aircraft stability during the resulting deceleration and descent to the engine relight envelope. To meet this requirement, a mix of both electric and hydraulic motor-driven pumps is used.

12.5.3.1 Fuel Transfers during a Typical Flight

Fuel management throughout the flight is under the control of the flight engineer. Figure 12.48 illustrates the transfers that take place during a typical flight with arrows indicating the direction of the fuel transfer.

Early in the flight, during the transonic acceleration, a fuel transfer from the forward trim tanks is initiated into the aft trim tank. Any fuel remaining in the forward trim tanks is transferred into the main transfer tank groups. This must be done as quickly as possible to avoid excessive fuel heating due to the geometry and location of the tanks. Even so, during the latter stage of emptying the forward trim tanks, fuel temperatures can approach 90–100 degrees C.

From an equipment perspective, this proved to be a particularly difficult challenge in fuel pump design, i.e. the need to pump high temperature fuel with minimal fuel head above the pump inlet.

Figure 12.49 shows actual fuel tank fuel temperature data from a typical flight illustrating how the operating envelope of Concorde drives fuel bulk temperatures well above the traditional upper limit of 55 degrees *C* associated with subsonic transport aircraft.

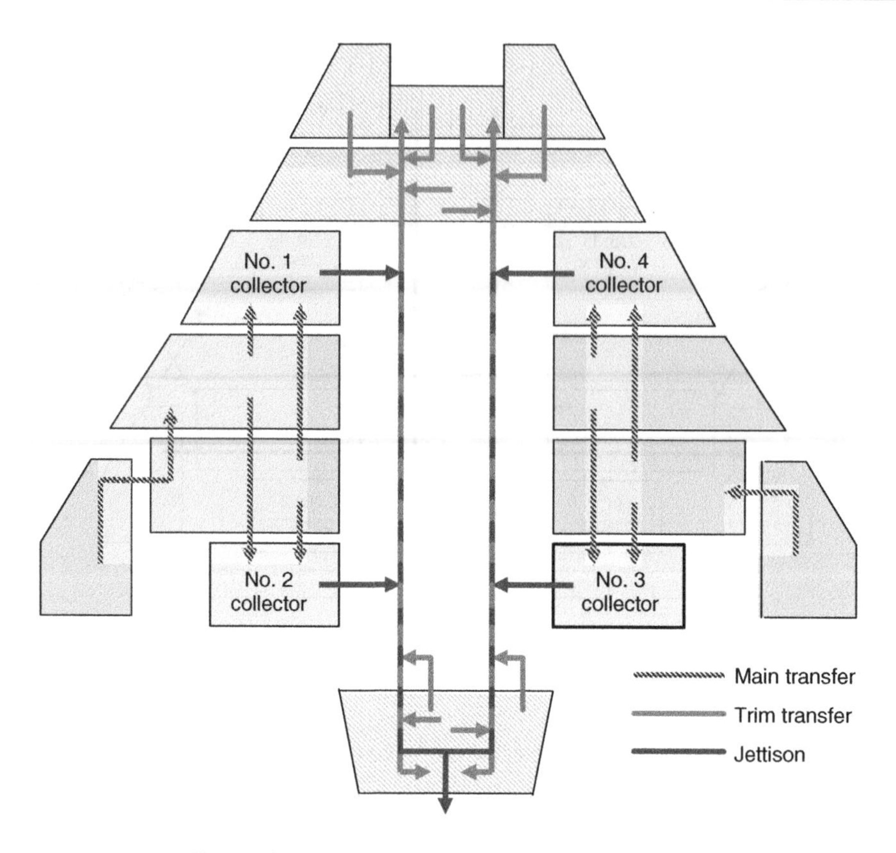

Figure 12.47 Transfer and jettison systems schematic.

As mentioned previously, there is very little movement of longitudinal CG from the optimal position during cruise and therefore no significant fore or aft transfers are necessary during this phase of the flight.

At the end of the cruise phase, fuel must be transferred forward as the aircraft decelerates to subsonic speed. This transfer is flight critical because the aircraft would be unstable when flying at subsonic speeds with a supersonic CG position. This high integrity function is therefore supported by providing sufficient redundancy to ensure successful forward transfer in the presence of multiple failures of either the pumping elements of the motive power sources.

12.5.3.2 Fuel Jettison

A fuel jettison system is provided to accommodate major failures occurring soon after take-off (e.g. and engine failure) by providing the ability to dump fuel overboard quickly and reduce the aircraft weight to below the design maximum landing weight limit. This system takes advantage of the large trim transfer pumping capacity. Again the dual fore and aft refuel/transfer galleries are used to move the fuel to the jettison outlet at the rear of the fuselage.

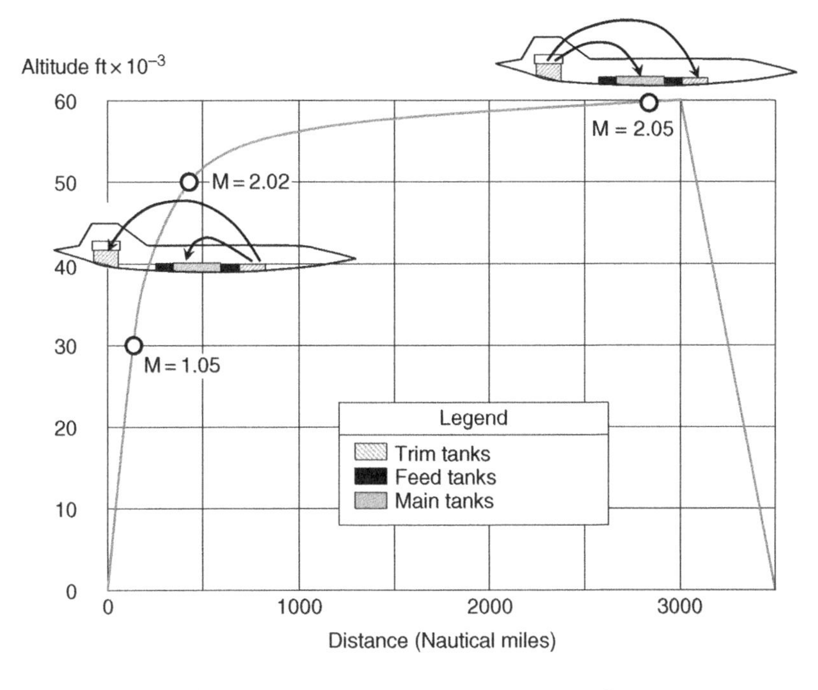

Figure 12.48 Mission profile showing transfers.

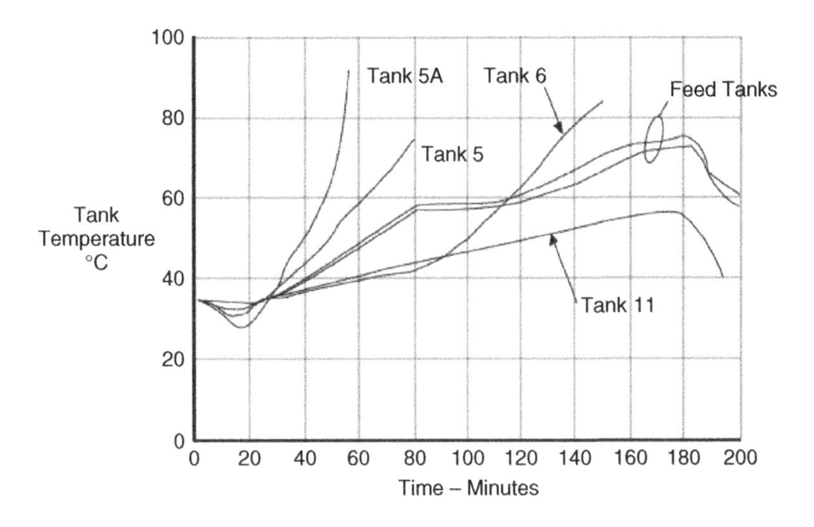

Figure 12.49 Fuel tank temperatures for a typical flight (courtesy of Airbus Industrie).

12.5.4 Fuel Feed

Figure 12.50 shows a high-level schematic of the feed system. Each engine is supplied from its own dedicated feed tank in accordance with traditional airworthiness regulation requirements; a fuel storage design arrangement considered standard in normal subsonic aircraft applications.

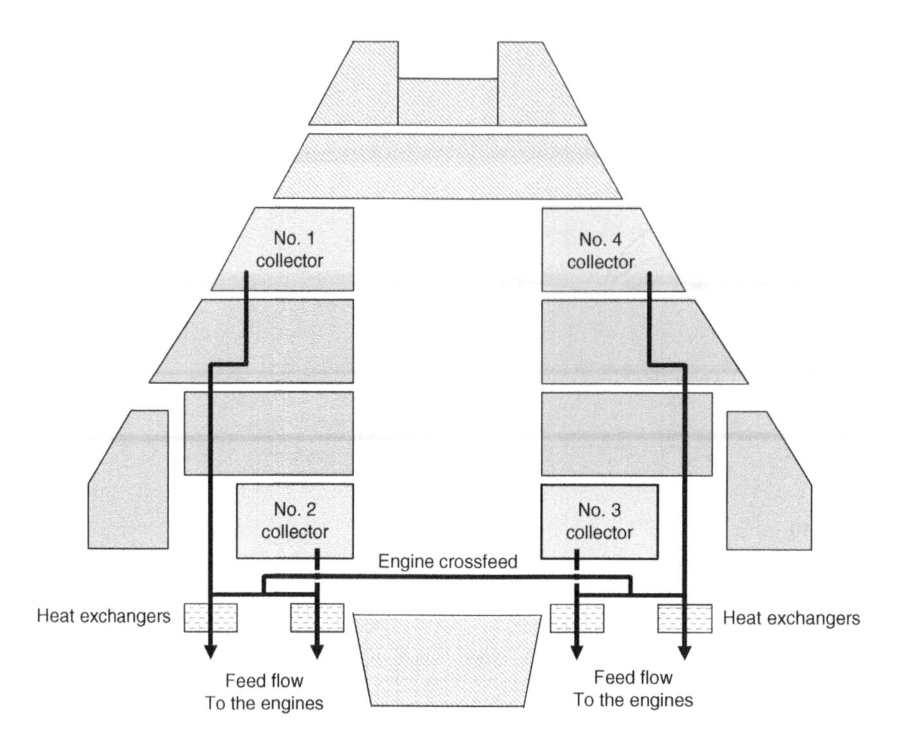

Figure 12.50 Feed system schematic.

The Concorde feed system differs substantially from that of traditional subsonic aircraft in the fuel temperatures that must be tolerated as a result of supersonic flight.

While subsonic aircraft of the same era used boost pumps designed to operate on both Jet A and Jet B fuels with temperatures ranging from the specified fuel freeze point (−48 to −60 degrees *C*) to an upper limit of +55 degrees *C*, Concorde's fuel pumps are designed to operate with fuel bulk temperatures in excess of +85 degrees *C*.

This design challenge is aggravated further by the fact that the fuel system must act as a heat sink to absorb surplus heat from other aircraft systems including air conditioning, hydraulics and engine oil.

To support these requirements a thermal management system comprising a number of heat exchangers together with a fuel recirculation system was deployed as an integral part of the feed system for each engine. This feed system and thermal management system arrangement is shown in Figure 12.51.

The requirement for operation during periods of negative g together with the very high fuel temperatures that can occur during supersonic operation, an accumulator was installed in the boost pump discharge line as shown in the figure.

Negative g operation on traditional subsonic aircraft is normally achieved by careful selection of pump location to ensure that the pump inlet does not become uncovered thus preventing air ingress into the feed line. For Concorde, in addition to ensuring that air is not supplied to the engine during these periods of negative g, it was also essential to ensure that during hot fuel conditions, that feed pressure would not fall so as to cause fuel vaporization.

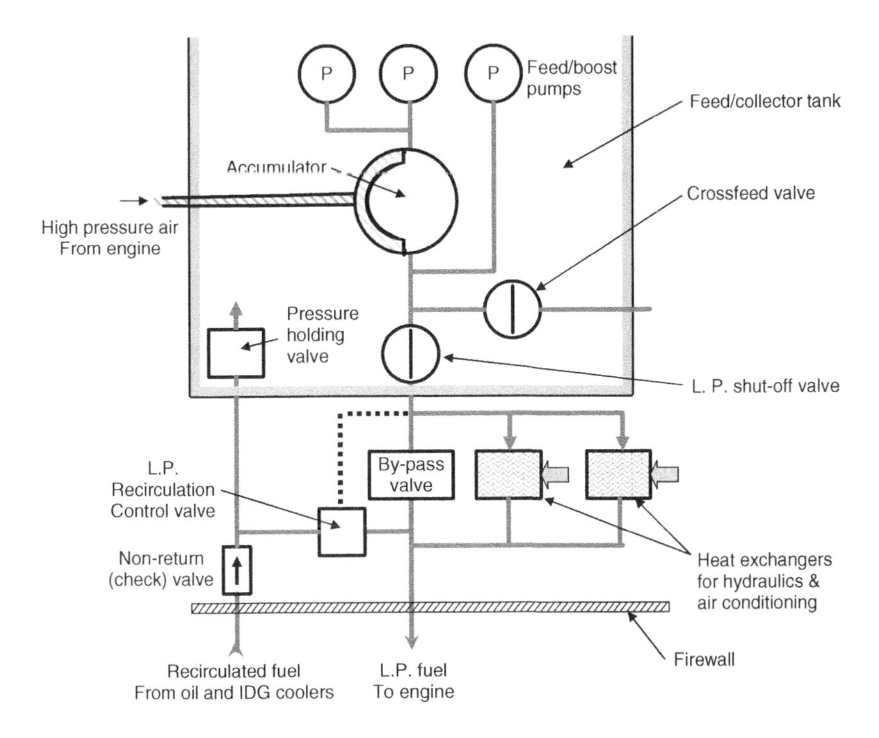

Figure 12.51 Feed system and thermal management schematic.

The accumulator arrangement accomplished this objective. During normal operation, the accumulator is full of fuel, however , if for any reason fuel pressure drops below a certain level, the accumulator will discharge its contents into the feed line thus preventing or minimizing any further loss in feed pressure or the formation of fuel vapor. The gas side of the accumulator is pressurized by engine bleed air.

Referring to Figure 12.51, each feed tank comprises three boost pumps to cover pump or power source failures. Each feed tank also contains a twin motor-driven shut-off valve that can isolate the fuel system from the engine, and a single motor-driven crossfeed valve to facilitate fuel feed from a failed engine's feed tank to the remaining good engines.

The thermal management function uses fuel from the feed tank to cool the following aircraft and engine systems:

- air conditioning systems
- hydraulic systems
- engine gearbox lubrication system
- Integrated Drive Generator (IDG) oil cooling system.

While there is adequate feed flow to meet the heat sink requirements during most of the operating regime, there are periods of low engine demand during the descent phase following supersonic cruise, when additional feed flow is required to meet the cooling systems needs.

The resulting thermal management system design has features not normally seen on other civil aircraft systems. A fuel recirculation system uses a mechanical control valve to maintain an approximately constant fuel flow through the heat exchangers irrespective of engine fuel demand. Excess fuel above the prevailing engine burn rate is returned to the feed tank. This return fuel can have a temperature in excess of +120 degrees *C* and therefore a pressure holding valve located in the feed tank maintains sufficient pressure to prevent boiling in the recirculation fuel line.

12.5.5 Vent System

The Concorde vent system is complicated because of the extensive operating flight envelope which involves operation at altitudes up to 60,000 ft at Mach numbers of up to Mach 2. As in all aircraft, the function of the vent system is to ensure that tank structural limits are maintained under all possible operating conditions including operation in the presence of system failures.

In the Concorde design, a main vent gallery runs throughout the fuselage and wings and exits in the tail cone of the aircraft. This location is well away from the engine inlets and therefore fuel vapors developed in the fuel tanks as a result of high skin temperatures, associated with high Mach number operation, can exit the aircraft safely. As in traditional subsonic aircraft, flame arrestors are located near each of two vent outlets, connected to the main vent gallery, to prevent flame propagation into the fuel system in the event of a direct lightning strike to the vent outlets that could ignite flammable fuel vapors. In this vent system design, no attempt is made to recover ram pressure at the main gallery outlet; however, some ram recovery for the vent system is provided for high altitude operation as will be described later.

All of the fuel tanks are connected to the main vent gallery using float-operated vent valves to prevent fuel from entering the vent system; the number of valves depending upon the tank geometry. In the main transfer tanks, additional pressure relief valves and drain valves are employed to ensure that safe vent system pressures are maintained under all operating conditions.

At the rear of the aircraft the vent gallery is connected to a surge (or scavenge) tank to collect any fuel overflow from the storage tanks. An externally mounted scavenge pump, activated by a sensor in the surge tank, pumps fuel to one of the collector tanks. An overboard overflow/pressure relief valve is opened during the refuel process to protect the tanks' structure from excessive pressures that could occur as a result of a refuel system failure. This valve is closed after refueling provided that the surge tank sensor is dry.

As is required in all aircraft, provision for fuel thermal expansion must be provided by ensuring a minimum of 2 % of tank volumetric capacity between the tank full quantity and vent overflow. In the case of Concorde, some tanks remain full for long periods during flight and the high skin temperatures can cause additional expansion. To accommodate this situation, additional pipes allow excess fuel to be expanded into a tank which is known by the fuel burn schedule to have sufficient ullage available.

Since operation at very high altitudes could result in fuel boiling, the main vent line is closed automatically as the aircraft climbs above 44,000 ft. Additional vent inlets located in the leading edge of the fin provide a nominal ram recovery of 1.5 psi above free airstream pressure. This allows satisfactory feed and transfer pump operation in the high altitude flight regime to provide:

- Satisfactory feed and transfer pump operation by avoiding the need to handle boiling fuel and, at the same time minimizing boil-off losses that would otherwise be as much as two tonnes during a typical mission without his provision.
- Greatly simplifying fuel pump design though the easement of the operating environmental requirements.

Concorde, being an aircraft designed and developed more than thirty years ago would today have to address the current fuel tank safety regulations, specifically SFAR 88 which provides for much more challenging constraints for in-tank electronics (i.e. gauging equipment). For Concorde, the potential for catastrophic fuel tank explosion is clearly significantly higher than for the traditional subsonic jets due to the very high recovery/skin temperatures that are inherent in any long-range supersonic transport aircraft mission and any aircraft designed today to serve a similar market would be required to provide an on-board fuel tank inerting system of some sort.

13

New and Future Technologies

Over the past 50 years we have seen enormous changes in avionics technologies beginning with the transistor and followed by exponential miniaturization in accordance with Moore's Law which predicts a doubling of memory capacity and computer throughput every two years. While there are important caveats to this law, for example, memory access times and software development productivity are not growing exponentially, we have nevertheless seen major advances in the design of avionics products which have greatly impacted fuel quantity gauging and fuel management systems.

Gauging and management systems for example are now able to achieve functional integrity levels in excess of 10^{-9} that are both affordable and effective through the application of dual-dual computer architectures with the addition of a third dissimilar channel.

The technology associated with in-tank gauging sensors has remained stubbornly consistent with the continued use of capacitance technology almost exclusively with relatively minor excursions into ultrasonics as covered in Chapter 7 and in Chapter 12 with the example of the Boeing 777. One of the main reasons for the conservatism seen in fuel gauging technology is the need for extremely high reliability due to the high cost of having to access the fuel tank for maintenance. The risk of making the wrong decision regarding in-tank sensor technology is that any problems that develop during the first few years of operation may have to be carried as an unplanned operational cost for the life of the program which is often longer than twenty years.

Like fuel gauging, fluid-mechanical technologies have not seen radical changes over this same period.

The following sections describe some of the recent fuel system technologies that are being investigated and speculate what newer technologies might be applied to future aircraft and the main drivers and obstacles involved.

13.1 Fuel Measurement and Management

13.1.1 Fuel Measurement

13.1.1.1 Basic Gauging Technology

The most important need to support a meaningful advancement in fuel gauging is for the development of improved in-tank gauging sensor technology. At the time of writing,

Aircraft Fuel Systems R. Langton, C. Clark, M. Hewitt, L. Richards

capacitance-based sensing still remains the technology of choice for aircraft gauging applications, with the industry adopting ac capacitance probe-based systems in the most recent new systems, for reasons explained previously. Over the many years of its use, this technology has been difficult to supersede, despite pervasive operational issues primarily associated with harness connectivity and water contamination. The key problem has been the development of a technology reliable enough to survive the extremely hostile environment of the tank for extended periods between major maintenance checks. The impact of SFAR 88, reference [10], has also added pressure to the industry to further improve in-tank sensing techniques with preferably non-intrusive or non-susceptible solutions.

Over the years, many alternative sensing techniques have been investigated which have included the use of pressure, radar, optics and ultrasonics techniques. Hybrid solutions involving combinations of these technologies have also been examined.

Ultrasonic gauging has been the only technology other than the traditional capacitance technology to see service in a production program. The Boeing 777 airliner and the Lockheed Martin F-22 Raptor military fighter are the only aircraft to adopt ultrasonics for in-tank fuel quantity sensing.

The apparent benefits of ultrasonics at the outset were seen to be the in the minimization of harness shielding and connector integrity-related problems that are magnified in capacitance systems by virtue of the fact that the system has to deal with very small capacitance signals in a hostile environment. In spite of such improvements in signal management the overall system benefits of the ultrasonic solution have clearly not been adequately fulfilled since all new aircraft starts in both military and commercial fields have reverted back to the traditional capacitance methodology. To emphasize this point, it should be noted that the Boeing 787 Dreamliner, the Airbus A380 super jumbo, and the new extra-wide body Airbus A350 have all chosen capacitance gauging as the preferred technology for fuel quantity measurement. On the military aircraft front, the latest fighter aircraft, the F-35 Lightning will also use capacitance gauging.

An ideal, clean sheet of paper approach would be to use a sensing technique utilizing a minimal number of passive (no electronics), non-intrusive sensors with no wiring interface. The increasing use of composite materials in aircraft construction allows the possibility of embedding non-intrusive sensors into the structure in manufacture, but maintenance difficulties associated with accessing faulty sensors have to be compensated for by the addition of redundant embedded sensors. The reliability of the sensor will be maximized if the sensor has little, or at best no electrical components within it. An additional consideration of the new sensor technology is its ease of interface with the signal conditioning within a nearby data concentrator or remote signal processor.

In the authors' opinion, the most promising technology that could be developed would be a combination of the use of light and micro-electric machines (MEMS). MEMS devices have the potential to be designed to measure pressure, temperature, density and acceleration when excited by light through an optical fiber. This new technology approach, which is currently being evaluated by specialist companies within the aerospace industry, offers a low recurring cost approach that could potentially provide an intrinsically safe and HIRF immune sensing solution that would be suitable, because of the small size of MEMS sensors, for embedding in composite structure and therefore be the ideal candidate technology to reliably operate in the harsh environment and meet today's stringent regulatory requirements.

Proprietary restrictions prevent a more detailed coverage of this topic.

13.1.1.2 Fuel Properties Measurement

A fundamental requirement for high accuracy capacitance gauging systems is the acquisition of fuel properties. Specifically this involves the simultaneous measurement of density, temperature and permittivity of the fuel on board.

During recent years, the approach to this requirement has been to use an integrated sensor package designated the Fuel Properties Measurement Unit (FPMU) comprising all three of these sensors through which a sample of the uplifted fuel is captured. Thus the critical characteristics of the residual fuel (from the last refuel) and the current fuel load can be determined.

Key issues with this approach relate to the quantities and location of these devices within the fuel tankage. The concerns here are the functional redundancy and the need to avoid contamination, particularly from water which can accumulate over time and adhere to in-tank equipment.

A new approach to the FPMU requirement is to install the unit within the refuel gallery of the aircraft. The challenge here is to maintain the parameter accuracies required under high flow conditions in both directions. This new 'In-line FPMU' approach will be used in the new Airbus A350 aircraft for the first time.

13.1.2 Fuel Management

Most issues related to the fuel management task are system-related since the task itself involves interfacing and interacting with other aircraft systems. One example of this is the issue of fuel uploading accuracy. In the normal process of refueling, the operator pre-selects the total fuel required and the fuel management system controls the refuel process by using the gauging system data to determine when to initiate closing of the various refuel valves. The challenge for the management system is to complete the auto-refuel process with the correct quantities in each tank and the total adding up to the pre-set fuel load. There are, however, many sources of error that can impact the performance of the management system, for example:

- variations in refuel pressure from location to location
- variations in refuel valve operating characteristics
- variations in dc line voltage.

The task of coping with these variables along with the very high refueling rates required to keep refuel times to a manageable level can lead to significant errors. If the error exceeds a predetermined limit, the refuel process is aborted leaving the operator to complete the refuel process manually. While this is not a common occurrence, it is nevertheless a source of aggravation to the operator and therefore the question that gets asked often is: 'With the technologies available to us today, why is not possible to deliver more precise fuel upload accuracy?'

The difficulty with this problem lies partly in the system interfaces within the aircraft that seem to preclude an easy solution. This is illustrated by Figure 13.1 which shows the interconnectivity between the refuel station, the gauging system, the fuel management and power management systems.

As indicated in the schematic, the fuel management system controls the pumps and valves by sending low level discrete commands to the power management system which then switches the load currents required. Pump and valve status (open/closed and high/low pressure) is typically fed back to the fuel management system directly as low-level discretes.

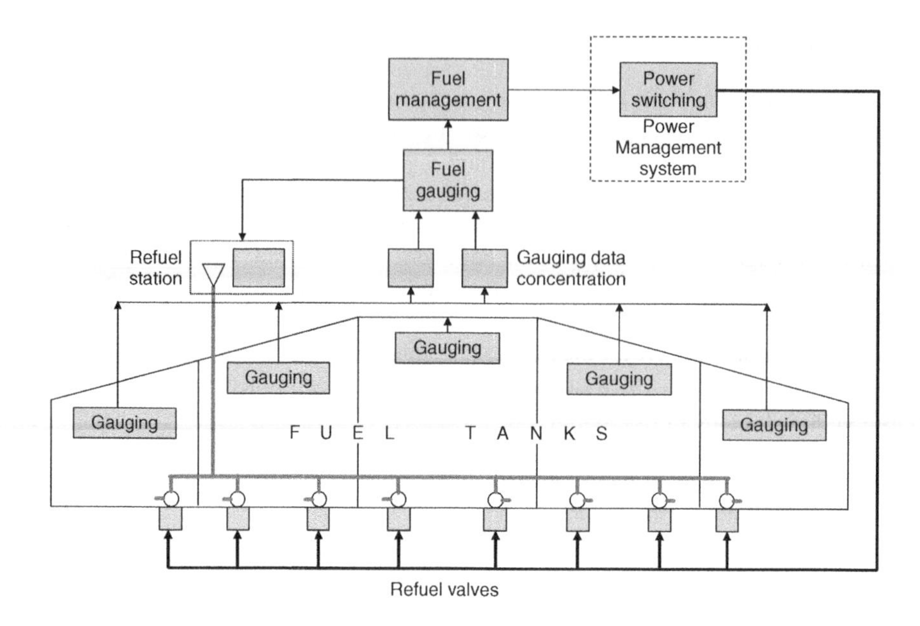

Figure 13.1 Refuel management system schematic.

The fundamental problem with this arrangement in commercial aircraft applications is that refueling rates can be extremely high in order to meet the refuel times required to support turn around time objectives. The ability to control shut-off valve closure selection times as the quantity approaches the target tank quantity is problematic. It is like trying to stop a speeding car at a specific point on the highway by anticipating when to stamp on the brakes but without any visual feedback as to where you are as the car slows to a stop.

There is an equally important issue that must be addressed in any solution and that is to ensure that the action of the valve closure must not induce unacceptable surge pressures in the refuel lines.

There are a number of commonly used techniques that come part way to addressing these problems that are used in today's systems:

- Provide two refuel shut-off valves mounted in series. Closing the first valve reduces the refuel line flow area to about 10 % of full flow. Subsequently the second valve is actuated to fully close the fuel flow into the tank. This is a 'Brute force' solution that is both expensive and unattractive technically.
- Use hydro-mechanical valve technology to provide variable rate closure capability (see Chapter 6) to eliminate surge pressures. This is effective as a surge pressure control measure; however, the target accuracy issue still remains.
- Use a refuel manifold approach to prevent surge pressures from entering into the refuel galleries within the aircraft. This approach is typically limited to aircraft with three or less fuel tanks and again, the target accuracy is not solved.

The technology is available today to provide modulating valves that communicate with the fuel management computer to give essentially perfect refuel accuracy together with surge pressure limitation however; this leads to a number of issues that must be resolved before such an approach is viable. See Section 13.2 below under 'Smart Valve Technology' for a more detailed discussion of the subject.

An intermediate approach is proposed here which is purely speculative that may offer a commercially viable alternative:

- Use a two-position valve to provide, say, 90 % closure upon initial selection and full closure following secondary selection.
- Use hydro-mechanical technology to provide a baseline surge free closure rate profile that remains available even if the secondary selection function is inoperative. (Alternatively some independent surge limiting means could be provided.)
- Communicate with the two-stage valve via wireless technology in order to simplify installation (no additional wiring).
- Wireless communications are available only during the refuel process (i.e. inhibited during flight) and hence have no functional safety impact.

13.2 Fluid Mechanical Equipment Technology

From a component perspective, the evolution of fluid mechanical equipment technology has been much slower than in avionics. In fact, much of the fluid-mechanical technology applied today is fundamentally that same as that used over fifty years ago. There has been a significant advancement in materials used and analytical capabilities that have resulted in optimized designs but for the most part, the technology in-service today is the same. That being said there are opportunities for technology advancement in the area of both pump and valve technology that are worth describing here. The following paragraphs capture some of the more notable technology development issues that are currently being addressed by the industry.

13.2.1 Fuel Valve Technology

There are a number of fundamental improvements that are being continuously pursued to improve in-service performance, cost, weight and reliability. These improvements in fuel valve technology are regarded as incremental rather than revolutionary improvements to a mature area of technology. Some examples of these incremental improvements include:

13.2.1.1 Surge Pressure and Overshoot Control

In current fuel valve usage both electro-mechanical and hydro-mechanical valves exhibit a relatively wide variation in flow and surge pressure control with variations in supply pressure and flow characteristics for typical product manufacturing tolerances. Motor-operated valves have variations in opening/closing time of more than two to one with variations in the electrical power supply and operating temperatures. Refuel control valves, for example, can present a wide variation in overshoot and surge pressure with variations in the functional characteristics of the ground refueling or aerial refueling tanker systems. Ideally, when commanded to close, there should be the same quantity of fuel passing through the valve regardless of variations in supply characteristics, electrical power, operating temperature or variations in the valve physical configuration.

Potential solutions that have been developed include:

- Velocity controlled valves using hydro-mechanical design techniques. Typically however, this is an open loop approach to improving surge pressure and overshoot performance and is therefore not particularly accurate or repeatable due to its dependence upon valve functional tolerances, environmental conditions and refuel pressure variations.
- A constant speed motor-operated valve that uses a voltage regulation technique to provide relatively constant opening/closing times by eliminating (or minimizing) the effects of aircraft power supply variations. This technique significantly reduces the functional variation in motor-actuator performance with minimal penalty in terms of cost, weight or reliability. This approach, however, only goes part way to solving the valve inconsistency–related issues since variations in other system factors (e.g. pressure variations) are not addressed.
- A third technique for improving overshoot control involves using the Fuel Management System to 'Learn' the functional characteristics of each control valve and to adjust the anticipatory valve selection process accordingly. It is also realistic to have the system compensate for variations in electrical supply voltage in the same way.

More radical techniques are also being considered including the use of valve designs that include closed loop control features. This is discussed later under the heading of 'Smart' pumps and valves. In either case the challenge is to avoid any significant increase in system cost or weight. System reliability in terms of functional integrity and availability must also not be negatively impacted.

13.2.1.2 Higher Operating Temperature Capability

Future aircraft advancements will likely include high speed, high altitude supersonic/hypersonic military and commercial aircraft. Operating temperatures will be very high as compared to today's aircraft. The fuel system equipment will need to be compatible with these much higher operating temperatures and will involve the use of new materials.

13.2.1.3 Valve Status Indication

The ability to provide reliable valve status information has been a perennial problem in aircraft fuel systems. Micro-switches used to identify valve status (open, closed or in transit) are in common use on motor-operated valves, however, they are often the weakest link in potential valve failure modes. This is a particularly challenging issue in long-range operations where moisture ingestion through the motor shaft dynamic seal is difficult to avoid.

Reliable valve status monitoring is also a weakness in current day hydro-mechanical valves. Most hydro-mechanical valves have very low operating force margins which makes it very difficult to incorporate any type of robust mechanical position indication switch.

While the benefits of valve status information in many of today's complex fuel systems greatly enhance system fault detection/accommodation and allow isolation of the failed

LRU for maintenance action, the addition of transducers may add a disproportional failure probability to the overall valve assembly since the least reliable element of the design may be the transducer itself.

13.2.1.4 Lower Weight Equipment

Lower weight equipment obviously results in lower system weight which improves overall aircraft efficiency. To date there has been only a limited use of light weight composite materials in the construction of aircraft fuel system equipment. Factors that have limited the use of lightweight composite materials in addition to 'that is how we have always done it' include availability of low cost stable materials, the relatively high cost of tooling, and the difficulty in providing adequate electrical bonding. The relatively small equipment cost benefits in the low quantities typical of the aerospace market has historically served to reduce interest in pursuing this approach to weight reduction. With the increasing cost of fuel, however, the importance of weight will drive industry to look much harder at the use of lower weight material in construction of fuel system equipment.

13.2.2 Revolutionary Fuel Pump and Valve Technology

The above discussion regarding fuel handling product technology development was focused on the incremental developments that are in vogue today. More revolutionary, longer term developments are also being addressed within the industry that offer more radical solutions to the fuel handling task and these are described in this section. Fuel pumps represent an area that has been driven to address more non-traditional designs over the past ten or fifteen years as a result of changes within the aircraft systems that interface directly with the fuel system; namely the power generation and distribution system. These drivers have impacted both commercial and military applications.

Advanced fuel pump technologies are addressed first below.

13.2.2.1 Smart Pump Technology

During the past 15 years aircraft fuel pump technology has developed rapidly primarily in response to changes in electrical power standards together with challenging new aircraft applications in both the military and commercial fields.

The first example was the increased use of wild frequency ac power replacing what was the 400Hz standard with frequencies varying from a low value of about 325Hz to maximum frequencies as high 800Hz. This allows the elimination of the variable speed/constant frequency devices specifically the Integrated Drive Generator (IDG) and the Variable Speed Constant Frequency (VSCF) converter; however the impact on electrically-powered equipment throughout the aircraft has been considerable.

The simplest solution for the motor-driven fuel pump is to use the slipping induction motor (see Chapter 6). The problem with this approach is the very low power factors that occur and for large power applications such as with the new A380 aircraft, the penalty in terms of generator capacity and wiring weight may be considered unacceptable. In this particular application, electronically-controlled brushless dc pumps were selected for the fuel system.

Motor controllers continue to advance with improved technologies and techniques for brushless dc motor control from open loop through scalar to vector control techniques for fast changing torque applications, reference [25]. A key technology area that has enabled the growth of 'Smart' pump applications is in power switching devices. New HEXFET and IGBT devices are capable of switching high inductive loads with minimal losses making them attractive and robust solutions for motor switching.

Figure 13.2 shows a high level schematic of a modern 'Smart' pump controller.

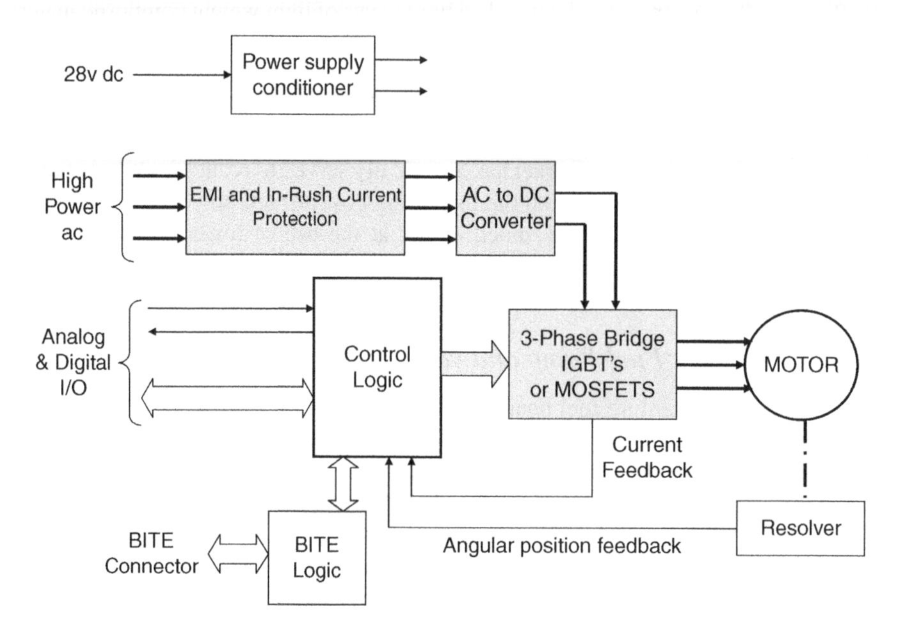

Figure 13.2 Smart pump controller high level schematic.

The control logic can be implemented in either software or in firmware, the former providing more flexibility in the accommodation of changes.

The smart pump controller must be designed to ensure that the influences of both conducted and radiated EMI are small and that the system is able to operate over the full performance range considering these effects. A particularly challenging requirement for the ac powered pump is to meet stringent waveform distortion requirements that occur as a result of the high power switching involved in the controlling function.

Figure 13.3 shows a high level schematic of the motor control system showing both position and current feedback loops.

To control the brushless dc motor, position sensors are employed to measure the rotation angle of the rotor relative to the stator. Hall devices and pulse encoders offer discrete information while a resolver provides continuous angular information and is often preferred. To enable fast response to torque changes, an inner current loop is employed using a precision shunt resistor. The time constant associated with this inner loop which is equal to the ratio of the stator inductance and resistance (L/R) is very small and therefore requires a high bandwidth controller.

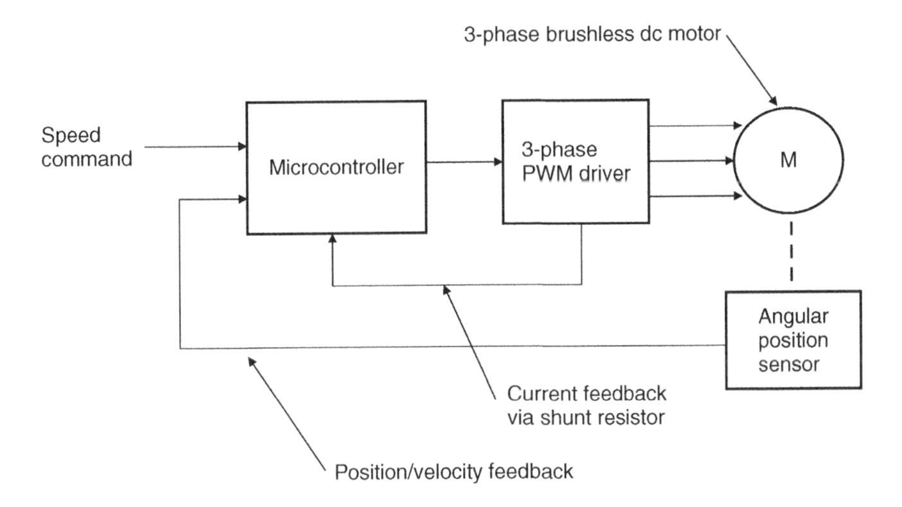

Figure 13.3 Brushless dc motor control concept.

The velocity control loop is associated with the much slower dynamics related to the motor and pump inertia.

More advanced control techniques have allowed elimination of the position sensor by employing a Kalman Filter estimator based on Modern Control Theory. Here estimates of speed and angle are obtained and used in the control algorithm.

These sensor-less techniques require substantially more bandwidth than the traditional sensor-based vector control.

The flexibility afforded by software is invaluable in motor control by its ability to eliminate the need for complex analog control solutions which are often driven by the need to manage uncertainties associated with the motor load and its dynamic characteristics.

These new motor control techniques are driving motor design to achieve very high efficiencies while at the same time requiring very tight control through the use of feedback control and Pulse-Width-Modulation (PWM) drivers. Software-based control is also a very challenging and technology-driven aspect of the industry.

The most recent commercial aircraft have moved on from the original standard 115 volts line-to-neutral ac power supply to 230 volts line-to-neutral generation systems which now appear to becoming a new standard for high power electrical loads.

In the military field, a new standard of 270 volts dc is now being utilized for high power usage including the fuel pump power source. Figure 13.4 shows a schematic of a new state-of-the-art double-ended fuel boost pump for military aircraft applications which can be compared with the earlier technology unit shown in Figure 5.5 from Chapter 5.

In this new version, the pumping elements comprise two inducers whose discharge becomes the input to the main impeller which is a radial impeller device. This approach takes advantage of the fact the inducer element is a fundamentally better suction device requiring a low NPSH. Having two small inducers in combination with a large radial impeller is a more effective design than the earlier dual element solution.

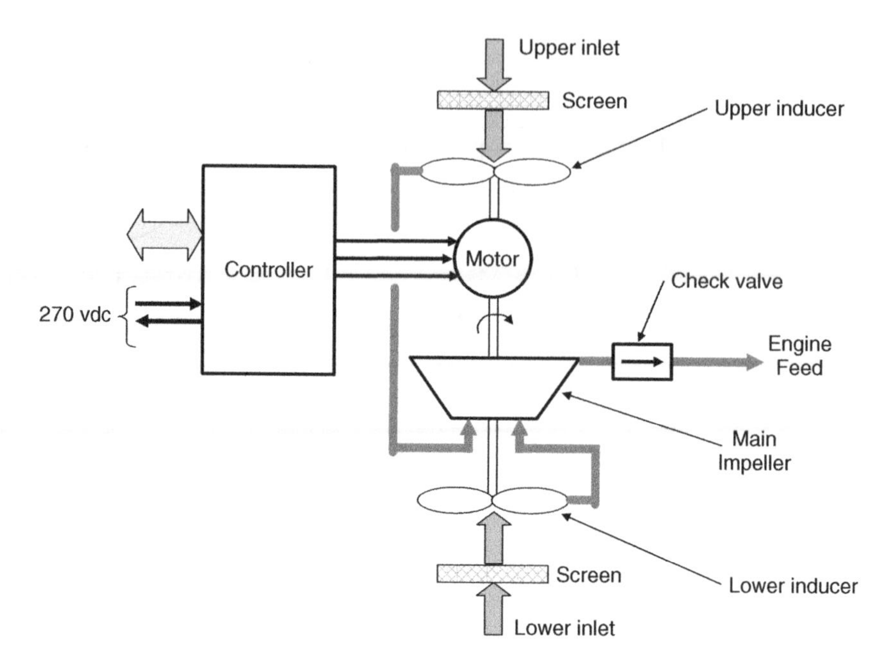

Figure 13.4 Smart boost pump schematic for a military application.

13.2.2.2 Smart Valve Technology

Smart valve technology has so far been confined to the development laboratories where 'smart' operational characteristics have been routinely demonstrated. There is nothing truly revolutionary about much of this work per se except that it is new to the fuel system application. The main issues here are cost and reliability. The traditional valves are extremely simple, low cost, low weight and are highly reliable. Adding smarts to this equipment typically involves the addition of sensors and intelligence (in the form of electronics).

Figures 13.5 and 13.6 are examples of a smart shut-off valve implementation where a predefined velocity profile can be downloaded and stored (and modified as necessary) within the valve itself.

The first figure is a top level schematic showing the introduction of a Digital Signal Processor (DSP) with on-board Non-Volatile Random Access Memory (NVRAM) to manage the control logic associated with the management of position and velocity control.

A Digital-To-Analog (DAC) converter provides visibility of control logic parameters for performance evaluation purposes and a serial digital interface provides the ability to download specific valve velocity profiles. An encoder is used to provide valve position information to the DSP.

The hardware implementation with today's digital technology is simple and low cost requiring only a handful of electronic components on a small circuit board.

The second figure shows the control logic that allows tight velocity control of the valve in both the opening and closing direction. This is accomplished via a slave datum which is a derived valve position command with an in-built velocity limit.

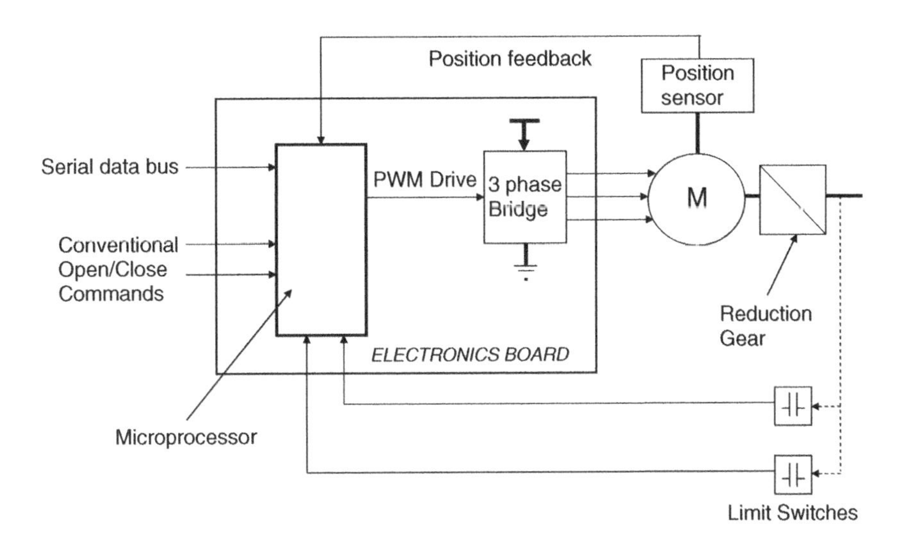

Figure 13.5 Top-level smart valve schematic.

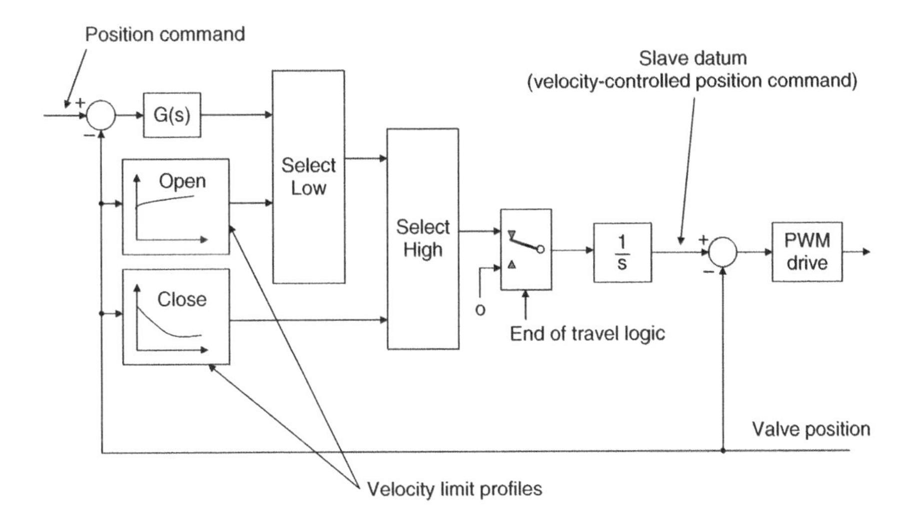

Figure 13.6 Smart valve control logic.

While the above example demonstrates the power of modern electronic technology as a means to enhance control valve functionality, there are a number of issues that must be addressed before such solutions become viable in real world applications.

The first and foremost is functional integrity. While the reliability of solid state electronics in a well designed package can be extremely high, the functional integrity of a refuel shut-off valve, for example, must be extremely high in order to meet demanding dispatchability requirements and to ensure that single failures can not result in loss of critical functionality, e.g. surge protection.

Adding functional redundancy to the above described design would result in a design solution that lacks credibility in cost, weight and reliability compared with the traditional solutions already available to fuel systems designers.

13.3 Aerial Refueling Operations

Current day aerial refueling operations require human functional coordination between the receiver and tanker aircraft. This task requires extreme skill from both the tanker and receiver aircraft aspects of the operation.

For the probe and drogue system the receiver aircraft must holds a tight position relative to the tanker throughout the aerial refueling operation while the tanker aircraft also holds its position and attitude. Automatic refueling hose tension control helps to keep the situation relatively stable. For the flying boom application, the receiver holds station while the boom operator 'Flies' the refueling boom into the receptacle.

The ability to perform these tasks efficiently and effectively becomes far more challenging when operating in bad weather or at night with the need to maintain radio silence for operational security reasons.

Coupling this situation with the burgeoning growth in Unmanned Aerial Vehicle (UAV) operations there is a fast growing need to develop a high degree of automation in the aerial refueling process. This would provide a major strategic solution for long-range UAV applications while providing long needed support to the traditional manned aircraft fleets.

While this technological challenge has been the focus of much discussion and little hard development activity over recent years, the critical needs of future UAVs may force the industry to address this key technological challenge through the application of extensive military research and development efforts in the relatively near term. The technological challenge involved in the development and qualification of an automated aerial refueling system represents both a major challenge and a major opportunity for the future aircraft fuel systems community.

References

1. SAE ARP 4754 Certification considerations for highly integrated or complex aircraft systems.
2. SAE ARP 4761 Guidelines and methods for the safety assessment process on civil airbourne systems equipment.
3. Design & Development of Aircraft Systems, Moir and Seabridge, John Wiley & Sons Copyright 2004.
4. Aircraft Systems, Third edition, Moir and Seabridge, John Wiley & Sons Copyright, 2008.
5. Military Standard MS24484 Relating to single point D1 refueling nozzles.
6. Jane's All The World's Aircraft, Copyright Jane's Information Group Ltd., 2004.
7. RTCA DO-178B Software considerations in airborne systems and equipment certification.
8. NATO Standard STANAG 3447 Aerial refuelling equipment dimensional and functional characteristics.
9. MIL STD 1553 Interface Standard for Digital Time-Division Command/Response Multiplex Data Bus.
10. SFAR 88 Fuel Tank System Fault Tolerance Evaluation Requirements.
11. I. Karassik, J. Messina, P. Cooper and C. Heald (2008) *Pump Handbook*, 4th edition, McGraw-Hill, pp 12.367–12.395.
12. MIL G 26988C Gage, Liquid quantity, Capacitor type transistorized, General specification for.
13. US Patent 4080562, Rubel et al, Apparatus for measuring capacitance or resistance and for testing a capacitance responsive gaging system. March 21st 1978.
14. US Patent 4147050, Rubel et al, Apparatus for testing a capacitance responsive gaging system.
15. FAR Part 25 Airworthiness Standards for Transport Category Airplanes.
16. Electrical Apparatus for Explosive Gas Atmospheres, ICE 600779-11, Section 6.
17. CENELEC document EN 50020. Electrical Apparatus for Potentially Explosive Atmospheres – Intrinsic Safety.
18. RTCA DO160 Environmental Conditions and Test Procedures for Airborne Equipment.
19. SAE AIR 1903 Aircraft Inerting System.
20. AIAA-81-1638, J.K.Klein, The F-16 Halon Tank Inerting System.
21. INCOSE Symposium, Brighton UK, 1999, "Collaborative Methods Applied to the Development and Certification of Complex Integrated Systems" by Roy Langton, George Jones, Steve O'Connor and Paul DiBella.
22. MIL F 8615 (and also SAE AIR 1408) Aerospace Fuel Systems Specifications and Standards.
23. MIL STD 810 Aircraft Environmental Test Standards.
24. SAE ARP 1401 Aircraft Fuel System and Component Icing Test.
25. Implementing Embedded Control for Brushless dc Motors, Yashvant Jani, Renesas Technology America, 28 January 2007.

Aircraft Fuel Systems R. Langton, C. Clark, M. Hewitt, L. Richards

Index

Aircraft Fuel Systems R. Langton, C. Clark, M. Hewitt, L. Richards

Printed and bound by CPI Group (UK) Ltd, Croydon, CR0 4YY

16/04/2025

14658831-0002